236
Advances in Polymer Science

Editorial Board:
A. Abe · A.-C. Albertsson · K. Dušek · W.H. de Jeu
H.-H. Kausch · S. Kobayashi · K.-S. Lee · L. Leibler
T.E. Long · I. Manners · M. Möller · E.M. Terentjev
M. Vicent · B. Voit · G. Wegner · U. Wiesner

Advances in Polymer Science

Recently Published and Forthcoming Volumes

High Solid Dispersions
Volume Editor: Cloitre, M.
Vol. 236, 2010

Silicon Polymers
Volume Editor: Muzafarov, A.M.
Vol. 235, 2011

Chemical Design of Responsive Microgels
Volume Editors: Pich, A., Richtering, W.
Vol. 234, 2010

Hybrid Latex Particles
Volume Editors: van Herk, A.M.,
Landfester, K.
Vol. 233, 2010

Biopolymers
Volume Editors: Abe, A., Dušek, K.,
Kobayashi, S.
Vol. 232, 2010

Polymer Materials
Volume Editors: Lee, K.-S., Kobayashi, S.
Vol. 231, 2010

Polymer Characterization
Volume Editors: Dušek, K., Joanny, J.-F.
Vol. 230, 2010

Modern Techniques for Nano- and Microreactors/-reactions
Volume Editor: Caruso, F.
Vol. 229, 2010

Complex Macromolecular Systems II
Volume Editors: Müller, A.H.E.,
Schmidt, H.-W.
Vol. 228, 2010

Complex Macromolecular Systems I
Volume Editors: Müller, A.H.E.,
Schmidt, H.-W.
Vol. 227, 2010

Shape-Memory Polymers
Volume Editor: Lendlein, A.
Vol. 226, 2010

Polymer Libraries
Volume Editors: Meier, M.A.R., Webster, D.C.
Vol. 225, 2010

Polymer Membranes/Biomembranes
Volume Editors: Meier, W.P., Knoll, W.
Vol. 224, 2010

Organic Electronics
Volume Editors: Meller, G., Grasser, T.
Vol. 223, 2010

Inclusion Polymers
Volume Editor: Wenz, G.
Vol. 222, 2009

Advanced Computer Simulation Approaches for Soft Matter Sciences III
Volume Editors: Holm, C., Kremer, K.
Vol. 221, 2009

Self-Assembled Nanomaterials II
Nanotubes
Volume Editor: Shimizu, T.
Vol. 220, 2008

Self-Assembled Nanomaterials I
Nanofibers
Volume Editor: Shimizu, T.
Vol. 219, 2008

Interfacial Processes and Molecular Aggregation of Surfactants
Volume Editor: Narayanan, R.
Vol. 218, 2008

New Frontiers in Polymer Synthesis
Volume Editor: Kobayashi, S.
Vol. 217, 2008

Polymers for Fuel Cells II
Volume Editor: Scherer, G.G.
Vol. 216, 2008

Polymers for Fuel Cells I
Volume Editor: Scherer, G.G.
Vol. 215, 2008

High Solid Dispersions

Volume Editor: Michel Cloitre

With contributions by

R. Besseling · R.T. Bonnecaze · M. Cloitre · M. Fuchs
G. Fytas · J.P. Gong · L. Isa · Y. Osada · W.C.K. Poon
A.B. Schofield · D. Vlassopoulos

Editor
Dr. Michel Cloitre
ESPCI ParisTech - Ecole Supérieure de Physique
 et Chimie Industrielles de la Ville de Paris
Matière Molle et Chimie (UMR CNRS-ESPCI ParisTech 7167)
10, rue Vauquelin
75005 Paris
France
michel.cloitre@espci.fr

ISSN 0065-3195 e-ISSN 1436-5030
ISBN 978-3-642-16381-4 e-ISBN 978-3-642-16382-1
DOI 10.1007/978-3-642-16382-1
Springer Heidelberg Dordrecht London New York

Library of Congress Control Number: 2010938607

© Springer-Verlag Berlin Heidelberg 2010
This work is subject to copyright. All rights are reserved, whether the whole or part of the material is concerned, specifically the rights of translation, reprinting, reuse of illustrations, recitation, broadcasting, reproduction on microfilm or in any other way, and storage in data banks. Duplication of this publication or parts thereof is permitted only under the provisions of the German Copyright Law of September 9, 1965, in its current version, and permission for use must always be obtained from Springer. Violations are liable to prosecution under the German Copyright Law.
The use of general descriptive names, registered names, trademarks, etc. in this publication does not imply, even in the absence of a specific statement, that such names are exempt from the relevant protective laws and regulations and therefore free for general use.

Cover design: WMXDesign GmbH, Heidelberg

Printed on acid-free paper

Springer is part of Springer Science+Business Media (www.springer.com)

Volume Editor

Dr. Michel Cloitre

ESPCI ParisTech - Ecole Supérieure de Physique
  et Chimie Industrielles de la Ville de Paris
Matière Molle et Chimie (UMR CNRS-ESPCI ParisTech 7167)
10, rue Vauquelin
75005 Paris
France
michel.cloitre@espci.fr

Editorial Board

Prof. Akihiro Abe

Professor Emeritus
Tokyo Institute of Technology
6-27-12 Hiyoshi-Honcho, Kohoku-ku
Yokohama 223-0062, Japan
aabe34@xc4.so-net.ne.jp

Prof. A.-C. Albertsson

Department of Polymer Technology
The Royal Institute of Technology
10044 Stockholm, Sweden
aila@polymer.kth.se

Prof. Karel Dušek

Institute of Macromolecular Chemistry
Czech Academy of Sciences
of the Czech Republic
Heyrovský Sq. 2
16206 Prague 6, Czech Republic
dusek@imc.cas.cz

Prof. Dr. Wim H. de Jeu

Polymer Science and Engineering
University of Massachusetts
120 Governors Drive
Amherst MA 01003, USA
dejeu@mail.pse.umass.edu

Prof. Hans-Henning Kausch

Ecole Polytechnique Fédérale de Lausanne
Science de Base
Station 6
1015 Lausanne, Switzerland
kausch.cully@bluewin.ch

Prof. Shiro Kobayashi

R & D Center for Bio-based Materials
Kyoto Institute of Technology
Matsugasaki, Sakyo-ku
Kyoto 606-8585, Japan
kobayash@kit.ac.jp

Prof. Kwang-Sup Lee

Department of Advanced Materials
Hannam University
561-6 Jeonmin-Dong
Yuseong-Gu 305-811
Daejeon, South Korea
kslee@hnu.kr

Prof. L. Leibler

Matière Molle et Chimie
Ecole Supérieure de Physique
et Chimie Industrielles (ESPCI)
10 rue Vauquelin
75231 Paris Cedex 05, France
ludwik.leibler@espci.fr

Prof. Timothy E. Long

Department of Chemistry
and Research Institute
Virginia Tech
2110 Hahn Hall (0344)
Blacksburg, VA 24061, USA
telong@vt.edu

Prof. Ian Manners

School of Chemistry
University of Bristol
Cantock's Close
BS8 1TS Bristol, UK
ian.manners@bristol.ac.uk

Prof. Martin Möller

Deutsches Wollforschungsinstitut
an der RWTH Aachen e.V.
Pauwelsstraße 8
52056 Aachen, Germany
moeller@dwi.rwth-aachen.de

Prof. E.M. Terentjev

Cavendish Laboratory
Madingley Road
Cambridge CB 3 OHE, UK
emt1000@cam.ac.uk

Prof. Dr. Maria Jesus Vicent

Centro de Investigacion Principe Felipe
Medicinal Chemistry Unit
Polymer Therapeutics Laboratory
Av. Autopista del Saler, 16
46012 Valencia, Spain
mjvicent@cipf.es

Prof. Brigitte Voit

Institut für Polymerforschung Dresden
Hohe Straße 6
01069 Dresden, Germany
voit@ipfdd.de

Prof. Gerhard Wegner

Max-Planck-Institut
für Polymerforschung
Ackermannweg 10
55128 Mainz, Germany
wegner@mpip-mainz.mpg.de

Prof. Ulrich Wiesner

Materials Science & Engineering
Cornell University
329 Bard Hall
Ithaca, NY 14853, USA
ubw1@cornell.edu

Advances in Polymer Sciences Also Available Electronically

Advances in Polymer Sciences is included in Springer's eBook package *Chemistry and Materials Science*. If a library does not opt for the whole package, the book series may be bought on a subscription basis. Also, all back volumes are available electronically.

For all customers who have a standing order to the print version of *Advances in Polymer Sciences*, we offer free access to the electronic volumes of the Series published in the current year via SpringerLink.

If you do not have access, you can still view the table of contents of each volume and the abstract of each article by going to the SpringerLink homepage, clicking on "Browse by Online Libraries", then "Chemical Sciences", and finally choose *Advances in Polymer Science*.

You will find information about the

– Editorial Board
– Aims and Scope
– Instructions for Authors
– Sample Contribution

at springer.com using the search function by typing in *Advances in Polymer Sciences*.

Color figures are published in full color in the electronic version on SpringerLink.

Aims and Scope

The series *Advances in Polymer Science* presents critical reviews of the present and future trends in polymer and biopolymer science including chemistry, physical chemistry, physics and material science. It is addressed to all scientists at universities and in industry who wish to keep abreast of advances in the topics covered.

Review articles for the topical volumes are invited by the volume editors. As a rule, single contributions are also specially commissioned. The editors and publishers will, however, always be pleased to receive suggestions and supplementary information. Papers are accepted for *Advances in Polymer Science* in English.

In references *Advances in Polymer Sciences* is abbreviated as *Adv Polym Sci* and is cited as a journal.

Special volumes are edited by well known guest editors who invite reputed authors for the review articles in their volumes.

Impact Factor in 2009: 4.600; Section "Polymer Science": Rank 4 of 73

Preface

High solid dispersions refer to a broad class of materials comprising colloidal or non-colloidal particles dispersed at very high volume fraction in a liquid phase. They are ubiquitous in daily life, in industry, in biology and in geophysics, encompassing a great variety of systems at the frontier between polymers, colloidal glasses and granular materials. Among these are personal care products, paints, drilling muds, biological tissues and various types of synthetic or natural slurries. In spite of their huge diversity in composition, high solid dispersions have an amorphous jammed structure in common, which is at the origin of remarkable properties. At high volume fraction, the individual motion of particles is dramatically reduced unless a sufficient force is applied to overcome steric constraints. A striking consequence is that high solid dispersions behave like solids at rest but yield and flow under large stresses. The existence and the nature of this solid–liquid transition, which can take a variety of forms, have perplexed engineers and scientist for a long time. Over the years, high solid dispersions have stimulated a lot of work both from an applied and fundamental perspective. On the applied side, controlling the flowability of high solid dispersions is central to the formulation of a large range of commercial products and to the processing of high performance materials such as coatings, ceramics and serigraphic inks. On the fundamental side, understanding the complex behaviour of high solid dispersions is an outstanding challenge for statistical and condensed-matter physics as well as for materials science, biology and geophysics. During the last decade, it has become clear that macroscopic descriptions of high solid dispersions have reached their limits, and that a microscopic approach linking the structure and the dynamics at different scales is highly desirable. The contributions collected in this volume review some results obtained recently in this direction.

The volume begins with a chapter by Vlassopoulos and Fytas who survey how macromolecular and colloidal chemistry can be used to design high solid colloidal dispersions exhibiting a rich variety of phase states and material properties. The authors describe two strategies. The first strategy uses well-defined colloidal particles with tunable interaction potentials much softer than the very short-range repulsion existing between solid particles. Soft particles, which can be as diverse as grafted hard spheres, multiarm star polymers, microgels, block copolymer micelles, dendrimers and dendritically branched polymers, display an extraordinary variety of architectures and topologies. The second strategy consists in engineering soft nanocomposites by blending different classes of colloidal particles or polymers. In both types of materials, interactions and transitions are manipulated by tuning the chemical (composition, architecture and topology) and physical (temperature, solvent, depletion, enthalpy–entropy balance) parameters. This in turn provides unique tools to tailor the dynamics and the rheology of soft particle dispersions.

As explained above, several aspects of the structure, local dynamics and macroscopic rheology of high solid dispersions exhibit strong analogies with the physics of glasses. In Chap. 2, Fuchs focuses on concentrated suspensions of colloidal, slightly polydisperse hard spheres interacting through excluded volume interactions, which constitute one of the simplest systems undergoing a glass transition. He discusses the universal scenario of the glass transition under shear and reviews recent extensions of Mode Coupling Theory that offers a microscopic theory of the linear and non-linear rheology of colloidal fluids and glasses under shear, starting from first principles.

In Chap. 3, Bonnecaze and Cloitre focus on high solid dispersions made of elastic and deformable particles packed together at volume fractions well above close-packing, they term soft glasses. Unlike hard sphere glasses, soft glasses interact through slowly varying repulsive forces of elastic origin that develop at the contacts between the particles. The solvent plays an important role in transmitting the elastic interactions through the glass. After reviewing some generic features of soft glasses, the authors show that many of their properties result from a subtle interplay between disorder and solvent-mediated elastohydrodynamic interactions, which thus constitute the two basic ingredients of a micromechanical description of soft glasses. The theory quantitatively accounts for near-equilibrium properties (statistical distribution, osmotic pressure, shear modulus), for the slip phenomena occurring when soft glasses are sheared near smooth surfaces, and for the non-linear rheology of soft glasses.

Generally, high solid dispersions do not flow homogeneously when they are sheared; instead the macroscopic deformation is localized in slip zones, shearbands or fractures. In Chap. 4, Isa, Besseling, Schofield and Poon review modern advances in fast confocal microscopy imaging and data analysis techniques, which enable time-resolved tracking of individual particles in Brownian and non-Brownian suspensions and glasses. After describing the sample environments and the experimental technique needed to perform this kind of experiments, they present several applications of fast confocal imaging as a unique tool to probe the flow response of hard sphere suspensions and glasses, giving emphasis to the relation between particle scale dynamics and non-linear rheological phenomena such as yielding, shear localization, wall slip and shear-induced ordering.

Chapter 5, by Gong and Osada, is devoted to biological tissues which constitute one particular class of high solid dispersions, consisting of water and various macromolecular components. Bio-tissues have exceptional mechanical properties such as low friction, high toughness, specific adhesion and shock-absorbance capacity. The authors discuss recent progress on the study and development of model synthetic soft and wet hydrogels as substitutes to natural bio-tissues. A strong emphasis is given to the rich and complex surface friction and lubrication properties of these materials, which are found to share common features with the slip properties of soft glasses reviewed in Chap. 3.

It was our intention to provide the Soft Matter community with a comprehensive review of some recent approaches on the rheology of polymer–colloid dispersions. We hope that the reader will feel that the different topics discussed in this volume, which each address a particular facet of high solid dispersions, complement each other and help to draw a bridge between microscopic phenomena and macroscopic rheology.

France *Michel Cloitre*

Contents

From Polymers to Colloids: Engineering the Dynamic Properties of Hairy Particles... 1
Dimitris Vlassopoulos and George Fytas

Nonlinear Rheological Properties of Dense Colloidal Dispersions Close to a Glass Transition Under Steady Shear 55
Matthias Fuchs

Micromechanics of Soft Particle Glasses 117
Roger T. Bonnecaze and Michel Cloitre

Quantitative Imaging of Concentrated Suspensions Under Flow.............163
Lucio Isa, Rut Besseling, Andrew B. Schofield, and Wilson C.K. Poon

Soft and Wet Materials: From Hydrogels to Biotissues 203
Jian Ping Gong and Yoshihito Osada

Index 247

Adv Polym Sci (2010) 236: 1–54
DOI:10.1007/12_2009_31
© Springer-Verlag Berlin Heidelberg 2009
Published online: 4 December 2009

From Polymers to Colloids: Engineering the Dynamic Properties of Hairy Particles

Dimitris Vlassopoulos and George Fytas

Abstract For many years, colloidal hard spheres and polymeric coils served as model limiting cases of soft matter behavior. Softening the potential of interactions has been an obvious possibility for altering properties and has been explored in great detail with sterically interacting charged colloids. As the behavior of such (and technologically relevant) systems is very complex, it is desired to isolate the role of interactions and use well-characterized systems. Today, it is possible to achieve this goal by taking advantage of the capabilities of macromolecular and colloidal chemistry. We show how to use well-defined soft sphere systems interacting via excluded volume repulsions to generate a rich variety of phase states and materials properties. This approach provides opportunities for bridging the gap between polymers and colloids in terms of property variation, and thus designing soft systems with desired properties. At the same time, previously unexplored aspects of important open problems such as glass transition and effects of solvent on the dynamics of correlated systems are addressed. Appropriate blending of the two main classes of soft matter and further directing their assembly and dynamic response has now become a challenging new task.

Keywords Colloidal dispersions · Colloidal glasses · Dynamics · Grafted particles · Hairy particles · Micelles · Nanoparticle-polymer hybrids · Phase diagrams · Polymers · Rheology · Soft colloids · Softness · Stars

Contents

1 Why Soft Colloids? 2
2 Model Soft Spheres 4
   2.1 Colloidal Star Polymers 4
   2.2 Block Copolymer Micelles 6

D. Vlassopoulos (✉) and G. Fytas
FORTH, Institute of Electronic Structure & Laser, and University of Crete,
Department of Materials Science & Technology, 71110 Heraklion, Crete, Greece
dvlasso@iesl.forth.gr

	2.3 Grafted Colloidal Particles	7
	2.4 Microgel Particles	8
	2.5 Selection of Model Systems	10
3	Tuning the Softness: From Polymers to Hard Spheres	11
4	Form, Structure and Diffusion Dynamics	12
	4.1 Comparing Different Systems	13
	4.2 Dynamics of Interacting Colloidal Star Polymers	16
	4.3 Dynamics of Core-Corona Systems: Block Copolymer Micelles and Grafted Particles	21
	4.4 Remarks on Crystallization	27
5	Vitrification	28
	5.1 Soft Colloids in the Glassy State	29
	5.2 Signatures of Transitions and Rheology Manipulation	31
6	Hybrid Systems and Other Emerging Applications	36
7	Conclusions and Outlook	43
References		46

1 Why Soft Colloids?

In the last two decades, with the evolution of soft condensed matter physics [1–8] the manipulation of materials properties has emerged as a theme of scientific and technological significance. The ultimate goal of such an activity is the rational design of materials with desired properties for particular applications. As a first approach, this challenge can be met by combining properties of different, well-understood classes of materials. Linear flexible polymers and hard sphere colloids represent two of the most studied, and thus better understood classes of soft materials [6, 9–11] (see Fig. 1). They also reflect two extreme cases in the spatial organization of a large number of molecular units: in the former case (a contiguous sequence of N monomers which are covalently bonded) flexible random coils and in the latter case solid, compact assemblies with a well-defined shape (here spherical). Moreover, due to this architectural disparity, there are different characteristic length scales: for polymers, the monomer size is typically 1 nm whereas a colloidal sphere can reach a diameter of 1 μm. These distinct features impart considerable differences in the structure and dynamics and consequently in the properties of these systems. For example, semidilute solutions (at concentration c above the overlap concentration c^*) of homopolymers exhibit cooperative concentration fluctuations which are controlled by the osmotic pressure of the system [10, 11]. This reflects weak interactions at monomeric scales, of $O(k_B T)$ or less, with a relevant correlation length ξ_c of $O(\text{nm})$ independent of N; very high concentrations are therefore needed for short-range ordering, if attainable at all. On the other hand, colloidal dispersions of solid particles with radius R in a host fluid exhibit a size-dependent dynamic behavior governed by collective thermal number density fluctuations, with correlation length of $O(\mu\text{m})$; the ordering occurs at relatively low number densities and is long-ranged [12, 13]. The stress is transmitted in both systems due to their entropy, but via different channels (chain elasticity and Brownian motion, respectively).

From Polymers to Colloids: Engineering the Dynamic Properties of Hairy Particles 3

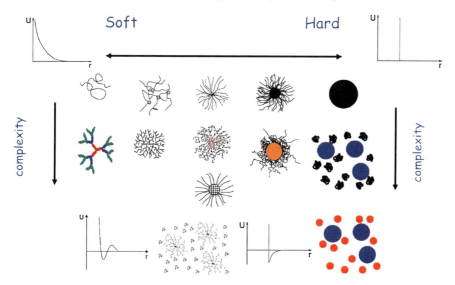

Fig. 1 Cartoon illustration of different systems which can be obtained using appropriate chemistry and span the gap between soft (limiting case being a polymer coil) and hard (limiting case being a hard particle) interactions. These interactions are characterized by the pair potentials U, which are plotted schematically against distance r from a reference particle. The *horizontal double arrow* indicates the possible ways to soften the hard sphere interactions via grafted particles, multiarm stars, microgels and eventually polymer coils, to the *left*. The *vertical arrows* indicate various possibilities of increasing complexity and/or changing interaction potential (as shown schematically), from *right*: mixing hard spheres with linear polymer coils or other spheres of smaller size, grafting particles with networks (crosslinking the grafted layer), block copolymer micelles whose core can be also crosslinked (thus imparting stability), dendrimers and dendritically branched polymers

Therefore, it becomes evident that the intermediate behavior between short-range polymeric and long-range colloidal interactions should be a rich area of research because of the great potential to combine polymeric with colloidal mesoscopic characteristics. A great deal of research effort has already been invested in this direction. The key is the interdisciplinarity via a synergy of synthesis, physical experiment, and theoretical rationalization or predictive power. One way of achieving optimum performance is by using chemical means to combine material properties, e.g., obtaining various macromolecular objects of more complex architecture. Some examples of the different possibilities are illustrated in Fig. 1, where starting from a colloidal hard sphere the interaction potential can be progressively softened with grafted hard particles, multiarm stars, microgels, and, eventually, polymer coils. Moreover, different architectures and complexity such as particles grafted with chemical networks, block copolymer micelles, dendrimers, and dendritically branched polymers can be achieved. Another way is simple mixing of, e.g., polymers and hard or soft particles (Fig. 1). This is a very efficient means to introduce attractions and represents an avenue to obtain a variety of colloidal gels [14–17] or glasses [16–20]. However, it will not be discussed here. Moreover, we emphasize that the methodology discussed here to tune the colloidal interactions from hard sphere to soft is of course not the only possibility. For example, recently a colloidal system was devised whose softness could be effectively tuned by varying the applied electric field, the solvent salt concentration, and the volume fraction of the particles [7].

2 Model Soft Spheres

To meet the above-mentioned challenge of rationally combining different features of different systems, the use of model systems becomes necessary. Whereas the discussed combinations would in principle lead to different material properties, there is a need to develop predictive power in order to coordinate different types of interactions in a single complex system, and hence correlate macroscopic properties to interactions. To do this with reasonable precision, one has to resort to model systems, i.e., systems of well-characterized features and behavior. In the particular case of interest with reference to Fig. 1, the challenge is to design, prepare, and use model spherical particles of tunable softness. Some popular examples are discussed below.

2.1 Colloidal Star Polymers

These are chemically homogeneous multiarm star polymers (usually homopolymers) with only excluded volume interactions [21, 22]. Due to their synthesis procedure (high vacuum anionic polymerization), they are stable and nearly monodisperse [23]. Their softness can be tuned at the synthesis level (number and size of arms) [23, 24] and/or by varying the temperature in different solvents [25, 26]. Moreover, these systems can be functionalized in various ways [27]. What made these systems truly ideal soft colloids were the breakthroughs in both theoretical description and synthesis. The former refers to the ability to describe analytically their internal structure [28] and their softness in terms of an effective interaction potential [24, 29].

The majority of the experimental studies have been carried out with multiarm 1,4-polybutadiene stars, which were synthesized by Roovers and co-workers via two distinct routes. (1) Using chlorosilane chemistry, central dendritic cores of spherical shape and different generations were synthesized, on which the desired number of polymeric arms were grafted [23, 30, 31]. With this approach regular stars with typical nominal functionality (i.e., number of arms, f) in the range 18–128 and nominal arm molar mass, M_a, in the range 10–80 kg mol^{-1} were synthesized. (2) Alternatively, a short 1,2-polybutadiene backbone chain was hydrosilylated with $HSi(CH_3)Cl_2$ yielding two coupling sites per monomer unit, which were substituted with 1,4-polybutadiene by addition of poly(butadienyl)lithium [32]. Such "irregular" stars (without truly spherical central core) with nominal f of 270 and M_a in the range 11–42 kg mol^{-1} were synthesized.

Several other efforts for synthesizing high-functionality star polymers have been reported in the literature. Stars of intermediate functionality (typically below 64)

[33–39] or high functionality (typically up to 250) [34] were prepared, whereas ultra-high functionalities (reaching nominal values of 6,400) were achieved using methologies similar to (2) above, leading to block copolymer stars (arborescent copolymers consisting of polyisoprene chains grafted on polystyrene hyperbranched polymers) [40]. Of course, there are always issues with the characterization of such complex systems, and different methods of preparation have their own merits and problems as well. For example, whereas the atom transfer radical polymerization methods typically yield stars of high polydispersity [35–37], recently an improved methodology drastically reduced this problem [38, 39]. Whereas there is no doubt that Roovers' stars are probably the best characterized, true model systems, the price one pays is that they are available in small amounts, and therefore the recent progress in the field as outlined above is very encouraging.

Due to their non-uniform monomer density distribution arising from their topology [28, 41], colloidal star polymers can be considered as effective core-corona particles with core radius $r_c \sim f^{1/2}$ [28] and softness defined as $s = L/(L+r_c)$, L being the corona thickness [42–44]. Typical value of s for the 128-arm stars is about 0.89 for M_a around 80 kg mol^{-1} [42]. Figure 2 illustrates a cartoon representation of a single such star in a good solvent along with the monomer density profile [28]. Note that the topology (geometric constraints due to curvature) is the only difference between a spherical brush [45] and a semidilute linear flexible polymer solution [10]: here, the blob size increases with the radial distance and three monomer density regimes can be observed [24, 28]: the inner melt-like core regime, the intermediate theta-like (coat or unswollen) regime where the blobs are ideal and only solvent can penetrate in a dense suspension, and the outer excluded volume

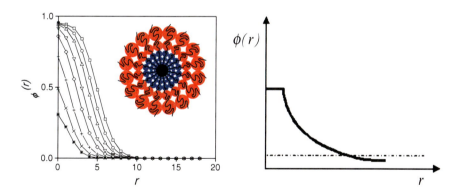

Fig. 2 *Left*: Simulated monomer density profile (distribution of intrastar density around the center of mass) for a melt of stars of varying arm number f (from *left star* to *right open square*: 2 (linear chain), 4, 8, 16, 24, 36, 48, 64) [41]. The low intrastar density at low functionalities indicates penetrability by other stars (to satisfy incompressibility condition), whereas at high functionality the star core is formed with constant density. *Inset*: Cartoon illustration of a multiarm star with the three areas in different colors: melt-like inner black core, theta-like intermediate blue unswollen and excluded-volume outer red swollen region. *Right*: The predicted Daoud–Cotton [28] monomer density distribution. The *horizontal dashed line* indicates the average solution concentration

(or swollen) regime where the blobs are swollen and star–star interpenetration can take place in dense suspensions. The Daoud–Cotton density profile was confirmed experimentally for polyisoprene and polybutadiene star polymers with functionalities in the range 8–129 [21, 46, 47] as well as by computer simulations [21, 24, 41].

Naturally, these star polymers (regular and irregular alike) are distinctly different from dendrimers: the latter represent another class of model colloidal particles [48], which however are not discussed in this review.

2.2 Block Copolymer Micelles

Block copolymers (such as diblocks and triblocks), when dispersed in a selective solvent for one block, self-assemble into supramolecular structures called micelles [49]. Depending on the block composition, the micelles can take different shapes, including spherical star-like structures (the core is usually larger in micelles compared to stars, and clearly separated from the corona due to the enthalpic repulsion of the two blocks); thus, micelles can be structurally similar to stars [50–52]. In fact, it was demonstrated [51–54] that the density profile of the star-like micelles in good solvent for the corona follows the Daoud–Cotton [28] scaling. However, in contrast to the star formation or other means of chemical grafting, the micelles are formed through a physical process. This has the advantage of simple and relatively inexpensive chemistry, and explains the fact that micelles represent by far the most studied soft sphere colloidal systems. On the other hand, the major disadvantage is the stability of the micelles due to the exchange kinetics of the chains participating in a micelle [55], especially in cases of temperature and concentration variation, and the attainability of equilibrium structures in the case of cores of glassy blocks. One way to overcome this important problem is by freezing the core below its glass transition temperature [56, 57] but there remain of course issues related to equilibrium.

Recently, a versatile class of poly(ethylene propylene)/poly(ethylene oxide) block copolymer micelles were introduced; they were stable due to a combination of high block incompatibility, kinetically frozen core, and high interfacial tension between core and solvent [53, 58]. Moreover, by using a co-solvent of varying composition, the aggregation number was controlled and soft spheres from star-like to micelle-like could be obtained. Another way is core stabilization via chemical crosslinking, say by UV radiation [59–64].

It is not our intention to review the vast field of block copolymer micelles and their nanotechnological applications. There are several relevant reviews for the interested reader [49, 65]. We mention only the versatility of these systems as model soft colloids and the various aspects that need careful consideration for designing appropriate systems.

The block composition, i.e., the relative size of the corona to the core, dictates to a great extent the softness of the micelles and therefore their properties and phase behavior [52, 66–71]. One particularly appealing feature of block copolymer micelles is their tunability [49, 72]. Parameters such as the pH strength and amount of added salt [73–75], the concentration [67, 76], the choice of solvent [77, 78], and the temperature variation [71, 79–83] are effective means for monitoring the micellar formation. It has been demonstrated (primarily via the use of rheological measurements) that by tuning the core/corona size ratio one can obtain a wide range of viscoelastic response, from polymeric to colloidal [71, 80–83]. The colloidal character is typically attributed to the hard core and the polymeric character to the softer corona. We note in particular the recent studies with triblock copolymer micelles based on poly(ethylene oxide) and poly(propylene oxide) middle blocks (the so-called Pluronics), in aqueous solutions [84]; these systems were proposed as models for investigating repulsive and attractive glass formation with soft spheres [16, 84]. Naturally, more complex structures (and possibly interactions) akin to those illustrated in Fig. 1 can be obtained from the micellization of architecturally complex block copolymers [85–87].

2.3 Grafted Colloidal Particles

From the ever expanding field of grafted polymers on surfaces, leading to the so-called polymer brushes [45, 88], we focus on grafted spherical particles. There is undoubtedly a wide range of possibilities for grafting different polymeric chains on different types of particles, organic and inorganic alike [89, 90]. Moreover, chemistry provides a lot of flexibility in manipulating particles properties, say via functionalization with a variety of groups. One may note that even hard sphere colloids consist of particles with a tiny grafted layer to impart stability by reducing the van der Waals attractions [13, 91]. Typically this layer is a small fraction of the particle diameter (not exceeding 10%), and an interesting question is what is the threshold beyond which the layer contribution becomes non-negligible as to soften the pair interaction potential and thus influence the properties [92–94].

Polymer-grafted silica (SiO_2) particles represent one of the most popular colloidal systems with tunable interactions. The chains of choice were mainly polystyrene, poly(dimethyl siloxane), poly(butyl methacrylate), and *n*-octadecyl or stearyl alcohol. Chain grafting provided the means to tailor the colloidal particle behavior from hard to soft, as well as introduce attractions in a controlled way by varying the temperature or adding non-absorbing polymer depletant [44, 95–112].

Poly(methylmethacrylate), PMMA, latex particles have also served as a model colloidal system for many years (mainly as hard spheres with hydroxy stearic acid (HSA) chains being the grafting choice [91, 113, 114]). Tunability was achieved by varying the core size and the size of the corona chains. The comparison between chemically grafted (stable) and end-adsorbed temperature-sensitive chains (usually surfactants) has shown that the adsorbed chain particles exhibit similar rheological behavior with chemically grafted particles [115].

Polystyrene (PS) latex particles were grafted with poly(ethylene oxide) or poly(ethylene glycol) chains, allowing temperature variation (and hence a respective variation of attractive interactions) in water, and thus property manipulation from

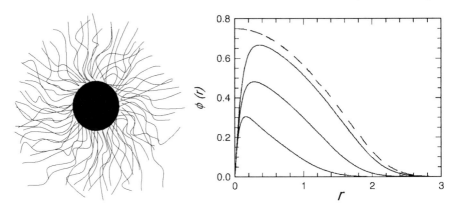

Fig. 3 Cartoon representation of an isolated grafted colloidal sphere (*black core*) along with the calculated monomer density profile $\phi(r)$ (for various core radii, r_c) [119]. The three *solid lines* correspond to values of the ratio of the overall particle radius R to the grafted polymer end-to-end distance of 0.5, 2 and 16, from *bottom* to *top*, respectively. The *dashed line* represents the self-consistent field theory calculation for the flat brush limit ($r_c \to \infty$) [119]. Note the depletion region next to the solid core surface ($r = 0$)

liquid to solid behavior [116, 117]. Recently, the limitation of aqueous environment for the emulsion polymerization preparation of particles has been overcome, and the synthesis of polyurethane-PMMA core-shell particles in nonaqueous emulsion has been reported [118].

The monomer density profile of the layer of grafted chains follows the usual radial decay and strongly depends on the grafting density. A representative example is shown in Fig. 3.

The synthetic flexibility allows combining features and interactions in this class of materials as well. For example, spherical polyelectrolyte brushes constitute another versatile grafted colloidal model with tunable properties [120–122]. As already mentioned, the grafted layer imparts softness and deformability of the colloidal particle. A feature that can make such particles similar to colloidal stars and micelles is the interpenetrability of the grafted chains when particles are brought into contact, typically at high volume fractions. On the other hand, it is possible to reduce or even eliminate this latter feature (and thus influence the monomer density distribution, softness and properties accordingly) by chemically crosslinking the grafted chains so that a rubbery network is grafted onto the solid particle [123]. This makes the grafted particle a microgel-like system and will be discussed in the next section.

2.4 Microgel Particles

Intramolecularly cross-linked polymeric latex particles that are swollen in a good solvent are known as microgels [124–130]. They can be viewed as sterically stabilized colloidal particles with an inhomogeneous density distribution of crosslinks

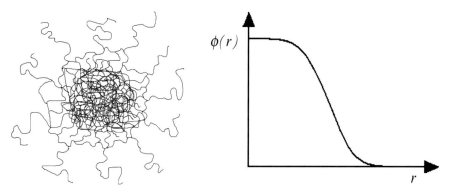

Fig. 4 *Left*: Cartoon illustration of a microgel particle, showing the heterogeneous distribution of crosslink density. *Right*: Corresponding radial monomer density profile $\phi(r)$

and free dangling chains at the surface. This structure suggests that their internal structure resembles that of core-corona systems, bearing analogies to stars and block copolymer micelles. However, there are important differences, in particular referring to the much reduced interpenetrability of the microgels because of the crosslinks [125, 130]. The core size and related degree of homogeneity of crosslink density depend on parameters such as temperature of the reaction and amount of initiator [124–128]. A typical crosslink density profile of a microgel particle [128, 131–133] is illustrated in Fig. 4. The achievable narrow particle size distribution [129] combined with the inherent steric stabilization of the particles make them tunable (e.g., by controlling the degree of crosslinking of the grafted layer) soft colloidal systems [134]. For example, by changing the solvent quality the microgel particles can swell or shrink and thus vary softness. Perhaps the most frequently studied system in this direction is the so-called PNIPAM particles, i.e., poly(N-isopropylacrylamide). This polymer undergoes low critical solution temperature transition, i.e., the particles swell at low temperatures and deswell upon heating [125, 128, 135–138].

Another possibility for controlling size and softness variation in colloidal particles is by grafting PNIPAM microgels with polymer chains, yielding high grafting densities and eventually attraction of the densely grafted microgels, as the temperature increases [136, 139]. This modification provides the means to modulate the size of both the shell and the core independently, typically by temperature, thus imparting more flexibility to the particle's response [139, 140]. Alternatively, PS latex or SiO_2 particles were grafted with a PNIPAM, yielding soft colloids whose size and softness can be varied at will with temperature, with important implications on the suspensions properties [125, 141–147]. In fact, it was found that, in the collapsed state, the dispersions behave similarly to hard-sphere suspensions (also forming crystals and glasses), whereas in the swollen state they behave like soft colloids [125, 140, 146–148]. The PNIPAM crosslinking density also contributed to the softness of the overall particle, with interesting implications on the rheological properties [125, 147–151].

The polyelectrolyte microgels have been established as model soft spheres as, in addition to the above features, their softness and properties can be tuned by altering the physico-chemical environment (pH, ionic strength, degree of ionization) [152–160]. The response varies from that of colloidal (polydisperse hard-sphere) suspensions and that of polymer gels and in this respect such microgels fit within the theme of Fig. 1 [157–160].

2.5 Selection of Model Systems

Given the above developments, one may wonder whether there is need for so many model soft colloidal systems and what are the criteria for selecting the appropriate system. It turns out that the first question can be answered in two ways. First, different systems, such as those described above, have relative merits and drawbacks and can thus serve as models for specific goals. Second, as already discussed in Sect. 1 in the context of Fig. 1, the field of soft colloids is very large and there are ample possibilities for tailoring properties at the molecular level and discovering new material functions. As to the choice of the particular system to study, this depends to a great extent on the question to be addressed. If the latter relates to applications, then the system that appears promising with respect to the expected practical result will be preferred. For example, nanoreactor applications can be realized with thermosensitive microgel particles, in which deposition and immobilization of metallic nanoparticles can take place with temperature variation [161, 162]. On the other hand, if the question relates to fundamental research, e.g., exploring the universality of behavior and the degree to which it holds, then the choice should be based on a synergy of chemistry, physical experiment, and guidance from theoretical predictions. We adopt the latter approach in this review, and therefore we elaborate a bit more by pointing to a few selection criteria below:

1. Well-characterized systems. This depends on the appropriate chemistry and subsequent characterization (typical issues here are the polydispersity, control of grafting density, reproducibility of procedure to obtain identical particles). One frequent problem here is that the price one pays for such systems is the availability of small amounts (sometimes only fractions of 1 g) of material. For example, multiarm star polymers are in many ways unique, clean, soft colloids [19, 23], but their nontrivial synthesis makes them not readily available. On the other hand, recent developments with block copolymer micelles from anionically synthesized polymers [54–58] and arborescent graft copolymer synthesis [40] appear to have adequately addressed this issue for making available different alternative star-like systems.
2. Proper description of the interaction potential. This is needed once we enter the non-dilute regime (where nearly all of the interesting and practically important phenomena occur), and here collaboration with theoreticians is much needed. For example, for the case of colloidal stars this link is well-established [19, 24].

3. System tunability. The interaction potential typically depends on external conditions (e.g., temperature, ionic strength) and internal system parameters (e.g., molar mass, relative core-to-corona size, degree of crosslinking). It is important to be able to access these parameters easily, in order to obtain maximum flexibility.
4. Simplicity. Many systems, especially of technological interest, are too complex to be easily understood [12, 134, 163–165]. It is important to isolate different contributions of the various parameters affecting the interactions and study them independently. To this end, systems such as those discussed in the previous sections offer this flexibility. They can be very simple (e.g., star polymers with only excluded volume type of interactions) and can become more complex with functionalization and other chemical modifications.
5. Experimental needs. Often there are requirements on the range of the overall sizes imposed by particular experimental tools (e.g., scattering), or when labeling (adding chromophores or partial deuteration) is needed. In addition, the system/solvent combination may create problems since it is nontrivial to use very volatile organic solvents for (often very long) rheological experiments. These are general considerations that, in the case of the systems discussed here with so many degrees of freedom, require very careful consideration.

3 Tuning the Softness: From Polymers to Hard Spheres

A quantitative description of the softness tunability, connecting structure to properties, is the pair interaction potential. There are several possibilities for extracting an empirical apparent potential, typically using rheological data (high-frequency plateau modulus or non-Newtonian viscosity) which are very sensitive to the interactions between particles [115, 147, 154, 163, 166–168]. However, for each system of a given softness the data fitting procedure yields an apparent potential and so there is no predictive power in the sense of tuning the potential with some parameter. For the different soft systems discussed, one wishes a tunable potential possibly extracted from first principles. In fact, this was performed in the last decade by Likos and co-workers for a class of model tunable soft colloids, the multiarm stars, using a coarse graining [169] methodology known as effective interactions approach [19, 24]. They proposed a tunable effective pair interaction potential $V(r)$ between stars [29], which is solely based on excluded volume constraints. The key tuning parameter was the functionality of the star (f). For high f values, the stars are virtually spherical objects exhibiting a Yukawa type of interaction at long distances, typical for sterically stabilized colloids, whereas at short distances they feel a logarithmic divergence, based on general scaling arguments by Witten and Pincus [170]. For $f \to \infty$, the stars reach the hard sphere behavior. On the other hand, for low f values the stars approach the limit of polymer coils. The potential is given by

$$\frac{V(r)}{k_B T} = \begin{cases} (5/18) f^{3/2} \left[-\ln(r/\sigma) + \left(1 + \sqrt{f}/2\right)^{-1} \right] & \text{for } r \leq \sigma \\ (5/18) f^{3/2} \left(1 + \sqrt{f}/2\right)^{-1} (\sigma/r) \exp\left[-\sqrt{f}(r-\sigma)/2\sigma \right] & \text{for } r > \sigma \end{cases}, \quad (1)$$

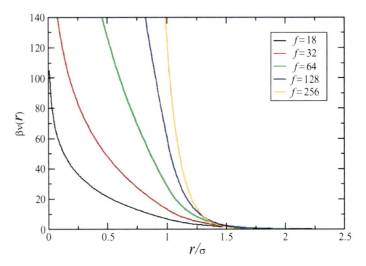

Fig. 5 Predicted star pair interaction potential normalized with the thermal energy unit $\beta = 1/k_B T$ vs the normalized center-to-center distance r/σ for different values of the functionality f [29]

where σ is the effective star diameter [29]. Figure 5 depicts this potential for different values of the functionality. It is evident that by tuning this single parameter f, the regime from the hard-sphere to the polymer-coil limits is spanned. Note that this potential is athermal and that the star–star interactions are purely repulsive (we call them excluded volume interactions). This strong variation of $V(r)$ with f is also reflected in the macroscopic properties of colloidal stars, as discussed in Sect. 4.3.

Interestingly (and not too surprisingly) this tunable potential also describes well star-like block copolymer micelles, i.e., micelles with very small core compared to the corona layer [29, 54, 58]. On the other hand, when the core becomes a significant fraction of the micellar particle, this potential fails to describe such systems (including grafted particles and microgels) successfully [171]. To our knowledge there is no other analytic form of pair interaction potential for other soft systems that provides this tunability with a single parameter, with the exception of ionic microgels [172]. This limitation currently represents another important advantage of the experimental star polymer systems. We also note that it is possible to tune the potential of other systems like charged stars and dendrimers [19, 48], although not in the straightforward and simple manner of the present repulsive stars. However, these types of systems and interactions are not considered in this review.

4 Form, Structure and Diffusion Dynamics

In this section we discuss the structural and dynamic features of soft colloids with emphasis on universal as well as distinct features in an effort to rationalize the experimental findings. Taking the paradigm of multiarm star polymers, the permanent attachment of many polymer chains on a spherical particle-like core results in an inhomogeneous monomer density profile as seen in Fig. 2. This leads to a distinct dynamic response, as compared to hard sphere colloids and flexible linear polymers. The dynamic structure factor $C(q,t)$ at different magnifications ($\sim q^{-1}$), selected by varying the magnitude of the scattering wavevector q, is the relevant physical quantity to quantify the behavior of the system at different volume fractions.

4.1 Comparing Different Systems

The shape and size of an individual mesoscopic object cannot be directly seen by real imaging. Instead, the form factor $P(q)$ in the reciprocal space is commonly utilized to quantify intra-particle spatial correlations. Following Fig. 1, the extreme soft systems (linear polymers) are characterized by the Debye form of $P(q)$ with maximum at $q = 0$ and power law scaling q^{-2} at high values of qR_g, with R_g being the polymer radius of gyration [173]. For the hard systems of Fig. 1, $P(q)$ exhibits characteristic interference oscillations at high qR with R being the hard sphere radius [173]. The form factor of the model soft colloidal particles presented in Sect. 2 will, as expected, display correlation lengths describing their inherent density profile $\phi(r)$ (see for example Fig. 2) and can be approximated by [54, 174]

$$P(q) = \left[\int_0^\infty 4\pi r^2 \phi(r) \frac{\sin(qr)}{qr} dr \right]^2 \bigg/ \int_0^\infty 4\pi r^2 \phi(r) dr. \qquad (2)$$

Figure 6 displays simulated form factors for a multiarm star polymer of varying functionality and a hard sphere [41]. The high-q asymptotic behavior, characteristic of the coil structure, is absent in the latter case. A handicap in the experimental determination of $P(q)$ is often the narrow-q range accessible by the scattering techniques that can be overcome through the combination of low-q light scattering and high-q X-ray and/or neutron scattering (utilized on the same system). Size and shape also determine the translational diffusion D_0 of the nanoparticles in dilute solution, and hence D_0 can prove the consistency of the scattering results.

The spatial correlations between different mesoscopic (and beyond) objects at sufficiently high volume fractions are manifested in the static structure factor $S(q) \sim I(q)/P(q)$, where $I(q)$ is the total scattering intensity pattern. Note that, whereas this relation between $S(q)$ and $I(q)$ is rather approximate above the overlap density [175], it serves as a guide to assess interacting colloidal systems. Whereas a concentrated solution of interpenetrated polymer chains is characterized by a trivial nanoscopic structure with a characteristic correlation length (mesh size) $\xi_c(\phi)$ (Sect. 1) that decreases with the volume fraction ϕ, an interacting multiarm star polymer forms a liquid-like [41, 176] or even crystalline [177, 178] order via the excluded volume interactions of the cores (Figs. 2 and 6). The structure factor (controlled by interactions, Fig. 5) affects the diffusion dynamics and the flow properties of the system.

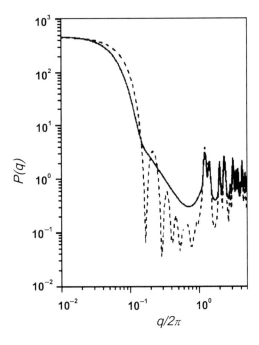

Fig. 6 Computer simulations results of the scattering wavevector dependence of the form factors $P(q)$ of a star with $f = 24$ arms (*solid line*) and a (compact lattice) hard sphere with nearly the same number of beads (*dotted line*). Taken from [41]

Taking the hard sphere colloids as a reference state, the mean-square displacement (MSD) in dilute suspensions is associated with the particle self-diffusion D_0, whereas at finite volume fractions the onset of interactions marks the alteration of the dynamics. The latter can be probed by the intermediate scattering function $C(q,t)$ which measures the spatiotemporal correlations $\phi_q(t)$ of the thermal volume fraction fluctuations [91]. Figure 7 depicts two representations (lower inset and main plot) of the non-exponential $C(q,t)$ for a nondilute hard sphere colloidal suspension for q near the peak of $S(q)$ (lower inset) [179]. In this case, the diffusive (q^2-dependent) initial rate (solid red line) reflects the collective diffusion coefficient $D_{\text{coll}}(q)$, which depends on the thermodynamics of the colloidal suspension (expressed by $S(q)$) and the hydrodynamic interactions. The measurement of $C(q,t)$ at low qR values yields the cooperative diffusion D_c which depends on the osmotic pressure ($\sim 1/S(qR \rightarrow 0)$) and hence increases with volume fraction. In the presence of size polydispersity, there is an additional incoherent contribution to $C(q,t)$ decaying through the self-diffusion coefficient $D_s(\phi)$ [42, 43, 91]. The latter can also be measured for monodisperse hard sphere suspensions at finite concentrations at qR corresponding to the first minimum of $S(q)$, i.e., when the interactions can be ignored [91]. These three different diffusion coefficients exhibit distinctly different dependence on q and ϕ. From these three transport quantities, $D_{\text{coll}}(q)$ is absent in monodisperse homopolymers, whereas D_s can hardly be measured in polydisperse homopolymers due to the vanishingly small contrast.

From Polymers to Colloids: Engineering the Dynamic Properties of Hairy Particles 15

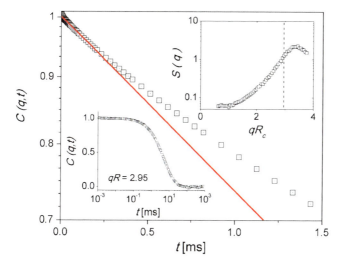

Fig. 7 Intermediate scattering function $C(q,t)$ for a suspension of PMMA hard spheres (radius $R = 118$ nm) recorded at $qR = 2.95$ and volume fraction $\phi = 0.42$ in a semi-log and log–log representation (*lower inset*). The static structure factor $S(q)$ is shown in the *upper inset* where the *vertical line* indicates the value $qR = 2.95$ at which the function $C(q,t)$ was recorded. The short-time collective diffusion is obtained from the initial plot of $C(q,t)$ (*red line*)

It is instructive to compare D_s for various soft systems such as polymers and colloids, as it shows sensitivity to the type of interactions. This is illustrated in Fig. 8 which depicts the normalized (to the extrapolated self-diffusion coefficient D_0 at zero concentration) inverse D_s for various systems (hard sphere suspensions, linear polymer solutions, star polymers, and block copolymer micelles) against the effective volume fraction, ϕ_{eff}. The latter is obtained in the usual manner for hard spheres (in reference to the random close packing) [91, 163], whereas for all other soft systems it is expressed as the ratio c/c_h^* of the mass concentration over the hydrodynamic overlap concentration (i.e., the c^* based on the hydrodynamic radius R_h rather than the radius of gyration, R_g). For example, Fig. 8 demonstrates the strong effect of softness on the volume fraction dependence of D_s. Alternatively, this plot can serve as a means to characterize different colloidal particles with respect to their softness or design systems interpolating between polymers and colloids. We shall not discuss here different possibilities for semi-empirical fitting of D_s (or viscosity) data [81, 180, 182]. We only note the different physics associated with the concentrated state of linear polymer (tube model) and hard sphere (long-time decay of density fluctuations). The representation of Fig. 8 is also considered appropriate for the zero-shear viscosities and allows obtaining master plots for direct comparison of different soft objects [183, 184]. However, the structure and interactions are manifested differently in the normalized viscosity vs ϕ_{eff} plot and the corresponding plot of Fig. 8 due to the effect of the internal (polymeric) structure. Only for the hard spheres does the volume fraction dependence of viscosity and self-diffusion coincide fully, as demonstrated in the figure.

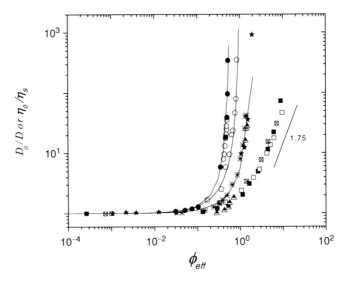

Fig. 8 Inverse self-diffusion coefficients $1/D$ (normalized to that at $c \to 0$, $1/D_0$) or zero-shear viscosities η_0 normalized to the solvent η_s (for hard spheres, where indicated) against the effective volume fraction for different soft systems: *open circles*: hard sphere PMMA particles (in cis-decalin) with small HSA layer and overall radius 247 nm, measured with photon correlation spectroscopy (PCS) [185]; *closed circles*: viscosity data for hard sphere poly(butyl acrylate – styrene) particles of diameter 254 nm with a layer of poly(ethylene oxide) of thickness 16 nm, suspended in water [115, 186]; *closed squares, open squares crossed with X* and *open squares*: linear PS (in good solvent acetophenone) of molar mass 34, 255, and 340 kg mol^{-1}, respectively, measured with fluorescence correlation spectroscopy (FCS) [187]; *open pentagons*: polystyrene – polyisoprene block copolymer micelles (PS-b-PI) with PS cores, in good PI-solvent decane (total molar mass 1.9×10^3 kg mol^{-1}, PS composition 0.26, $f = 1,470$), measured with PCS [188]; *stars, half-filled up triangles* and *filled stars*: multiarm polybutadiene stars 12807, 12814, and 12880 (with nominally $f = 128$ arms and nominal arm molar masses 7, 14, and 80 kg mol^{-1}), respectively, measured by pulsed field gradient (PFG)-NMR in good solvent d-toluene [43, 182, 189]. The lines through the hard spheres, micelles and stars data are empirical fits (Krieger-Dougherty and Doolittle equations [12, 92, 186]) yielding maximum packing fractions of about 0.7, 1, and 4, respectively (the latter value indicating strong interpenetration). The line with slope 1.75 represents scaling prediction for linear coils

4.2 Dynamics of Interacting Colloidal Star Polymers

For a multiarm star polymer with f arms, the Daoud–Cotton model [28] (Fig. 2) implies that the radius of the star scales as $R \sim N_a^\nu f^{(1-\nu)/2}$, where $\nu \approx 3/5$ is the Flory exponent and N_a is the arm degree of polymerization. The concentration profile scales with the radial distance r as $\phi(r) \sim (r_c/r)^{(3-1/\nu)}$ for $r > r_c$, whereas $\phi(r) = 1$ for $r < r_c$. As mentioned in Sect. 2, scattering measurements in both stars and block copolymer micelles have demonstrated that the extracted form factor conforms to the Daoud–Cotton density profile, even for micelles as big as 450 nm in radius [188].

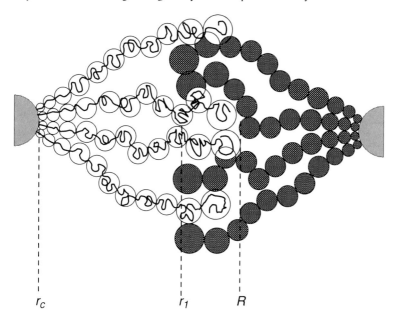

Fig. 9 Schematic representation of (sections of) multi-arm star polymers in semidilute solution in good solvent. The three different length scales, the radius of the star R, the coat r_1, hard core r_c are indicated. The *open circles* denote the correlation length (ξ_c, blob size)

In the semidilute regime, the stars are densely packed on the R-scale (Fig. 9) and some interpenetration takes place. Each star consists of a core with monomer density $\phi(r) = 1$ and a coat of radius r_1 where the arms are stretched (due to geometrical reasons) and $\phi(r) \sim (r_c/r)^{(3-1/\nu)}$ decreases with distance reaching the bulk (mean) solution concentration ϕ at $r \sim r_1$ (limit of arm interpenetration in the semidilute solution) [43]. Mass conservation implies that $\phi = \phi_1(1 - c\upsilon_1) + c\upsilon_{co}$, where $\phi_1 = (r_1/r)^{(3-1/\nu)}$, $c = \phi/(N_a f \upsilon)$ is the number concentration of stars, υ is the monomer volume, $\upsilon_1 = (4\pi/3)\, r_1^3$ is the coat volume, and υ_{co} denotes the volume occupied by the total number of monomers in the coat region. The term $(1 - c\upsilon_1)$ represents the fraction of the total volume with uniform monomer concentration ϕ_1, whereas the term $c\upsilon_{co}$ reflects the coat contribution to the mean volume fraction. The crossover ϕ^* to the semidilute regime follows from the above relation for $r_1 = R$ (star radius) yielding $\phi^* = 3\nu^{3\nu}(\upsilon_c/N_a f \upsilon)^{3\nu-1} \sim f^{2/5} N_a^{-4/5}$ where $\upsilon_c = (4\pi/3)\, r_c^3 \sim f^{3/2}$ is the core volume. It follows that for similar molar masses, multiarm stars display significantly lower ϕ^* than linear chains in the same solvent.

The curvature-induced spatial inhomogeneous density distribution modifies the free energy per unit volume in the semidilute regime, $F = F_{id} + F_{int}$, which in turn alters the static and dynamic properties of the system. The free energy includes the ideal mixing free energy contribution, $F_{id} = c\ln\phi$ and the interaction term, $F_{int} = (1 - c\upsilon_1)F(\phi_1) + cF_{co}$ which consists of the excluded-volume monomer interactions and the arm stretching contributions. In the homogeneous ϕ_1 region, the interpenetrating arms resemble linear chains in the semidilute regime and hence, $F(\phi_1) = (k/b^3)\phi_1^{3\nu/(3\nu-1)}$ where the numerical constant reflects the chain rigidity. The term $F_{co} = (k'/b^3)\upsilon_c[\ln(r_1/r_c)]$ is the free energy of a coat region with $k' > k$ being a numerical factor accounting for the (stronger) stretching of the arms inside the coat region. The final result is [43]

$$F_{int} \approx F(\phi)\left\{1 + [3\nu(k'/k)/(3\nu-1)\ln(1/\phi) - 3\nu - 1](\upsilon_c/N_a f\upsilon)\phi^{-1/(3\nu-1)}\right\}, \quad (3)$$

with $F(\phi) = (k/b^3)\phi^{3\nu/(3\nu-1)}$ being the free energy of linear chains in semidilute solution at volume fraction ϕ. Note that the uniform concentration in the coat region is $\phi_1 \approx \phi[1 - (3\nu - 1)(\upsilon_c/N_a f\upsilon)\phi^{-1/(3\nu-1)}] < \phi$ and the correction term (due to the star architecture) is $\upsilon_c/N_a f\upsilon \sim f^{1/2}$. Similarly, the correction term in (3) is larger than F_{id} by a factor of $\upsilon_c \sim f^{3/2}$ and hence $F \approx F_{int}$. The computed osmotic modulus $K = \upsilon\phi^2(\partial^2 F/\partial\phi^2)$ of the star solutions is lower than the corresponding semidilute linear homopolymer solutions because of the lower contribution of the monomers in the coat region. This is another architectural effect that reduces the excluded-volume interactions of the star solution.

The exclusion of "foreign" arms (belonging to different stars) from the coat region of a given star in nondilute solutions is clearly displayed in the simulated monomer density profile of star polymer melts, as depicted in Fig. 2 [41]. Whereas at low functionalities there is considerable star interpenetration, for $f > 24$ the interpenetration becomes increasingly prohibited. This excluded volume effect causes a liquid-like ordering as observed in solutions above ϕ^* by small angle neutron scattering (SANS) [26, 190] or in melts by small angle X-ray scattering (SAXS) [176, 191]. The spacing of the structure, d, which is the most probable separation between star cores distance between the cores was found to scale as $d \sim (fN_a)^{1/3}$ in agreement with the simulation of the pair correlation functions of the star centers [41].

The static scattering intensity $I(q \to 0)/c$ as a function of concentration c is a sensitive index of the (static) crossover concentration c^*, as shown in Fig. 10 [43, 192].

The values of c^*, which were obtained from the peak in the experimental $I(q=0)/c$ vs multi-arm star concentration c, support the scaling prediction for the stars as shown in the inset of Fig. 10. Above c^*, the decrease of $I(q=0)/c$ with concentration is reminiscent of the overlapping linear polymer physical networks [10, 193] with a slope close to $(c/c^*)^{-1/(3\nu-1)} \sim c^{-1.3}$. The star architecture seems responsible for the steeper drop of $I(q=0)/c \sim c/K$ due to the stronger, relative to homopolymers, concentration dependence of the osmotic modulus, i.e., $I(q=0)/c \sim c^{-1/(3\nu-1)}[1 + (3\nu-1)/(3\nu)^{3\nu/(3\nu-1)}(k'/k)(c/c^*)^{-1/(3\nu-1)}]$. The suppressed thermal concentration fluctuations in semidilute polymer solutions relax via the cooperative diffusion [10, 193] with the diffusion coefficient, $D_c = s\phi(\partial\mu/\partial\phi)$ where s is the sedimentation coefficient $s = \phi\xi^2$ and $\mu = \upsilon(\partial F/\partial\phi)$ is the chemical potential. For semidilute star solutions, s is locally non-uniform due the different $\xi(\phi)$, yielding $D_c = D_0 c^{\nu/(3\nu-1)}[1 - (3\nu-1)/(3\nu)^{3\nu/(3\nu-1)}(k'/k - 1)(c/c^*)^{-1/(3\nu-1)}]$ [43].

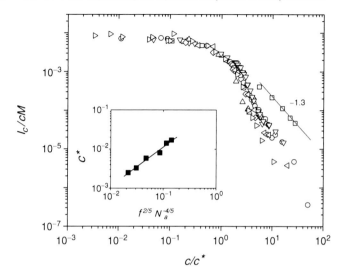

Fig. 10 Normalized static scattering intensity vs reduced concentration c/c^* for different polybutadiene multiarm stars with f arms (nominal values 64–128) and arm molar masses M_a (nominal values in the range 7–80 kg mol^{-1}: (*circle*) 6407 (i.e., $f = 64$, $M_a = 7,000$ g mol^{-1}), (*triangle*) 12807, (*inverted triangle*) 12814, (*diamond*) 12828, (*left pointing triangle*) 12856, (*right pointing triangle*) 1280, and a linear polybutadiene of 165 kg mol^{-1} (*open square*) [42, 43, 189, 192]. The slope of -1.3 represents the scaling for the linear homopolymer. The *inset* depicts the conformity of c^* to the scaling prediction ($c^* \sim f^{2/5} N_a^{-4/5}$)

The star architecture effects are more important for $I(q \to 0)$ than for D_c because the ratio of the corresponding correction terms, $k'/(k'-k)$, is large when $k' \sim k$. Nevertheless, the experimental $D_c(c/c^*)$ reveals a stronger speed-up of D_c with concentration in multiarm stars compared to the semidilute linear polymer solutions. The hard core contribution to the osmotic pressure is essentially hidden in the inhomogeneous density profile and the thermodynamic properties of the star solutions are primarily determined by their polymeric character.

The particle character of the star is more pronounced in the intermediate scattering function $C(q,t)$, which is more complex than the polymeric counterpart. Figure 11 depicts the $C(q,t)$ at $q = 0.035$ nm^{-1} ($qR_g = 1.5$) for the largest polybutadiene star ($R_g = 42$ nm) in cyclohexane solution with $c = 0.016$ g mL^{-1} ($c/c^* = 6.8$) at 20°C [43, 189]. The dynamics extend over 4 decades in time and the underlying processes are best resolved by the inverse Laplace transformation (ILT) which yields a continuum spectrum of relaxation times $L(\ln(\tau))$ [193]. At this concentration, well above c^*, the fast diffusive process is identified with the cooperative diffusion based on the increasing D_c (and decreasing intensity) associated with first peak of $L(\ln(\tau))$. The third slowest process represents the relaxation of thermal number density fluctuations and is associated with the self-diffusion of the stars as verified by independent PFG–NMR measurements of the true self-diffusion coefficient D_s in the same solutions [43, 189]. Self-diffusion is seen by light scattering due to inherent functionality polydispersity of the stars that causes an incoherent

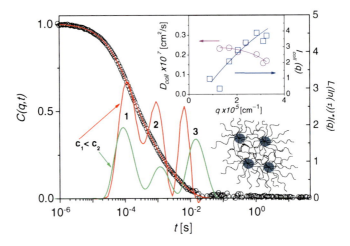

Fig. 11 The dynamic structure factor $C(q,t)$ of polybutadiene star 12880 (nominally $f = 128$, $M_a = 80 \, \text{kg mol}^{-1}$) in cyclohexane at $c_1 = 0.016 \, \text{g mL}^{-1}$ and $q = 0.035 \, \text{nm}^{-1}$, along with the fit (*solid line*) from the ILT analysis. The corresponding relaxation distribution function $L(\ln(\tau))$ (shown here for c_1 and $c_2 = 0.023 \, \text{g mL}^{-1}$) embraces the cooperative diffusion (**1**), the collective apparent diffusion (**2**), and the self-diffusion (**3**). The slowing-down of the middle structural mode (**2**) and the increase of its intensity with q are shown in the *upper inset* whereas the lower cartoon illustrates the liquid-like ordering [43, 189]. The core regions are drawn out of scale (larger) for clarity

scattering in analogy to size polydispersity in colloidal dispersions [91, 194] and composition polydispersity in diblock copolymers [195]. In the present case, the finite functionality polydispersity (ε) of the stars can cause density fluctuations due to the exchange of two effective populations, namely small $f(1-\varepsilon)$ and large $f(1+\varepsilon)$ stars. In the limit of small polydispersity $\varepsilon \ll 1$, there is no phase separation and the process is driven by the ideal gas entropy of mixing of the two types of stars and hence the intensity and decay rate of the slow process are [43] $I_p \sim (f^2 \varepsilon^2 / N_a) \phi^{[(3\nu-3)/(3\nu-1)]} \sim \phi^{-1/6}$ and $\Gamma_p \approx D_s q^2$, respectively. Note that, in contrast to cooperative diffusion, $D_s(c)$ decreases with increasing concentration following the viscosity of the system [42, 43].

The intermediate mode is associated with the correlations between the star positions leading to liquid-like ordering. It describes the thermal relaxation of the structure exhibiting distinct characteristics (upper inset to Fig. 11); its intensity I_{coll} increases with q and the collective diffusion coefficient D_{coll} (expressed as Γ_{coll}/q^2, Γ_{coll} being corresponding relaxation rate) decreases with q since the fluctuations with wavelength of the order of the inter-star distance become more probable and hence long-lived [195]. This is reflected on the colloid-like behavior observed in dense suspensions of hard spheres [196] and is absent in homopolymer solutions. Above c^*, the star positions are correlated in space and the cores immersed in a sea of entangled arms (Fig. 9) undergo a collective diffusion controlled by both hydrodynamic $H(q)$ and thermodynamic $S(q)$ interactions; for a hard-sphere colloidal suspension, $D_{\text{coll}} = D_0 H(q)/S(q)$ [196]. The collective structural rearrangements, which reflect the colloidal nature of the high-functionality stars, are also manifested in the viscoelastic response. As the functionality increases, in addition to the arm retraction (a mechanism that governs the terminal, i.e., flow, behavior of a star polymer), an additional slow relaxation process is observed and assigned to the structural rearrangements of the liquid like ordered stars [197]. The shift of the terminal flow to lower frequencies with increasing functionality is a consequence of the increasing colloidal contribution due to the star topology. A controlled manipulation of the latter at the synthesis level allows tailoring of the mechanical response [42].

There is a host of other intriguing phenomena associated with the structure and dynamics of stars, which we only list here. The inhomogeneous monomer density distribution in Fig. 2 is responsible for temperature and/or solvency variation in analogy to polymer brushes attached on a flat solid surface [198]. In fact, multiarm star solutions display a reversible thermoresponsive vitrification (see also Sect. 5) which, in contrast to polymer solutions, occurs upon heating rather than on cooling [199]. Another effect is the organization of multiarm stars in filaments induced by weak laser light due to action of electrostrictive forces [200]. This effect was recently attributed [201] to local concentration fluctuations which provide localized-intensity dependent refractive index variations. Hence, the structure factor specific to the particular material plays a crucial role in the pattern formation.

4.3 Dynamics of Core-Corona Systems: Block Copolymer Micelles and Grafted Particles

We argue that the above features of star dynamics are generic for soft systems of the core-shell type for which stars serve as prototype. Support for this comes from the dynamic light scattering (DLS) investigation of large block copolymer micelles, where all three relaxation modes, i.e., cooperative, structural and self-diffusion are observed [188]. In particular, the star model discussed above applies to core-shell particles with a small spherical core relative to the chain (shell) dimensions. For a surface number density $\sigma^* = f/(4\pi r_c^2)$ the polymer layer thickness under good solvent conditions is $L \sim N^\nu \sigma^{*(1-\nu)/2} r_c^{(1-\nu)} = N^{3/5} \sigma^{*1/5} r_c^{2/5}$ where r_c stands for the size of the micellar core [202–204]. For comparison, the size of a swollen brush layer on a flat interface scales as $L \sim \upsilon^{1/3} N \sigma^{*1/3}$ with υ being the excluded-volume parameter. This is a clear effect of the interface curvature which changes the configurational statistics of the anchored chains and hence the layer structure and interactions [203]. For intermediate curvatures, the expression $R^{5/3} - R_c^{5/3} = 8R_c^{2/3} N (4\pi \sigma *)^{1/3} / (3\upsilon 4^{1/\nu})$ was proposed [204] for the overall particle radius R in good solvents ($\nu \sim 3/5$). However, for an arbitrary r_c/L ratio, self-consistent-field-theory (SCFT) calculations are more reliable than these scaling predictions [1195].

The form and the structure factor of the PEP–PEO star-like block-copolymer micelles in aqueous solutions were thoroughly investigated by SANS [53–55, 205].

The experimental form factor $P(q)$ shown in Fig. 12a can be expressed as $P(q) = [b_c F_c(q,r_c) + b_s F_s(q,r_c,R)]^2$, where b_c, b_s are the contrast factors for the core (c) and shell (s) with core radius r_c and overall micelle radius R, whereas $F_c(q)$, $F_s(q)$ are the scattering amplitudes of the core and shell, respectively. Under core contrast conditions ($b_s \sim 0$), the expected first minimum for the compact sphere $F_c(q)$ at high q values falls outside the q-range, whereas under shell contrast conditions the power-law behavior arising from blob (swollen PEO shell) scattering is observed. Hence, the dual colloid-polymer character of the particle is clearly reflected in Fig. 12a.

The star-like micellar interactions lead to liquid-like ordering at volume fraction $\phi \approx \phi^*$ as seen in the $S(q)$ of Fig. 12b ($\phi = 0.072$) [55]. Except for the very low-q region, all main features of $S(q)$, i.e., peak positions and heights, are identical in multiarm stars of the same f. Above ϕ^*, the transition to a bcc polycrystalline structure is evidenced by the Bragg peaks, the height (exceeding 2.8) of the first peak of $S(q)$ and the 2D SANS pattern [54].

In contrast to colloidal stars, the dynamic structure $C(q,t)$ of other core-corona systems is essentially unexplored because of the lack of well-characterized systems amenable to DLS experiments. The prerequisites for such studies include sufficiently large particles (for light scattering), stability (in many micelles there is exchange kinetics of constituent chains or non-equilibrium trapped core configurations), large variety of grafting densities and grafted particles uncontaminated from remaining free polymer chains. The few systems investigated so far are diblock copolymer micelles with either glassy core [188] or fixed core by crosslinking [206] and silica particles chemically grafted with poly(dimethyl siloxane) chains [101].

The low-q range of $C(q,t)$ reveals the cooperative diffusion and self-diffusion of the core-shell micelles. Whereas the features of the former (as discussed above) appear to be generic, the latter is very much influenced by the internal structure of the system. In particular, stable block copolymer micelles with rubbery (chemically crosslinked) cores were shown to be responsive depending on the solvent used. For a highly selective solvent (good for the grafted hairs and poor for the core), a slowing-down of the self-diffusion with the (effective) volume fraction was observed (Fig. 13) akin to that reported for other colloidal particles of varying interactions (Fig. 8). On the other hand, it was found that a mutually good solvent for core and corona yielded core swelling, and the core would shrink in turn at high volume fractions as a result of the enhanced osmotic pressure and the associated expulsion of solvent from the core. This shrinking, which we call osmotic deswelling, had an impact on the reduced slowing-down of the self-diffusion coefficient (Fig. 13) with volume fraction and was confirmed by Brownian dynamic simulations [206]. Remarkably, the inverse relative (to the solvent) zero-shear viscosity is identical to the reduced (to that at $c \to 0$) self-diffusion when plotted against the effective volume fraction [206]. Note that similar osmotic deswelling behavior (with consequences on tuning the rheology) has been reported recently for polyelectrolyte microgel particles [156, 207].

The high-q region in the vicinity of the peak position q^* of $S(q)$ is dominated by the collective diffusion with a weaker contribution of the arm relaxation. However, in sharp contrast to the respective situation with hard sphere colloidal suspensions

From Polymers to Colloids: Engineering the Dynamic Properties of Hairy Particles

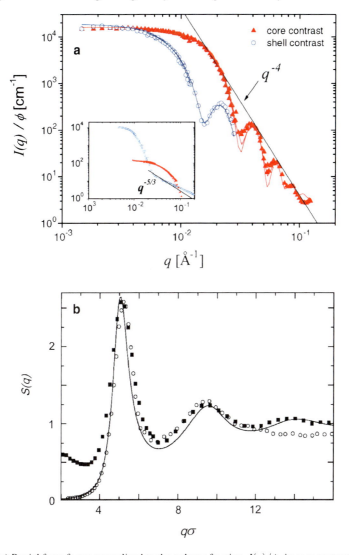

Fig. 12 (a) Partial form factor normalized to the volume fraction, $I(q)/\phi$, in core contrast (*closed symbols*) and shell contrast (*open symbols*) for h-PEP-dh-PEO in d-aqueous solutions. Data are shown for two different micellar systems. Main plot: $M_{PEP} = 4.2\,\mathrm{kg\,mol^{-1}}$, $M_{PEO} = 5.8\,\mathrm{kg\,mol^{-1}}$, $f = 1{,}600$ and $\phi = 5 \times 10^{-4}$. *Solid lines* represent predictions and the limiting slope q^{-4} [54, 55]. *Inset*: $M_{PEP} = 1.2\,\mathrm{kg\,mol^{-1}}$, $M_{PEO} = 21.5\,\mathrm{kg\,mol^{-1}}$, $f = 130$ and $\phi = 10^{-3}$ [54, 55]. The $I \sim q^{-5/3}$ law in the high-q region is typical of polymer chains in good solvent [173] and arises from the swelling of PEO blobs in the shell, i.e., blob scattering [205]. (**b**) Structure factor $S(q)$ of the same PEP–PEO suspensions as those in the *inset* of (**a**) but in a water/DMF mixture with fraction of DMF $\chi_{DMF} = 0.5$ (yielding $f = 63$), compared to a respective multiarm polybutadiene star ($f = 64$) in d-tolune, at $\phi/\phi^* \approx 1$. The *solid line* is the theoretical prediction of $S(q)$ based on the Likos potential, (1). Taken from [55]

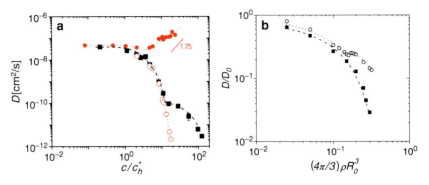

Fig. 13 (a) Dependence of the diffusion coefficients of a poly (2-cinnamoylethyl acrylate)/poly (butyl methacrylate) core/shell diblock copolymer micelle (with crosslinked core, $f = 39$ arms and hydrodynamic radius in mutually good solvent toluene $R_h = 48$ nm) on the effective hydrodynamic volume fraction c/c_h^* (based on R_h). Data are shown for the cooperative (*red solid symbols*) and self-diffusion coefficients in two different solvents (toluene (*black solid symbols*) and cyclohexane (*red open symbols*), the latter being poor for the core). Lines are drawn to guide the eye. The slope of 1.75 is the cooperative diffusion scaling prediction for linear polymers [10]. (b) Brownian dynamics simulations of the volume fraction dependence of the self-diffusion coefficient of a core-shell particle (normalized to that at infinite dilution), depicted with *solid squares*. The case of osmotic deswelling is accounted for by reducing the core size with volume fraction (*open circles*). The data confirm the trend of apparent self-diffusion speed-up observed experimentally [206]. Lines are drawn to guide the eye

[196], D_{coll} does not exhibit the thermodynamic slowing-down at q^*. This finding, observed with block copolymer micelles [188], was attributed to the softness of these particles (being the only difference from the hard-sphere particles of [196]) which apparently lead to peculiar, not yet fully understood, non-local $H(q)$ interactions. The fact that softness was the controlling parameter for this effect was demonstrated by reducing the softness of the particles (from the ultrasoft micelles of [188]) and observing the emergence of a weak slowing-down (weaker compared to hard spheres) at q^*. Well-characterized hybrid particles consisting of a silica core ($r_c = 250$ nm) grafted with poly(dimethyl siloxane) (PDMS) brush (thickness $L = 20$ nm) were recently synthesized and utilized for this purpose [101].

The short-time collective diffusion coefficient $D_{\text{coll}}(q)$ in concentrated suspension was measured by a combination of standard single and two-color dynamic light scattering in order to suppress the effect of turbidity (which was at moderate levels). The data revealed a weak slowing-down $D_{\text{coll}}(q)$ around the peak of $S(q)$ compared to hard sphere systems as seen in Fig. 14. These nearly-monodisperse particles with small corona can undergo crystallization by sedimentation, in accordance with recently reported results with PMMA-coated silica particles having $r_c = 65$ nm and L varying between 85 and 355 nm [204]. Despite these intriguing observations, however, and in view of the rich parameter space (volume fraction, solvency, r_c/L, grafting density) as well as synthetic challenges with grafted particles (Sect. 2.3), the elucidation of the dynamics of dense suspensions of core-shell particles is still in its infancy.

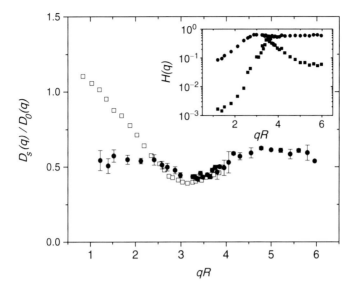

Fig. 14 Normalized D_{coll}/D_0 of PDMS coated silica suspension with $\phi = 0.3$ in a symmetric mixture of toluene and heptane (*solid circles*) along with hard sphere suspension (*open squares*) at similar volume fraction. The hydrodynamic interactions expressed in $H(q)$ for the two systems (*solid squares* for the hard sphere suspension) are shown in the inset [101]. This system is crystallized by sedimentation as seen in the photograph

The recent synthesis of model PMMA-grafted SiO$_2$ nanoparticles with the flexibility of tuning grafting density and r_c/L [112] provided a means to continue the investigations of $C(q,t)$ along the already discussed path of parameter space. Their dynamic response should display common and distinct features compared with the established equilibrium dynamics of hard sphere colloids. The similarities should include the three aforementioned diffusion coefficients which are, however, expected to be quantitatively different because of the significant alteration of the interaction potential. In addition, the curvature-dependent brush-like nature of the polymeric shell should be manifested in the osmotic pressure of the suspension and the associated dynamics of the total density fluctuations.

The dynamic structure factor $C(q,t)$ of two such PMMA-grafted SiO$_2$ particles (code DP150 with $N_{\text{PMMA}} = 150$, $f = 1{,}056$ and code DP760 with $N_{\text{PMMA}} = 770$, $f = 628$, both grafted on the same silica core with $r_c = 10$ nm) was very recently measured by PCS over a broad concentration range [205]. Three different mechanisms contributed to the relaxation of $\phi_q(t)$ for $q < q^*$, and the diffusion coefficients are plotted against volume fraction in Fig. 15. In the dilute regime ($\phi < 5 \cdot 10^{-3}$), $C(q,t)$ exhibited a single-exponential decay with self-diffusion coefficient D_0. Above this concentration, the onset of deviation from the single-exponential shape of $C(q,t)$ was detected and the normalized collective short-time diffusion coefficient $D(q \to 0)/D_0$ decreased with q (cf Fig. 14) and with ϕ (inset to Fig. 15) for both systems. For the present good-solvent conditions (for the PMMA chains) the decrease of $D(q \to 0)$ reflects the increase of the friction coefficient

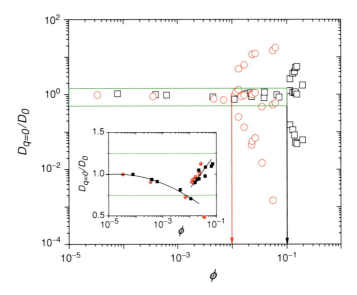

Fig. 15 Normalized diffusion coefficients $D(q \to 0)/D_0$ as functions of the volume fraction ϕ for PMMA-grafted silica particles DP 150 (*open square*) and DP 760 (*open circle*) in CCL$_4$ (good solvent for PMMA) at 20°C [208]. The different crossover concentrations in the dynamics are indicated by *solid* and *vertical arrows*. The inset shows in magnification the area bounded by the rectangle around a normalized diffusion coefficient value of 1 in the main plot, using the same colors and symbols (*solid*) for the two systems. Lines are drawn to guide the eye

with concentration. Above about $\phi = 0.01$, $D(q \to 0)$ reverses its concentration dependence following the sudden increase of the osmotic pressure, i.e., the strong decrease of the intensity occurring at about $\phi^* \sim 5 \times 10^{-3}$ and 0.01 for the large (DP760) and small (DP150) particles, respectively. This dominance of the thermodynamic contribution evolves from D_0 and is not the only change around ϕ^*. As seen in Fig. 15, two additional diffusive processes appear in the large DP760 at $\phi \geq \phi^{**} \approx 0.01$, whereas for the smaller DP150 this dynamic splitting occurs at much higher $\phi^{**} \sim 0.1$. The fastest diffusive process associated with relatively weak scattering intensity speeds-up with increasing concentration and hence it relates to the polymeric (PMMA) cooperative diffusion which is expectedly absent in hard sphere colloids but present in multiarm stars (Figs. 10 and 11). The slowest diffusion process that exhibits strong slowing-down with concentration is associated to the particle self-diffusion, as already discussed for the hard-sphere colloids, multiarm stars and micelles. The inversion of the concentration dependence of the $D(q \to 0)$ was also witnessed in the scattering intensity. According to Fig. 15 (see vertical arrows), the bifurcation of $D(q \to 0)$ into cooperative and self-diffusion of the core-shell particles occurs close to ϕ^* for DP760 but at a different $\phi^{**} \sim 0.1$ for the DP150 system. The presence of the fastest diffusion that speeds-up with concentration appears at about the bifurcation concentrations. Notably both dynamic and intensity data can be represented (not shown) in a master plot against ϕ^{**}. Whereas this behavior reflects the dual polymeric-colloidal nature of the particles response, a theoretical rationalization of the dynamic results of Fig. 15 remains a challenge.

4.4 Remarks on Crystallization

Well-characterized systems (hard or soft particles) are nearly monodisperse. This implies that at high volume fractions (typically 0.5 for hard sphere systems [13]) colloidal crystallization is observed. This has indeed been the case of hard spheres [13, 91] and a number of soft sphere systems such as grafted particles with relatively small brush length [204, 209], block copolymer micelles [67, 210–212], and microgels [213–215]. We note that a typical way to suppress crystallization and transit from a fluid to the glass state is by introducing size polydispersity into the system [91]. However, for virtually monodisperse star polymers crystallization remains a grand experimental challenge. It turns out that crystallization is indeed expected for the stars at $c \sim c^*$ as a result of the concentration-induced enhanced osmotic pressure which overbalances the elastic energy of the stretched arms [177]. At higher concentrations the crystal melts and the semidilute polymer solution behavior is recovered. Moreover, based on the potential of (1) [29] various crystalline phases were predicted (Fig. 16) [178]. However, there is only one reported direct experimental observation of star crystallization [216] for nominal values of $f = 128$, $N_a = 130$, after gentle pre-shearing [217] (shear can of course influence spatial organization of

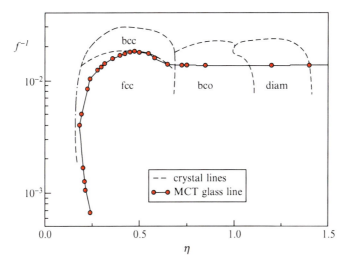

Fig. 16 Calculated star phase diagram (inverse functionality f^{-1} again packing fraction η) [178]. The different regimes of the formed crystal phases (bcc, fcc, bco, diam) are indicated by *dashed curves*. The calculated MCT glass line [224] is also shown (*red closed circles* and line through them)

colloidal systems in many ways [218–220]). There is also indication of approach to crystallization based on indirect experimental information from PFG–NMR [182].

It is thus evident that stars do not crystallize as "easily" as, say, micelles (which have a similar form). There are many possible reasons for this [221]: significant fluctuations of the outer blobs of the stars (as well of the overall star position) with amplitude exceeding the Linderman criterion limit, weak metastable state of the stars, and difference between stars and micelles in relation to the presence of a well-defined, solid core in the latter. Indeed, the difference in core size and shape is important and can explain why star-like micelles could more easily crystallize [54, 58], whereas it also affects particles mechanical response and deformability [222]. We note that a hybrid class of block copolymer stars, i.e., multiarm stars with diblock arms [184] crystallized in selective solvents for the outer blocks (conditions for stable micelle formation) [223]. With regard to the metastability issue, it appears that stars at high packing fractions get trapped into glassy states, and the original phase diagram of Fig. 16 was complemented with the glass lines calculated from mode coupling theory (MCT) analysis [224]. Indeed, this glassy state has been observed in stars [22, 225] and star-like micelles [54, 58]. But what would be the fate of a star glass? Recently, molecular dynamics (MD) simulations demonstrated that long annealing of a (thermally induced) star glass leads to a crystal [226]. Recent experimental data appear to support this important finding [227]. In fact, Fig. 17 depicts the radial distribution function of a multiarm ($f = 128$) star in the glassy state (a) and the eventual crystalline state (b) after long aging.

5 Vitrification

It is evident from the above discussion that, upon increasing their volume fraction, colloidal systems tend to reach a frozen state, either glassy or crystalline. The glassy state is a non-equilibrium state, not well understood yet. In fact, it is known that the glass transition represents one of the outstanding challenges in condensed matter physics [18, 228–232]. Whereas the phenomenology of hard spheres colloidal glasses is well advanced [18] and various versions of the MCT serve as a good framework to describe hard colloidal glasses [18, 231, 233–236], little attention has been paid in the area of the soft colloids. To this end, the knowledge acquired with hard sphere systems [18] serves as a guide for any further development. In particular, hard-sphere-like colloidal glasses are characterized by repulsive interactions, non-ergodicity and nearly time-independent MSD [237], enhanced, nearly frequency-independent storage modulus [238], spatial and dynamic heterogeneity [239, 240], and very slow dynamics and aging [241, 242]. We attempt a discussion in general terms of the glass transition of soft colloids, using the same qualitative concept, that of caging [18].

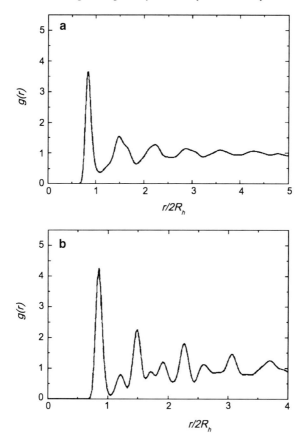

Fig. 17 Radial distribution function of a multiarm star (with nominal $f = 128$ and $M_a = 56\,\mathrm{kg\,mol^{-1}}$) star in n-decane at $50\,^\circ\mathrm{C}$ ($R_h = 42.4\,\mathrm{nm}$, $c/c_h^* = 1.17$) in the glassy state (**a**) and the crystalline state (**b**) after long aging [226]

5.1 Soft Colloids in the Glassy State

Dense systems are intimately related to packing (geometrical) constraints by their neighbors. Starting from the limiting cases of linear flexible polymers and hard sphere colloids (cf Fig. 1), we sketch the high volume fraction regime attempting at a generic qualitative description [233]. For long chain polymers (above the entanglement limit) the topological constraints of the neighbors (entanglements) in the concentrated regime (or in the melt) confine a test chain within a so-called tube [11, 243] as illustrated in Fig. 18a. The dynamics of this dense system is equivalent to the escape of the chain from its tube, the so-called reptation [11, 244]. Due to the very large number of chains, this is treated within the mean-field approach. On the other hand, when colloidal hard spheres at $\phi \approx 0.58$ form a glass, a given particle is constrained by a small number (not exceeding 12) of neighbors which restrict and eventually arrest its macroscopic motion (on the scale of its size), forming

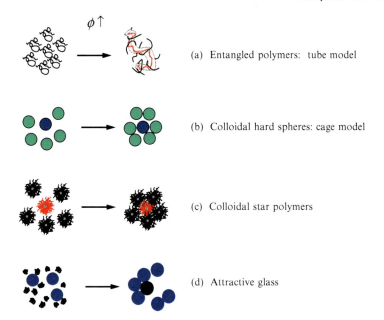

Fig. 18 Schematic representation of crowded soft systems: (**a**) entangled polymers, (**b**) repulsive colloidal hard spheres, (**c**) colloidal star polymers, and (**d**) attractive hard spheres. The former are described by the tube model for entanglements, whereas the latter three by the general cage model for colloidal glasses

an effective cage [91, 245, 246]. This is illustrated in Fig. 18b. Local motions of the particle within its cage (resembling rattling) are of course possible (the so-called β-relaxation). The escape of the particle from its cage (the so-called α-relaxation) is the determining factor for the flow of the glass. In this case a mean-field description is not possible, and different approaches have been proposed, the mode coupling theory being the most successful to date [18]. Thus, topological constraints at the scale of the whole (end-to-end distance) system are discussed in the context of entanglements in polymers and glass transition in colloids [233]. In the intermediate case of soft colloids, one can consider a modified cage picture. The terminal motion of the dense star suspension signaling macroscopic flow is that of the center of mass of the star [42, 197, 225]. The star is constrained within the cage of its neighbors; however, at the same time, due to their inherent structure (Fig. 2) the stars interpenetrate, forming effectively some engagements of the outer blobs (Fig. 9). The star has to disengage from its neighbors (a process absent in hard spheres) and move its center of mass (cage escape) for macroscopic flow to occur [225]. This is illustrated in Fig. 18c, and this interaction of the outer blobs of the stars has consequences on their properties. For completeness, we discuss the cage concept as applied to attractive colloidal glasses, i.e., the re-entrant glassy state induced in colloid/non-absorbing polymer mixtures at large polymer concentration (depletion strength) [16, 17, 247]. In that case, illustrated in Fig. 18d, a caged particle is bonded to some neighbors. This bears analogies to the star glass cage, where the interstar engagement can well be considered as a type of transient bonding. Note, however, that despite the interesting qualitative similarities, the physics of the two systems (attractive glasses from hard sphere colloid/polymer mixtures and star glasses) is not quite the same.

5.2 Signatures of Transitions and Rheology Manipulation

Here we focus on the main signatures of colloidal glassy states as extracted from studies of dynamics. A dynamically arrested state (no matter whether it is a glass or a gel) is characterized by non-ergodicity [91, 237, 247]. In Fig. 19a we depict a classic set of experimental intermediate scattering functions from DLS measurements with colloidal hard spheres over a range of volume fractions from the dilute to the glassy regime, along with the respective MCT lines [237]. Vitrification is accompanied by a non-relaxing (to zero) $C(q,t)$, which is a clear sign of non-ergodicity. We also show in Fig. 19b the experimental $C(q,t)$ of colloidal stars [248], exhibiting a qualitatively identical behavior.

In accordance with the $C(q,t)$, the MSD of a glassy suspension exhibits a plateau in its time-dependence [18, 249, 250]. We show in Fig. 20 a specific example from the calculated MSD (from MD simulations) of a colloidal star suspension, where the necessary increase of the volume fraction to enter into the vitrified regime is mediated by temperature [251] (the star size, and thus volume fraction, increases with temperature [25, 26]).

A very rich rheological response characterizes colloidal glasses and can serve as a means of identifying and/or distinguishing their different kinds [22, 252–255].

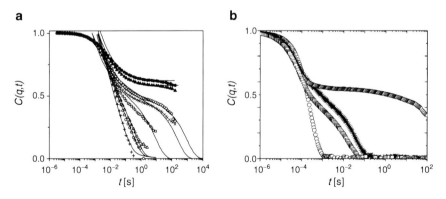

Fig. 19 (a) Intermediate scattering function $C(q,t)$ of hard sphere PMMA ($R_h = 205$ nm) suspension (in cis-decalin) at $qR = 2.68$, from the dilute to the glassy regimes (volume fraction: (+) 0.494, (*open circle*) 0.528, (*open triangle*) 0.535, (*open square*) 0.558, (×) 0.567, (*diamond*) 0.574, (*solid triangle*) 0.581, (*solid circle*) 0.587). *Solid lines* represent MCT fits. Taken from [237]. (b) Respective response of a colloidal polybutadiene star suspension (in good solvent cyclohexane, at 20°C) at different values of the effective hydrodynamic volume fraction c/c_h^*: (*open circle*) 0.06, (*open square*) 0.72, (*asterisk*) 1.02, (*open triangle*) 1.48. Taken from [248]

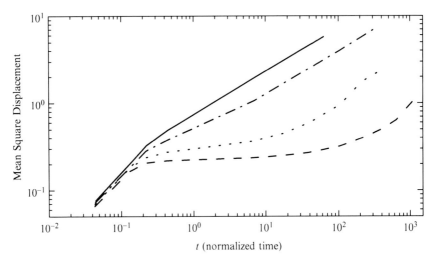

Fig. 20 Calculated star (nominal $f = 128$, $M_a = 56 \text{kg mol}^{-1}$) mean-square displacement as a function of time for different temperatures: $T = 35°\text{C}$ (*solid line*) with estimated effective hydrodynamic volume fraction $c/c_h^* \approx 0.895$, 45°C (*dashed-dotted line*) with $c/c_h^* \approx 1.033$, 50°C (*dotted line*) $c/c_h^* \approx 1.168$, and 55°C (*dashed line*) $c/c_h^* \approx 1.332$. The observed emergence of plateau (for 50 and 55°C) represents evidence of caging effects with increasing temperature. Data are taken from [251]

Here we limit the discussion to the linear response. Small amplitude oscillatory shear measurements (or dynamic frequency sweeps) provide information on the dynamics of these systems. We show such data for a hard sphere silica suspension (in ethylene glycol) at various volume fractions [238] in Fig. 21a, b and for a glassy microgel particle (PNIPAM-coated PS latex) aqueous suspension [256] in Fig. 21c. We note that, as the volume fraction increases, the character of the suspension in Fig. 21a, b expectedly changes from Newtonian liquid to viscoelastic liquid and eventually to a viscoelastic solid (or glass). The latter exhibits a nearly frequency-independent storage modulus (G', Fig. 21a) and a frequency-dependent and much lower in value loss modulus (G'', Fig. 21b). Mode-coupling theory can successfully describe the linear viscoelastic spectrum. Based on this, the minimum in G'' is assigned to the β-relaxation time, i.e., the in-cage local motion of the particle, whereas the low-frequency crossover of G' and G'' (usually not reached experimentally) to the α-relaxation, i.e., the macroscopic flow. Not surprisingly, the same general features hold for soft colloids, as shown in Fig. 21c. There are of course differences as discussed in the framework of the cage picture, which are reflected in the values of moduli, extent of G' plateau and whether it is a true plateau or a weak power law, and the low-frequency response. Recently, an effort was made to extend the rheological data to longer times. This is done with colloidal star polymers [225]. As seen from the extracted frequency-dependent viscoelastic moduli in Fig. 22a, the near plateau G' and minimum in G'' features of a colloidal glass are preserved. In addition however, there is a low-frequency G'' peak which is attributed

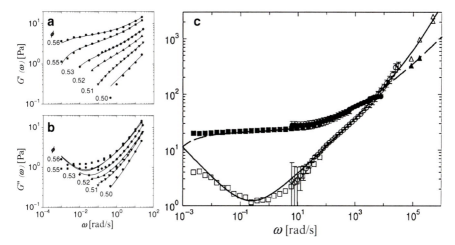

Fig. 21 (**a, b**) The frequency dependencies of the storage G' (**a**) and loss G'' (**b**) moduli for different volume fractions of uncoated silica hard spheres ($R_h = 210$ nm) dispersed in ethylene glycol [238]. The *solid lines* represent MCT predictions. (**c**) Respective data (G': *solid square*; G'': *open square*) for an aqueous glassy microgel suspension (PNIPAM-coated PS latex particles, overall radius 105.3 nm and effective volume fraction 0.585 at 10°C), along with the MCT lines [256]. The minimum of G'' marks the inverse β-relaxation time

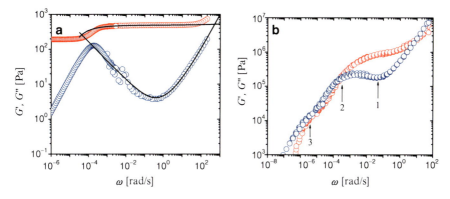

Fig. 22 (**a**) Frequency-dependent linear viscoelastic moduli (G': *red circles*; G'': *blue squares*) of a colloidal star (with nominal $f = 128$, $M_a = 80$ kg mol^{-1}) suspension in the glassy state in squalene (at $c/c_h^* = 2.1$) at 25°C, along with the MCT fit (*solid lines*) [225]. (**b**) Respective data for a star polybutadiene melt (with nominal $f = 128$, $M_a = 14$ kg mol^{-1}) at reference temperature $T_{\text{ref}} = -94$°C [176, 197], depicting characteristic relaxation times as indicated: (1) rattling time (of outer parts of arms); (2) arm disengagement time; (3) time for star overall motion (center of mass)

to the star arm disengagement (cf Fig. 18), but the macroscopic flow occurs at even lower frequencies. This intermediate (peak) arm relaxation is also observed in star melts (Fig. 22b) where excluded volume interactions on the scale of the star size dictate the system's response [197] and is characteristic of the softness of these systems (imparted by the arms). It can also be considered as a kind of debonding, and the arm viscoelasticity clearly plays a role in detecting it. In the case of attractive glasses such a mode has not been clearly distinguished in the linear viscoelastic spectrum [257].

Remarkably, MCT describes well the key glassy features around the β-relaxation for both G' and G'' without essentially any adjustable parameters: the plateau G'_p and minimum G''_β values are taken from the experiments, whereas the additional mode coupling parameters (usually referred to as a', b', and B) assume constant values for all test systems [225, 256]. It is also worth noting that MCT describes the pre-transitional approach to the glass for hard spheres [238], and it was successfully applied well within the glassy state for ultrasoft colloids [225].

These results confirm the high sensitivity of rheology to transitions, and at the same time challenge us to find ways to tailor the rheology of soft colloids. This can have profound consequences in a wide range of technologies where glassy materials with substantial thermodynamic and kinetic stability are needed (e.g., [258]). In general, one can in many ways enhance the viscoelastic properties and create a stronger glass or do the opposite and fluidize a glass. The former can be easily performed by adding material and the latter by applying load and causing yielding of the glass [158, 160, 186, 257, 259–262]. Here, we do not discuss these possibilities, but instead briefly mention the not-so-obvious effects of thermodynamic forces in achieving vitrification and melting.

When soft colloids (such as stars, block copolymer micelles, or particles sterically stabilized with grafted chains) are suspended in solvents of intermediate quality at high concentrations, an increase of temperature leads to an increase of their effective volume fraction, which in turn can yield vitrification [26, 190, 199]. A representative example is depicted in Fig. 23a for a star with nominal $f = 128$ and

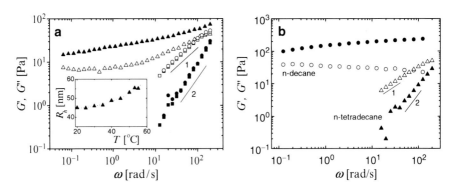

Fig. 23 (a) Frequency-dependent linear viscoelastic moduli (G': *closed symbols*; G'': *open symbols*) of a colloidal star with nominal values $f = 128$ arms and $M_a = 80$ kgmol^{-1} at a concentration 5 wt% in *n*-tetradecane and different temperatures (*circles*: 40°C, *squares*: 50°C, *triangles*: 55°C). A liquid-to-solid transition is marked between 50 and 55°C. Lines with slopes 1 and 2 indicate terminal behavior of G″ and G′, respectively. *Inset*: The temperature dependence of the hydrodynamic radius R_h of the same star, indicating swelling. (b) Respective moduli for the same system at 40°C in two different solvents, *n*-decane (*circles*, solid-like behavior) and *n*-tetracane (*triangles*, liquid-like behavior) [26]

$N_a = 1,480$ in n-tetradecane, where the rheology changes from Newtonian liquid to a glass with increasing temperature. This is due to the swelling of the star with temperature (demonstrated in the inset of Fig. 23a), which reflects a volume fraction increase. Such an effect was also explained using a temperature-dependent [25] star potential and performing MD simulations [251]. The MSD vs time behavior exhibits a long-time plateau at high temperatures (cf Fig. 20). This opens the route for rheology manipulation via selection of different solvents, as the varying quality imparts different size and thus volume fraction. For example, Fig. 23b shows the same star at the same temperature in two different solvents. Whereas in n-tetradecane the suspension remains in the fluid state, in n-decane (better quality) it is already vitrified.

Another example of rheology manipulation concerns mixtures involving soft spheres. Here the starting point is the knowledge from colloid-polymer mixtures and in particular the depletion effects of non-absorbing polymers [15, 17, 263, 264] and the recent discovery of re-entrant attractive glasses [16, 247]. The question of interest is what happens when softness comes into play. Mixtures involving model soft spheres and in particular star polymers were recently investigated in detail [20, 248, 265, 266]. In one type of mixtures the additive was a small (compared to the star) linear homopolymer with the same chemistry. The other type involved small stars. Whereas the findings have been recently summarized [22] we only mention here the key results of relevance to rheology tailoring. With the original big star suspension in the glassy state, it was found that addition of a small additive, linear or star, affected the viscoelastic properties a great deal via depletion mechanism. By tuning the strength (additive concentration) and range (size ratio) of depletion, the rheology changed at wish: from glass to fluid to re-entrant glass. A host of different glassy states were discovered which are still under investigation [267], and which are not seen in mixtures involving hard spheres [268]. Of particular interest is the recently discovered asymmetric caging of binary mixtures with small size difference and large fraction of added small component, akin to sheared hard sphere glasses [267]. The softness of interactions appears responsible for these phenomena, as it has a distinct signature on the depletion interactions [19, 248]. For example, the range of size ratios is much larger compared to colloid/polymer mixtures, and more types of multiple glassy states are observed. We show here two characteristic examples of depletion-induced rheology alteration, for a star/linear polymer (Fig. 24a) and a star/star mixture (Fig. 24b).

In closing this section we note that aging is an inherent problem in colloidal glasses, hard and soft alike. For soft colloids there is only fragmental evidence on this important parameter [269, 270], suggesting the scaling of viscoelastic properties with the so-called waiting (i.e., aging) time, and more work is currently underway. Related to aging and shear rejuvenation (i.e., shear-induced melting or rheological homogenization) [157, 225] is the problem of thixotropy in these systems [271]. Despite the ample rheological literature on the subject, systematic thixotropy investigations with well-defined protocols and well-behaving systems (not necessarily model systems, i.e., systems of unknown interaction potential) began only recently [272, 273].

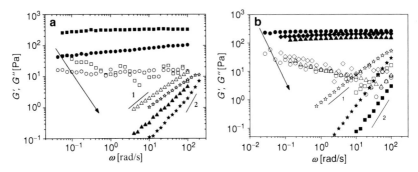

Fig. 24 (a) Star/linear polymer mixtures: effects of adding small homopolymer chains in melting a big star colloidal glass. Dynamic frequency sweeps of a polybutadiene star glass (with nominal $f = 270$ and $M_a = 43\,\mathrm{kg\,mol^{-1}}$, $c = 2.5\,\mathrm{wt\%}$ in good solvent toluene, yielding an effective hydrodynamic volume fraction of 1.4) at 10°C and several mixtures with linear homopolymer of molecular mass $22.6\,\mathrm{kg\,mol^{-1}}$ and different concentrations, C_{lin}: (*filled circles, open circles*) 0.6 wt%, (*filled triangles, open triangles*) 1.53 wt%, (*filled stars, open stars*) 2.51 wt% (*solid symbols*: G'; *open symbols*: G''). [19, 265]. *Solid lines* with slopes 1 and 2 represent the terminal scaling for G'' and G', respectively. *Arrow* indicates the effect of added linear chains in inducing glass melting. (b) Respective rheological data of asymmetric binary star mixtures in toluene at 20°C [20, 266]. G' (*bold symbols*) and loss G'' (*open symbols*) moduli as function of frequency for different mixtures consisting of a big star 12880 at $c_{12880} = 2.55\,\mathrm{wt\%}$ in the glassy state and a small star 3210 (nominally $f = 32$ and $M_a = 10\,\mathrm{kg\,mol^{-1}}$) at different concentrations: c_{3210}: 0 wt% (*circles*), 0.066 wt% (*diamonds*), 0.21 wt% (*triangles*), 0.50 wt% (*stars*), 0.82 wt% (*squares*). It is evident that adding small stars to a big stars glass leads to fluidization (with increasing c_{3210}), as indicated by the *arrow*

6 Hybrid Systems and Other Emerging Applications

Extensive research in the past decade has focused on the structure-property relations of bulk polymer nanocomposites, motivated by the perspectives of exploiting the particular properties of nanoscale materials to achieve polymer composite materials with new functionalities and greatly enhanced strength. Research on clay-polymer hybrid materials [274] was initiated after the first invention by researchers at Toyota [275] (mica in nylon). The focus shifted later to particles with well-defined shape, size, and surface modification [276–282]. Nevertheless, contradictory results on the improvement of mechanical properties have been obtained in different laboratories, and thus many questions about the mechanism of enhancement in nanoparticle-filled polymer materials remain open [278–284]. In these prior studies the importance of the dispersed state of the particle fillers as well as of the interfacial interactions between the particle filler and polymer matrix components were recognized [285]. If nanoparticles are segregated, the optimum performance is that of traditional fillers. Traditionally, the dispersion of particles in polymers has proven difficult due to entropy loss of the polymer chains in the absence of an offset by favored enthalpic interactions. The understanding of and control over the interactions might have important implications in structural engineering of the composites via a directed spatial particle distribution [286], as well as in many biomedical applications since the nanoparticle-mediated cellular response can be size-dependent [287].

The dispersion of nanoparticles in polymer matrices has shown to be related to the subtle interplay between enthalpic and entropic interactions that depend on the chemistry and architecture of both polymer-coated particle and matrix polymer. Owing to the vast parameter space which might affect the still missing structure-property relationship, both theoretical and experimental studies are focused on the size of the two mixture constituents and the surface functionalization of the nanoparticles, in addition to their volume fraction. For nanoparticle-filled homopolymers, miscibility is achieved when $R < R_g$ where R denotes the effective particle radius (i.e., the sum of core radius and shell thickness) [288], in agreement with theoretical considerations [289, 290]. The polymer can accommodate the small particles without substantial entropy penalty resulting from the chain stretching, in order to avoid the impenetrable particles, but expels the large particles due to substantial conformational entropy penalty. The anticipated increase in R_g (entropically unfavorable) should be compensated by a negative interaction parameter (χ) in addition to the usual favorable entropy of mixing ($\Delta S_{mix} > 0$, consistent with homogeneous dispersion). This enthalpy term ΔH is χN for chains of N monomers and $\Delta H = 4\pi R^2 \chi$ for the hybrid (polymer/nanoparticle) system.

The effect of the well-dispersed nanoparticle on the polymer dimension is controversial since all possible situations (increase, decrease, or no change at all) on R_g have already been reported in the literature [291, 292]. Based on a very recent SANS study [293], the dispute is attributed to the ill-defined equilibrium morphology of the investigated dispersions. Even if the disordered state is thermodynamically stable, it might be kinetically inaccessible unless a suitable processing strategy is adopted. To assist access to that physical state in a mixture of polystyrene and chemically identical nanoparticles (PS microgels), its solution in a common solvent (toluene) was dripped into a non-solvent (methanol), inducing rapid precipitation. The scattering wavevector dependence of the SANS intensity, $I(q)$, from a PS/nanoparticle with 2% dPS of similar molar mass (M_w) revealed miscibility for $R < R_g$ and immiscibility for $R > R_g$. For the well-dispersed mixture, the d-PS was found to swell by about 15% in the presence nanoparticles with $\phi = 0.1$.

Computer simulations treating melt densities properly need to include nanoparticle mobility and account for mixing entropy and enthalpic interactions in order to predict a nanoparticle-induced polymer swelling [294]. One limitation of the computational studies is the neglect of the polymeric surfactant shell bound to the particle surface [276]. Earlier work [295–299] on the stabilization of grafted colloids in polymer melts suggested that compatibilization requires the molecular weight of the surface-bound polymer to exceed the molecular weight of the chemically identical matrix polymer. Moreover, attractive interactions between the grafted polymer chains and the mobile melt chains can induce structural changes (e.g., swelling) to the brush and thus enable the interpenetration of the melt free chains [109, 296, 300, 301]. However, experimental tests of the effect of grafting density, size, and shape of the particles and molecular weight of both grafted and free chains on the physical state of the hybrid mixtures have been rare [288].

Polymer architectural complexity is also a key feature in the morphology of the hybrid mixture since, in addition to the basic thermodynamic forces, microphase separation of polymers can be utilized to direct the spatial distribution of the nanoparticles and thereby dictate the properties of the composite material. For example, the self-assembly of diblock copolymers provides host microenvironments for the nanoparticles and tuning their final location (e.g., interfacial vs selective-layer arrangement) can influence the materials optimal performance [302]. From the experience with the homopolymers/nanoparticle blends, however, the block copolymer morphologies are not always expected to template the particle spatial distribution; alternatively, the orientation and the morphology of the polymer microdomains can be modified by the nanoparticle characteristics [286, 303]. As already discussed, the two main parameters that are considered important for the structure of particle-block copolymer blends are the conformational entropy and the enthalpic contribution. They are controlled by the size of the nanoparticles and the nanoparticle ligand (grafted) chains. Because of the entropic penalty, large particles are expelled from the diblock copolymers (polymers have to stretch around the impenetrable particle) and reside in the center of either block's domain (depending on the nanoparticle ligand chains), whereas smaller particles are more uniformly dispersed [298, 303–307].

On the experimental side, this size exclusion has been observed in the case for diblock copolymers [308] in which nanoparticles of different size but same nanoparticle ligand chains assume distinct locations within the microstructure (Fig. 25) and for particle interpenetration into polymer brushes [309].

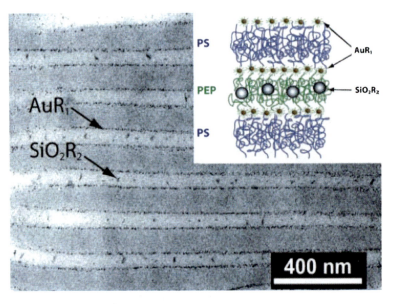

Fig. 25 Transmission electron micrographs (TEM) of a ternary nanocomposite of PS-poly(ethyl propylene) (PEP) diblock copolymer with two types of nanoparticle-ligand systems: AuR_1- and SiO_2R_2-functionalized (R_1,R_2 are alkyl groups) nanoparticles of total volume fraction 0.02. The former appear along the interface of the lamellar microdomains, whereas the latter reside in the center of PEP microphases. Schematically, the nanoparticle distribution is shown in the *inset*. Taken from [308]

The spatial distribution of particles can be also driven by enthalpic interactions through appropriate surface functionalization defining the grafting density σ^* and the polymer ligand molecular weight [302, 305]. A crude estimate of the critical grafting density σ_c^*, below which the core-shell nanoparticles (of particle core radius r_c and attached chain radius of gyration R_g) are directed to the interface of a lamellar diblock, suggests $\sigma_c^* \sim \left[(r_c + R_g)/r_c R_g\right]^2$, in good agreement with experiment [305]. Therefore, the right combination of such thermodynamic forces can guide the particle placement within a diblock copolymer matrix. This ability can be utilized to create "gradient materials" [308, 310] by blending diblock copolymers with bidisperse particles and compositionally heterogeneous structured silica hybrids through the right choice of the size of silica core-shell nanoparticles and the size of the hydrophilic block of the block copolymer [311]. Conversely, the nanoparticles can alter the morphology of the diblock copolymers as already mentioned. This is due to the stretching of the matrix chains (caused by the particle excluded volume) and the reduction of the interfacial energy for particle localization at the interface between the incompatible blocks, which change the free energy of the block copolymer, leading to a transition from one microphase to another. Directing gold nanoparticles functionalized by thiol-terminated PS to the interface between PS and poly (2-vinylpyridine)(P2VP) microdomains, a transition from a lamellar to bicontinuous morphology of the PS-P2VP copolymer was observed [312], in agreement with theoretical predictions [313]. This microphase transition can also be induced by increasing the particle volume fraction for both symmetric (lamellar to hexagonally packed cylinders) and asymmetric (cylinders to lamellae) block copolymers. These phase transformations are captured by advanced computer simulations that combine the appropriate methods to correctly describe both the polymeric and colloidal nature of the composites [307]. In a similar context, a symmetric diblock sandwiched between two solid surfaces changes its lamellar orientation from parallel to perpendicular to the surfaces after blending with nanoparticles [286, 314]. More complexity, but also more possibilities for tunability, can be imparted under the influence of external fields, such as flow (e.g., [218, 315]).

Studies on the thermodynamic properties of mixtures of nanoparticles and binary polymer blends and the effect of particle characteristics (size, volume fraction, surface functionality) on the phase behavior of the composite are relatively few. It is known that phase separation kinetics and morphology are affected by the preferential wetting of a solid surface by one component of the mixture. Therefore, inclusion of particles drastically affects the pattern evolution of phase separation, driving the particles into the more wetting phase. Since the coupling between wetting and phase separation is dynamic, the pattern evolution depends strongly on the particle mobility. The volume fraction of particles is a crucial parameter for the resulting morphology since imbalanced coverage of the particles causes attractive depletion forces. Theoretically, this wetting-induced depletion leads to an unusual re-entrant morphological transformation, where the more wetting phase is altered from network to droplet to network with increasing particle concentration [316]. With regard to the influence of the size of the nanoparticles on the miscibility of binary polymer blends, theoretical calculations have shown that a decrease of the size can drive a phase separating system into the miscible one-phase region [317].

The nanoparticle/polymer mixtures (of respective size ratio R/R_g) display new material properties not inherent to the constituent components, while many more are expected to emerge. An entropy-driven effect is the migration of nanoparticles with $R \sim R_g$ into cracks [318]. This suggests ways for designing self-healing systems to restore their original mechanical strength. In contrast, smaller particles remain in the polymer matrix without a substantial entropic penalty. The optical properties of diblock copolymer/nanoparticle blend depend sensitively on the materials microstructure. The absorbance spectra reveal a significant decrease in the transparency with a red shift in the case of interfacial particle segregation as opposed to the selective-layer uniform particle dispersion [302]. The optical turbidity can be suppressed through the control of the polymer coatings, which allow adjusting the light scattering of the embedded nanoparticles. Because of the large surface area per particle volume, the polymer ligands play a significant role in determining the properties of the nanoparticles. Thus the effective refractive index of SiO_2-PS core-shell nanoparticles can be tuned by the appropriate choice of the grafting density and degree of polymerization of the grafted chains in the shell [112]. Promising photonic and phononic properties of assembled mixtures of polymers and nanoparticles are essentially unexplored [319–321].

The main research challenge remains the relation of the morphology of the nanoparticle-filled polymers to their thermomechanical performance, which is of paramount importance for the use of particle-based nanocomposites in technological applications. In general, mixing rigid fillers with polymers increases the elastic modulus of the material [274, 275]. In the case of the nanofillers, the induced changes are not well-understood, primarily due to their poor dispersion. There have been relatively few theoretical and experimental studies of the thermomechanical behavior of hybrid composites with well-defined spatial distribution of nanoparticles. Self consistent field theory (SCFT) was applied to study the elastic properties of symmetric block copolymers with spherical particles ($R \sim R_g$, $\phi = 0.15$) sequestered in the lamellar phase of one block [322]. While the shear modulus was unaffected, the tensile modulus of the material decreased, due primarily to the swelling of the lamellar block. The composite's elastic constants reflect, in principle, volume (packing) and/or energetics (high strength bonds); their influence on the dynamics and glass transition of the polymeric system due to the addition of nanoparticles is essentially unexplored. Computer simulations of the segmental dynamics of a polymer melt surrounding a nanoscopic filler suggest close similarity to the behavior of ultrathin films in the sense that the shift of glass transition temperature, T_g, is fully correlated with the polymer filler interactions [323]. An attempt to quantify this analogy experimentally relies on well-characterized SiO_2-PS core-shell particles and particle spatial distribution [324]. For the entropic mixtures of CdSe-PS core-shell with linear PS, both the Young modulus E and the T_g were found to decrease with increasing nanoparticle fraction. The trivial plasticization effect due to the short ligand chains was excluded and the weakening effect was attributed to a possible formation of a depletion region around the nanoparticle [325]. Notably, when enthalpic (cohesive) interactions become a significant factor, an increase of T_g and a slowing-down of the segmental dynamics have been reported [326].

The matrix transport properties such as viscosity are severely affected by the presence of nanoparticles, but in an unanticipated manner. A strong decrease of the zero-shear viscosity was observed in entropic mixtures of PS nanogel particles ($R \leq R_g$) with PS homopolymers [276]. This highly unusual reduction was tentatively ascribed to a disruption of the polymer chain packing that leads to an increase of the free volume. However, there was no accompanying decrease in the T_g of the blend. It remains to be seen whether new physical phenomena (in addition to the classical hydrodynamic interactions) become relevant for nanosized filler additives. In [276], the PS nanoparticles were produced by dripping a solution of PS chains with different amount of pendant crosslinking groups into a hot non-solvent to activate the crosslinking process, and deformability cannot be excluded. On the other hand, the studies with core-shell particles also point to an additional problem for the interpretation of the thermomechanical properties of filled polymers, i.e., the detachment of ligands during the thermal annealing cycle that is necessary to relieve stress from the spin-cast films. This cleavage is a consequence of the weak covalent bonding of most ligand-particle chemistries and effectively results in plasticization of the matrix polymer with low molecular weight ligand molecules. The use of core-shell particles, however, offers structural flexibility and enables the control of the density profile, i.e., reduced crowding with increasing distance from the particle center via grafting density and ligand chain molecular weight. This will form the basis for the hypothesis that the grafted-chain density is an important parameter for the thermomechanical properties of polymer nanocomposite mixtures.

Knowledge of the bulk mechanical behavior of the particle-filled hybrids necessitates access to the elastic properties of the structured particles. Forces which can be negligible in macroscopic systems (such as surface tension, depletion, and confinement effects) often become significant at the nanoscale. Therefore, the behavior of the same materials in nanoscopic systems can deviate considerably from the bulk [327]. The nanomechanical properties in structured materials are a precondition to realize specific needs in structural engineering applications. Conventional dynamic rheological measurements of the macroscopic systems are usually not sufficient as they cannot extend to the needed high frequencies. The latter relate to the vibrational modes confined in the individual particles as a result of the elastic motion at the nanoscale. These modes should sensitively depend on the geometrical, architectural, interfacial and mechanical characteristics of the nanoparticles. It turns out that there is a paucity of appropriate non-destructive experimental techniques to probe these particle vibrations (sometimes called "music of spheres" [327]) since both high-frequency resolution and sensitivity are required to measure a few eigenfrequencies. Raman scattering [328] has been utilized to detect few eigenfrequencies of nanoparticles with dimension below 10 nm whereas Brillouin light scattering (BLS) [193, 329] and optical pulse-probe techniques [330] can probe respectively the spontaneous and stimulated vibrations confined in sub-micrometer particles. In the latter technique, the excited acoustic oscillations are observed in the form of modulations of the transient reflectivity of the probe laser and hence the particles must possess good reflectance (e.g., gold). In BLS, laser light is scattered inelastically by the density fluctuations (phonons) associated with these particle localized modes at thermal equilibrium and there are no such limitations. BLS is basically a coherent scattering phenomenon [193, 331] and therefore capable to record the phonon dispersion $\omega(k)$, where ω is the phonon frequency and k the scattering wavevector. For non-transparent media, where a well-defined k is not possible due to multiple scattering effects, BLS can be applied only to reveal localized, k-independent modes confined in space. Figure 26 shows two characteristic BLS spectra [332] for bare silica spheres with $R = 90$ nm and a silica-PMMA core-shell particles with the same core and PMMA thickness of 112 nm. The spectra can be well represented by up to nine Lorentzian line shapes as shown by the solid lines in the Stokes side of the spectra. The peak position of these spectral lines yields the resonance frequencies of the particle eigenmodes. While for the bare silica only two resonance frequencies were resolved, the BLS spectrum of the core-shell system (with about double radius) is much richer. These frequencies relate to the spheroidal (i,l) modes with i being the i-th mode of the l-th spherical harmonic, and depend on the mass density ρ and two sound velocities c_l and c_t with longitudinal and transverse polarization, respectively, of the core and shell materials. Hence, the Poison ratio $\sigma = \left(c_l^2 - 2c_t^2\right)\left(2c_l^2 - c_t^2\right)$, the Young modulus $E = \rho c_l^2 (1+\sigma)/(1-2\sigma)$, and the shear modulus $G = \rho c_t^2$ can be obtained [332]. These values which are not a priori known for such systems and length scales can be obtained from the comparison of the experimental eigenfrequencies with the theoretically computed density-of-state spectra. Each resonance mode appearing at frequency $f(i,l)$ matches the corresponding experimental frequency by adjusting the materials elastic constants.

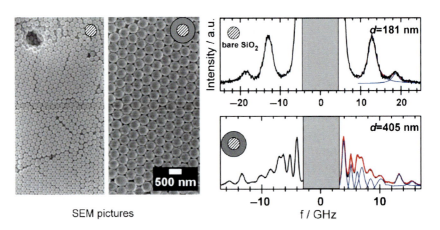

Fig. 26 *Left*: Scanning electron microscopy (SEM) pictures of silica ($R = 90$ nm) and silica-PMMA core-shell particles ($R = 202$ nm). *Right*: BLS eigenmode spectra of the bare silica core (*top*) and core-shell particle (*bottom*). The various *solid lines* on the Stokes-side (positive frequencies) indicate the representation by a sum of up to nine Lorentzian line shapes. Taken from [332]

The feasibility of BLS to measure the high frequency moduli was first demonstrated in dilute suspensions of giant core-shell micelles [333].

Another promising use of colloidal particles in polymeric systems is related to the manipulation and control of the properties of biphasic systems. Recently, colloidal particles were used effectively as compatibilizers in polymer blends [334, 335], a technologically important phenomenon similar to the Pickering emulsions, i.e., particle-stabilized emulsions [336]. An important related development was the possible stabilization and arrest of the bicontinuous interface in a binary liquid demixing via spinodal decomposition using particles that are neutrally wetted by both liquids [337, 338]. The resulting new class of soft solids, named bijels (bicontinuous interfacially jammed emulsions gel) could have many potential applications, for example, in microfluidics and as micro-reaction media. Three-dimensional samples of bijel in which the interfaces are stabilized by essentially a single layer of particles were recently produced and the mean interfacial separation in these bijels was tuned [338, 339]. We believe that this is a very exciting field and the various possibilities offered will be explored in great detail in the near future.

These examples constitute another proof of the reachness of soft hairy particles and the ample possibilities one has to engineer their properties.

7 Conclusions and Outlook

The main message of this review is the fact that softness imparts an incredibly large range of flexibility for tuning the properties of colloidal systems. Through various examples involving the use of well-characterized model systems, it has been demonstrated that it is possible to manipulate interactions and transitions in soft colloids with the chemical (grafting, core-to-corona size ratio) and physical (temperature, solvent, depletion, fraction of enthalpic contribution) parameters of the systems. This, in turn, provides the key for tailoring their dynamics and rheology. Of course, the use of other external fields such as flow and electric field can induce microstructural changes and are associated with a host of interesting phenomena (such as ordering, shear banding, shear thickening). These aspects are not addressed in this review, but the interested reader is referred to recent publications [218, 340–351]. In fact, directed colloidal self-assembly under non-equilibrium conditions (e.g., external fields) is an emerging promising field of research with potentially important technological implications [352].

The significant recent advances in both colloidal and polymer chemistries have enabled the successful fabrication of complex, defect-free nanostructures following a bottom-up approach [353]. Two recent related examples are mentioned to stress the point. Recently, triblock copolymer with divalent counter-ions in mixed solvents led to the formation of particles with tunable internal structure mimicking lipid anphiphiles for potential use in drug delivery. The mechanism of formation involves either nanophase separation within the triblock copolymer nanoparticle upon addition of water or microemulsion formation similar to that in lipid systems

[354]. The second example is the formation of patchy nanoparticles. Bottom-up approaches depend on the availability of well-defined building blocks and their directed self-assembly into the predesigned structure of interest. The self-assembly of soft particles is controlled by the strength, size, and distribution of attractive and repulsive forces at their surface where the interaction with their surroundings (being other building blocks or solvent) occurs [355–357]. The strength and direction of these forces on the surface is referred to as patchiness. Soft patchy nanoparticles were studied and materialized with binary AB and BC diblock copolymers where the patchiness was driven by the immiscibility of the A and B blocks at the particle surface, or the assembly of colloidal particles with prebuilt patchiness and sometimes formation of anistropic objects [358–361]. Soft patchy nanoparticles can be found in proteins and viruses, whereas non-biological soft patchy particles are very rare. Polyhedral structures were formed by colloidal spheres using capillary forces [362–364], whereas in a controlled guest-host chemistry the degree of random patchiness can be adjusted by the guest-host ratio during the self-assembly [365].

In the context of the newly emerging decorated colloids and engineering of effective interparticle potentials, a promising development is the covalent attachment of DNA oligonucleotide on solid cores. This was realized more than 10 years ago with the assembly of DNA-based gold nanoparticles upon addition of linking DNA duplex [366], and this field has received considerable interest in the colloid chemistry community [367]. The choice of the attached DNA sequences and the DNA linking molecules can direct different micrometer-sized crystalline gold nanoparticle structures [368]. Aside the many promising applications of these oligonucleotide-decorated nanoparticles, they exemplify patchy colloids with selective interactions and certain valency, i.e., number of cites that prevent multiple bonding on the same site simultaneously. These unique features can allow for unprecedented modifications of the phase diagram of this new class of colloids. For example, the region in the $T - \phi$ plane in which clustering and gelation is expected, shifts towards lower ϕ and/or lower T with decreasing valence (patchiness), in contrast to the hard spheres colloidal suspensions when the range of added attractive interactions is decreased [369, 370]. This initial theoretical work and the new possibilities these new materials promise will soon trigger the missing experimental investigations of their structural and dynamic properties in a rich parameter space, which in turn will motivate additional theoretical efforts [371].

It has been known for more than two decades that light can manipulate (move, trap, and guide) particles [372]. Electrostrictive forces hold transparent particles larger than 100 nm in the brightest part of the beam; this effect led to the prolific method of optical tweezers. Efficient optical trapping of silver and gold nanoparticles has been extended to particles diameters down to 20 nm which. due to the strong scattering by surface plasmons, allow for a concurrent optical visualization [373]. The potential antimicrobial properties and the ease of surface functionalization make these metal nanoparticles an obvious choice for nanomarkers in medical applications.

The guiding of the particles in the focus of the light source changes the refractive index in the colloidal suspensions and this induced Kerr effect can create self-focusing that counterbalances the natural diffraction and enhance light transmission. The optical beam is therefore self-trapped by this induced waveguide, creating the so-called optical spatial solitons (OSS) predicted and observed in various materials [374, 375] including colloidal solutions [201, 376]. Due to the nonlinearities, modulational instabilities further lead to the break-up of single soliton (filament) propagation into self-ordered OSS (multi-filaments) arrays in those media [377]. The variety of nonlinearities and the richness of colloid science along with the elimination of the beam spatial spreading could provide ways for new investigations of OSS with impact in optics and photonics [378].

The recent progress in describing the structure of soft colloids (e.g., the star pair interaction potential) and the complexity of the chemistry and resulting internal structure of the hairy particles available necessitate the development of predictive tools for gaining a thorough understanding of their performance and designing novel systems. In particular, for dense colloids mode couling theory in different versions remains the reference framework. Recent approaches [236, 379] with more microscopic information appear promising. On the other hand, a number of alternative or complementary coarse-grained methodologies have recently emerged, notably the soft glassy rheology (SGR) model [380–382], the responsive particle dynamics (RPD) mesoscopic model [383–386], and the multiple-particle-collision dynamics (MPC) [387, 388] or stochastic rotation dynamics [389] models that account for hydrodynamic interactions mediated by the solvent (as alternatives to Stokesian dynamics [390]). These approaches hold the premise of a rigorous understanding (based on knowledge of structure) and control of complex phenomena such as shear-induced melting and/or deformation, aging, thixotropy and shear banding in a wide range of soft materials [225, 257, 270–273, 340–346, 377, 384, 391–396]. To this end, novel high-resolution rheo-physical experiments will be indispensable [397–400]. At the same time, our understanding of complex phenomena in the bulk such as yielding, could be enhanced greatly by parallel model studies in confined space. This was recently demonstrated with the rheological study of flocculated two-dimensional suspensions [401]. This study was possible thanks to the emergence of interfacial rheometry instrumentation, which holds the promise for further advances in this direction [402–404]. As a result of that work, possible mechanisms for cluster break-up under flow were suggested [401, 405].

This is a very exciting area of research with many potential cutting-edge technological applications, and we can anticipate several new developments and applications. Pandora's box has been just opened.

Acknowledgments We are indebted to many long-term collaborators and colleagues, especially J. Roovers, C. N. Likos, N. Hadjichristidis, and A. N. Semenov. Our first investigations on soft colloids were undertaken in collaboration with our late friend Tadeusz Pakula more than 10 years ago. We acknowledge the contributions of our current and former students and in particular M. Kapnistos and E. Stiakakis. We are also thankful to J. Mewis for critical reading of the manuscript and helpful comments. Financial support from the EU (RTN-HUSC HPRN-CT-2000-00017 and NoE-Softcomp NMP3-CT-2004-502235) and the Greek General Secretariat for Research and Technology (PENED) is gratefully acknowledged.

References

1. Jones RAL (2002) Soft condensed matter. Oxford University Press, New York, NY
2. Hamley IW (2000) Introduction to soft matter. Wiley, New York, NY
3. Daoud M, Williams CE (eds) (1999) Soft matter physics. Springer, Berlin
4. Cates ME, Evans MR (eds) (2000) Soft and fragile matter. Nonequilibrium dynamics, metastability and flow. Institute of Physics Publishing, Bristol
5. Witten TA (2004) Structured fluids. Polymers, colloids, surfactants. Oxford University Press, New York, NY
6. McLeish TCB (ed) (1997) Theoretical challenges in the dynamics of complex fluids. NATO ASI E339. Kluwer, Amsterdam
7. Yethiraj A, van Blaaderen A (2003) Nature 421:513
8. Liu AJ, Nagel SR (1998) Nature 396:21
9. Rubinstein M, Colby RH (2003) Polymer physics. Oxford University Press, New York, NY
10. DeGennes, PG (1979) Scaling concepts in polymer physics. Cornell University Press, Ithaca, NY
11. Doi M, Edwards SF (1986) The theory of polymer dynamics. Oxford Science, New York, NY
12. Russel WB, Saville DA, Schowalter WR (1989) Colloidal dispersions. Cambridge University Press, New York, NY
13. Pusey PN, van Megen W (1986) Nature 320:340
14. Zaccarelli E (2007) J Phys Condens Matter 19:323101
15. Poon WCK (2002) J Phys Condens Matter 14:R859
16. Pham KN, Puertas AM, Bergenholtz J, Egelhaaf SU, Mousaid A, Pusey PN, Schofield AB, Cates ME, Fuchs M, Poon WCK (2002) Science 296:104
17. Sciortino F (2002) Nat Mater 1:145
18. Sciortino F, Tartaglia P (2005) Adv Phys 54:471
19. Likos CN (2006) Soft Matter 2:478
20. Zaccarelli E, Mayer C, Asteriadi A, Likos CN, Sciortino F, Roovers J, Iatrou H, Hadjichristidis N, Tartaglia P, Löwen H, Vlassopoulos D (2005) Phys Rev Lett 95:268301
21. Grest GS, Fetters JL, Huang JS, Richter D (1996) Adv Chem Phys XCIV:67
22. Vlassopoulos D (2004) J Polym Sci B 42:2931
23. Roovers J, Zhou L-L, Toporowski PM, van der Zwan M, Iatrou H, Hadjichristidis N (1993) Macromolecules 26:4324
24. Likos CN (2001) Phys Rep 348:267
25. Likos CN, Löwen H, Poppe A, Willner L, Roovers J, Cubitt B, Richter D (1998) Phys Rev E 80:6299
26. Stiakakis E, Vlassopoulos D, Loppinet B, Roovers J, Meier G (2002) Phys Rev E 66:051804
27. Charalabidis D, Pitsikalis M, Hadjichristidis N (2002) Macromol Chem Phys 203:2132
28. Daoud M, Cotton JP (1982) J Phys (Paris) 43:531
29. Likos CN, Löwen H, Watzlawek M, Abbas B, Jucknischke O, Allgaier J, Richter D (1998) Phys Rev Lett 80:4450
30. Zhou L-L, Hadjichristidis N, Toporowski PM, Roovers J (1992) Rubber Chem Tech 65:303
31. Toporowski PM, Roovers J (1986) J Polym Sci Polym Chem Ed 24:3009
32. Roovers J, Toporowski P, Martin J (1989) Macromolecules 22:1897
33. Yamamoto T (2003) Studies of long time relaxation of multi-arm star polymers. Ph.D Thesis, Kyoto University
34. Ishizu K, Ono T, Uchida S (1997) J Colloid Interface Sci 192:189
35. Matyjaszewski K, Miller PJ, Pyun J, Kickelbick G, Diamanti S (1999) Macromolecules 32:6526
36. Xia J, Zhang X, Matyjaszewski K (1999) Macromolecules 32:4482
37. Gao H, Matyjaszewski K (2006) Macromolecules 39:3154
38. Gao H, Ohno S, Matyjaszewski K (2006) J Am Chem Soc 128:15111
39. Gao H, Matyjaszewski K (2007) Macromolecules 40:399
40. Teertstra SJ, Gauthier M (2007) Macromolecules 40:1657

41. Pakula T (1998) Comp Theor Polym Sci 8:21
42. Vlassopoulos D, Fytas G, Pakula T, Roovers J (2001) J Phys Condens Matter 13:R855
43. Semenov AN, Vlassopoulos D, Fytas G, Vlachos G, Fleischer G, Roovers J (1999) Langmuir 15:358
44. Mewis J, Frith WJ, Strivens TA, Russel WB (1989) AIChE J 35:415
45. Halperin A, Tirrell M, Lodge TP (1992) Adv Polym Sci 100:31
46. Wilner L, Jucknischke O, Ricther D, Roovers J, Zhou L-L, Toporowski PM, Fetters LJ, Huang JS, Lin MY, Hadjichristidis N (1994) Macromolecules 27:3821
47. Richter D, Farago B, Fetters LJ, Huang JS, Ewen B (1990) Macromolecules 23:1845
48. Ballauff M, Likos CN (2004) Angew Chem Int Ed 43:2998
49. Gohy J-F (2005) Adv Polym Sci 190:65
50. Halperin A (1987) Macromolecules 20:2943
51. Förster S, Wenz E, Lindner P (1996) Phys Rev Lett 77:95
52. Gast AP (1996) Langmuir 12:4060
53. Lund R, Willner L, Stellbrink J, Radulescu A, Richter D (2004) Macromolecules 37:9984
54. Laurati M, Stellbrink J, Lund R, Willner L, Ricther D, Zaccarelli E, Richter D (2007) Phys Rev E 76:041503
55. Lund R, Willner L, Stellbrink J, Lindner P, Richter D (2006) Phys Rev Lett 96:068302
56. Bang J, Viswanathan K, Lodge TP, Park MJ, Char K (2004) J Chem Phys 121:11489
57. Lodge TP, Bang J, Park MJ, Char K (2004) Phys Rev Lett 92:145501
58. Laurati M, Stellbrink J, Lund R, Willner L, Ricther D, Zaccarelli E (2005) Phys Rev Lett 94:995504
59. Zilliox J-G, Rempp P, Parrod J (1968) J Polym Sci C 22:145
60. Liu GJ (1999) In: Nalwa HS (ed) Handbook of nanostructured materials and nanotechnology, vol 5. Academic, New York, NY
61. O'Reilly RK, Hawker CJ, Wooley KL (2006) Chem Soc Rev 35:1068
62. Liu GJ (1998) Curr Opin Colloid Interface Sci 3:200
63. Baek K-Y, Kamigaito M, Sawamoto M (2002) J Polym Sci A 40:633
64. Gurr PA, Qiao GG, Solomon DH, Harton SE, Spontak RJ (2003) Macromolecules 36:5650
65. Park C, Yoon J, Thomas EL (2003) Polymer 44:6725
66. Ruzette AV, Leibler L (2005) Nat Mater 4:19
67. McConnell GA, Gast AP, Huang JC, Smith SD (1993) Phys Rev Lett 71:2102
68. Laflèche F, Durand D, Nicolai T (2003) Macromolecules 36:1331
69. Pham QT, Russel WB, Thibeault JC, Lau W (1999) Macromolecules 32:2996
70. Pham QT, Russel WB, Thibeault JC, Lau W (1999) Macromolecules 32:5139
71. Bhatia SR, Mourchid A, Joanicot A (2001). Curr Opin Colloid Interface Sci 6:471
72. Discher DE, Kamien RD (2004) Nature 430:519
73. Lee AS, Bütün V, Vamvakaki M, Armes S, Pople JA, Gast AP (2002) Macromolecules 35:8540
74. Kimerling AS, Rochefort WE, Bhatia SR (2006) Ind Eng Chem Res 45:6885
75. Bütün V, Liu S, Weaver JVM, Bories-Azeau X, Cai Y, Armes SP (2006) React Funct Polym 66:157
76. McConnell GA, Gast AP (1997) Macromolecules 30:435
77. Alexandridis P, Athanassiou V, Hatton TA (1995) Langmuir 11:2442
78. Lin YN, Alexandridis P (2002) Langmuir 18:4220
79. Yang L, Alexandridis P, Steytler DC, Kositza MJ, Holzwarth JF (2000) Langmuir 16:8555
80. Sato T, Watanabe H, Osaki K (2000) Macromolecules 33:1686
81. Watanabe H (1999) J Non-Newtonian Fluid Mech 82:315
82. Watanabe H (1997) Acta Polym 48:215
83. Watanabe H, Matsumiya Y (2005) Macromolecules 38:3808
84. Chen SH, Chen W-R, Mallamace F (2003) Science 300:619
85. Mountrichas G, Mpiri M, Pispas S (2005) Macromolecules 38:940
86. Pispas S, Hadjichristidis N, Mays JW (1996) Macromolecules 29:7378
87. Cheng G, Hua F, Melnichenko YB, Hong K, Mays JW, Hammouda B, Wignall GD (2008) Macromolecules 41:4824

88. Advincula RC, Brittain WJ, Caster KC, Rühe J (2004) Polymer brushes: synthesis, characterization, applications. Wiley InterScience, New York, NY
89. Vincent B (1993) Chem Eng Sci 48:429
90. Edmondson S, Osborne VL, Huck WTS (2004) Chem Rev 33:14
91. Pusey PN (1991) In: Hansen JP, Levesque D, Zinn-Justin J (eds) Liquids, freezing and glass transition. North Holland, Amsterdam
92. Cheng Z, Phan S-E, Russel WB (2002) Phys Rev E 65:041405
93. Krishnamurthy L (2005) Microstructure and rheology of polymer-colloid mixtures. Ph.D Thesis, University of Delaware
94. Krishnamurthy L, Wagner NJ (2005) J Rheol 49:475
95. de Kruif CG, Rouw PW, Briels WJ, Duits MHG, Vrij A, May RP (1989) Langmuir 5:422
96. de Kruif CG, Briels WJ, May RP, Vrij A (1988) Langmuir 4:668
97. Grant MC, Russel WB (1993) Phys Rev E 47:2606
98. Chen M, Russel WB (1991) J Colloid Interface Sci 141:564
99. Vrij A, Philipse AP (1996) In: Pelizzetti E (ed) Fine particles science and technology. Kluwer, Amsterdam
100. Milling A, Vincent B, Emmett S, Jones A (1991) Colloids Surf 57:185
101. Petekidis G, Gapinski J, Seymour P, van Duijneveldt JS, Vlassopoulos D, Fytas G (2004) Phys Rev E 69:042401
102. Verduin H, de Gans BJ, Dhont JKG (1996) Langmuir 12:2947
103. Verduin H, Dhont JKG (1995) J Colloid Interface Sci 172:425
104. Rueb CJ, Zukoski CF (1997) J Rheol 41:197
105. Nommensen PA, Duits MHG, van den Ende D, Mellema J (1999) Phys Rev E 59:3147
106. Nommensen PA, Duits MHG, van den Ende D, Mellema J (2000) Langmuir 16:1902
107. Cawdery N, Milling A, Vincent B (1994) Colloids Surf A Physicochem Eng Asp 86:239
108. Mewis J, Biebaut G (2001) J Rheol 45:799
109. Green DL, Mewis J (2006) Langmuir 22:9546
110. Castaing JC, Allain C, Auroy P, Auvray L (1999) Eur Phys J B 10:61
111. Pyun J, Matyjaszewski K (2001) Chem Mater 13:3436
112. Bombalski L, Dong H, Listak J, Matyjaszewski K, Bockstaller MR (2007) Adv Mater 19:4486
113. Genz U, Dagguano B, Mewis J, Klein R (1994) Langmuir 10:2206
114. Mewis J, D'Haene DP (1993) Makromol Chem, Macromol Symp 68:213
115. Raynaud L, Ernst B, Verge C, Mewis J (1996) J Colloid Interface Sci 181:11
116. Shay JS, Raghavan SR, Khan SA (2001) J Rheol 45:913
117. Zackrisson M, Strasdner A, Schurtenberger P, Bergenholtz J (2006) Phys Rev E 73:011408
118. Haschick R, Mueller K, Klapper M, Muellen M (2008) Macromolecules 41:5077
119. Kim JU, Matsen MW (2008) Macromolecules 41:4435
120. Guo X, Ballauff M (2000) Langmuir 16:8719
121. Ballauff M (2003) Macromol Chem Phys 204:220
122. Ballauff M, Borisov O (2006) Curr Opin Colloid Interface Sci 11:316
123. Dingenouts N, Norhausen C, Ballauff M (1998) Macromolecules 31:8912
124. Murray MJ, Snowden MJ (1995) Adv Colloid Interface Sci 54:73
125. Saunders BR, Vincent B (1999) Adv Colloid Interface Sci 80:1
126. Pelton RH, Chibante P (1986) Colloids Surf 20:247
127. Pelton RH (2000) Adv Colloid Interface Sci 85:1
128. Stieger M, Richtering W, Pedersen JS, Lindner P (2004) J Chem Phys 120:6197
129. Bartsch E, Antonietti M, Schupp W, Sillescu H (1992) J Chem Phys 97:3950
130. Eckert T, Bartsch E (2004) J Phys Condens Matter 16:S4937
131. Saunders BR (2004) Langmuir 20:3925
132. Mason TG, Lin MY (2005) Phys Rev E 71:040801(R)
133. Saito R, Ishizu K (1997) Polymer 38:225
134. Wolfe MS (1992) Progr Org Coating 20:487
135. Heskins M, Guillet JE (1968) J Macromol Sci Chem 8:1444
136. Hu T, Wu C (1999) Phys Rev Lett 83:4105

137. Wu C, Qiu X (1998) Phys Rev Lett 80:620
138. Yang H, Yan X, Cheng R (2002) Macromol Rapid Commun 23:1037
139. Berndt I, Pedersen JS, Richtering W (2005) J Am Chem Soc 127:9372
140. Seelenmeyer S, Deike I, Rosenfeldt S, Norhausen C, Dingenouts N, Ballauff M, Narayanan T, Lindner P (2001) J Chem Phys 114:10471
141. Ballauff M (2003) Macromol Chem Phys 204:220
142. Makino K, Yamanoto S, Fujimoto K, Kawaguchi H, Oshima H (1994) J Colloid Interface Sci 166:251
143. Senff H, Richtering W (2000) Colloid Polym Sci 278:830
144. Berndt I, Pedersen JS, Richtering W (2006) Angew Chem 45:1737
145. Berndt I, Popescu C, Wortmann F-J, Richtering W (2006) Angew Chem 45:1081
146. Kang M, Wellert S, Pastoria-Santos I, Lapp A, Liz-Marzan LM, Hellweg T (2008) Phys Chem Chem Phys 10:6708
147. Senff H, Richtering W (1999) J Chem Phys 111:1705
148. Crassous JJ, Ballauff M, Drechsler M, Schmidt J, Talmon Y (2006) Langmuir 22:2403
149. Senff H, Richtering W, Norhausen C, Weiss A, Ballauff M (1999) Langmuir 15:102
150. Berndt I, Richtering W (2003) Macromolecules 36:8780
151. Crassous JJ, Siebenburger M, Ballauff M, Drechsler M, Henrich O, Fuchs M (2006) J Chem Phys 125:204906
152. Rodriquez BE, Wolfe MS, Fryd M (1994) Macromolecules 27:6642
153. Buscall R (1994) Colloids Surf A Physicochem Eng Asp 83:33
154. Paulin SE, Ackerson BJ, Wolfe MS (1996) J Colloid Interface Sci 178:251
155. Paulin SE, Ackerson BJ, Wolfe MS (1998) Phys Rev E 55:5812
156. Borrega R, Cloitre M, Betremieux I, Ernst B, Leibler L (1999) Europhys Lett 47:729
157. Cloitre M, Borrega R, Leibler L (2000) Phys Rev Lett 85:4819
158. Meeker SP, Bonnecaze RT, Cloitre M (2004) Phys Rev Lett 92:198302
159. Cloitre M, Borrega R, Monti F, Leibler L (2003) Phys Rev Lett 90:068303
160. Meeker SP, Bonnecaze RT, Cloitre M (2004) J Rheol 48:1295
161. Lu Y, Mei Y, Ballauff M, Drechsler M (2006) J Phys Chem B 110:3930
162. Lu Y, Mei Y, Ballauff M, Drechsler M (2006) Angew Chem Int Ed Engl 45:813
163. Mewis J, Vermant J (2000) Progr Org Coating 40:111
164. Ottewill RH, Rennie AR (1998) Modern aspects of colloidal dispersions. Results from the DTI colloid technology programme. Kluwer, Amsterdam
165. Smay JE, Cesarano J, Lewis JA (2002) Langmuir 18:5429
166. Buscall R, Goodwin JW, Hawkins MW, Ottewill RH (1982) J Chem Soc Faraday Trans 78:2889
167. Wagner NJ (1993) J Colloid Interface Sci 161:169
168. Buscall R (1995) In: Goodwin JW, Buscall R (eds) Colloidal polymer particles. Academic, New York, NY
169. van Megen W, Snook I (1984) Adv Colloid Interface Sci 21:119
170. Witten TA, Pincus PA (1986) Macromolecules 19:2509
171. Cloitre M (2008) Private communication
172. Denton AR (2003) Phys Rev E 67:011804
173. Benoit H, Higgins J (1994) Polymers and neutron scattering. Oxford University Press, New York, NY
174. Harreis HM, Likos CN, Ballauff M (2003) J Chem Phys 118:1979
175. Götze IO, Likos CN (2005) J Phys Condens Matter 17:S1777
176. Pakula T, Vlassopoulos D, Fytas G, Roovers J (1998) Macromolecules 31:8931
177. Witten TA, Pincus PA, Cates ME (1986) Europhys Lett 2:137
178. Watzlawek M, Likos CN, Löwen H (1999) Phys Rev Lett 82:5289
179. Voudouris P, Loppinet B, Petekidis G (2008) Phys Rev E 77:051402
180. Bartsch E, Frenz V, Möller S, Sillescu H (1993) Physica A 201:363
181. Cheng Z, Zhu J, Chaikin PM, Phan S-E, Ruseel WB (2002) Phys Rev E 65:041405
182. Fleischer G, Fytas G, Vlassopoulos D, Roovers J (2000) Physica A 280:266
183. Vlassopoulos D, Fytas G, Pispas S, Hadjichristidis N (2001) Physica B 296:184

184. Roovers J (1994) Macromolecules 27:5359
185. Sergè PN, Meeker SP, Pusey PN, Poon WCK (1995) Phys Rev Lett 75:958
186. Petekidis G, Vlassopoulos D, Pusey PN (2004) J Phys Condens Matter 16:S3955
187. Cherdhirankorn T, Best A, Koynov K, Peneva K, Muellen K, Fytas G (2009) J Phys Chem B 113:3355
188. Sigel R, Pispas S, Vlassopoulos D, Hadjichristidis N, Fytas G (1999) Phys Rev Lett 83:4666
189. Seghrouchni R, Petekidis G, Vlassopoulos D, Fytas G, Semenov AN, Roovers J, Fleischer G (1998) Europhys Lett 42:271
190. Loppinet B, Stiakakis E, Vlassopoulos D, Fytas G, Roovers J (2001) Macromolecules 34:8216
191. Vlassopoulos D, Pakula T, Roovers J (2002) Condens Matter Phys 5:105
192. Roovers J, Toporowski PM, Douglas J (1995) Macromolecules 28:7064
193. Brown W (ed) (1993) Dynamic light scattering. The method and some applications. Oxford University Press, New York, NY
194. Pusey PN (1985) In: Pecora R (ed) Dynamic light scattering. Plenum Press, New York, NY
195. Boudenne N, Anastasiadis SH, Fytas G, Xenidou M, Hadjichristidis N, Semenov AN, Fleischer G (1996) Phys Rev Lett 77:506
196. Sergè PN, Pusey PN 1996 Phys Rev Lett 77:771
197. Kapnistos M, Semenov AN, Vlassopoulos D, Roovers J (1999) J Chem Phys 111:1753
198. Yakubov GE, Loppinet B, Zhang H, Rühe J, Sigel R, Fytas G (2004) Phys Rev Lett 92:115501
199. Kapnistos M, Vlassopoulos D, Fytas G, Mortensen K, Fleischer G, Roovers J (2000) Phys Rev Lett 85:4072
200. Sigel R, Fytas G, Vainos N, Pispas S, Hadjichristidis N (2002) Science 297:67
201. Conti C, Ruocco G, Trillo S (2006) Phys Rev Lett 97:123903
202. Förster S, Zisenis M, Wenz E, Antonietti M (1996) J Chem Phys 104:9956
203. Lin EK, Gast AP (1996) Macromolecules 29:390
204. Ohno K, Morinaga T, Takeno S, Tsujii Y, Fukuda T (2007) Macromolecules 40:9143
205. Stellbrink J, Rother G, Laurati B, Lund R, Willner L, Richter D (2004) J Phys Condens Matter 16:S3821
206. Loppinet B, Fytas G, Vlassopoulos D, Likos CN, Meier G, Liu G (2005) Macromol Chem Phys 206:163
207. Cloitre M, Borrega R, Monti F, Leibler L (2003) C R Phys 4:221
208. Voudouris P, Fytas G, Matyjaszewski K, Bockstaller M (2009) Macromolecules 42:2721
209. Yoshinaga K, Chiyoba M, Ishiki H, Okubo T (2002) Colloids Surf A 204:285
210. Mortensen K, Brown W, Norden B (1992) Phys Rev Lett 68:2340
211. Watanabe H, Kotaka T, Hashimoto T, Shibayama M, Kawai H (1982) J Rheol 26:153
212. Castelletto V, Caillet C, Hamley IW, Yang Z (2002) Phys Rev E 65:050601(R)
213. Lyon LA, Debord JD, Debord SB, Jones CD, McGrath JG, Serpe MJ (2004) J Phys Chem B 108:19099
214. Alsayed AM, Islam MF, Zhang J, Collings PJ, Yodh AG (2005) Science 309:1207
215. Meng Z, Cho JK, Debord S, Breedveld V, Lyon LA (2007) J Phys Chem B 111:6992
216. Jucknischke O (1995) Untersuchung der Struktur von Sternpolymeren in Lösung mit Neutronenkleinwinkelstreuung. Ph.D Thesis, Westfaelischen Wilhelms-Universitaet Muenster
217. Mortensen K (2007) Private communication
218. Vermant J (2001) Curr Opin Colloid Interface Sci 6:489
219. Butler P (1999) Curr Opin Colloid Interface Sci 4:214
220. Hamley IW (2000) Curr Opin Colloid Interface Sci 5:341
221. Vlassopoulos D, Fytas G, Fleischer G, Pakula T, Roovers J (1999) Faraday Discuss 112:225
222. Watanabe H, Matsumiya H, Ishida S, Takigawa T, Yamamoto T, Vlassopoulos D, Roovers J (2005) Macromolecules 38:7404
223. Makrakis S, Loppinet B, Petekidis G, Vlassopoulos G, Roovers J (2008) in preparation
224. Foffi G, Sciortino F, Tartaglia P, Zaccarelli E, Lo Verso F, Reatto L, Dawson KA, Likos CN (2003) Phys Rev Lett 90:238301
225. Helgeson ME, Wagner NJ, Vlassopoulos D (2007) J Rheol 51:297
226. Rissanou A, Yiannourakou M, Economou IG, Bitsanis IA (2006) J Chem Phys 124:044905

227. Wilk A, Stiakakis E, Petekidis G, Vlassopoulos D (2009) submitted to Phys Rev Lett
228. Anderson PW (1995) Science 267:1615
229. Debenedetti PG, Stillinger FH (2001) Nature 410:259
230. Ediger MD, Angell CA, Nagel SR (1996) J Phys Chem 100:13200
231. Götze W (1991) In: Hansen JP, Levesque D, Zinn-Justin J (eds) Liquids, freezing and glass transition. North-Holland, Amsterdam
232. Angell CA, Ngai KL, McKenna GB, McMillan PF, Martin SW (2005) J Appl Phys 88:3113
233. Cates ME (2003) Ann Henri Poincare 4(Suppl 2):S647
234. Schweizer KS (2007) Curr Opin Colloid Interface Sci 12:297
235. Mayer P, Miyazaki K, Reichman DR (2006) Phys Rev Lett 97:095702
236. Schilling R (2000) J Phys Condens Matter 12:6311
237. van Megen W, Underwood SM (1994) Phys Rev E 49:4206
238. Mason TG, Weitz DA (1995) Phys Rev Lett 75:2770
239. Weeks ER, Weitz DA (2002) Phys Rev Lett 89:095704
240. Chen DT, Weeks ER, Crocker JC, Islam MF, Verma R, Gruber J, Levine AJ, Lubensky TC, Yodh AG (2003) Phys Rev Lett 90:108301
241. Cipelletti L, Ramos L (2005) J Phys Condens Matter 17:R253
242. Purnomo EH, van den Ende D, Mellema J, Mugele F (2007) Phys Rev E 76:021404
243. Edwards SF (1967) Proc Phys Soc 92:9
244. de Gennes PG (1971) J Chem Phys 55:572
245. Cohen EG, de Schepper IM (1991) J Stat Phys 63:241
246. Frenkel J (1946) Kinetic theory of liquids. The Clarendon Press, Oxford
247. Pham K, Egelhaaf SU, Pusey PN, Poon WCK (2004) Phys Rev E 69:011503
248. Stiakakis E, Petekidis G, Vlassopoulos D, Likos CN, Iatrou H, Hadjichristidis N, Roovers J (2005) Europhys Lett 72:664
249. Puertas AM, Fuchs M, Cates ME (2002) Phys Rev Lett 88:098301
250. Salaniwal S, Kumar SK, Douglas JF (2002) Phys Rev Lett 89:258301
251. Rissanou AN, Vlassopoulos D, Bitsanis IA (2005) Phys Rev E 71:011402
252. Coussot P (2007) Soft Matter 3:528
253. Stokes JR, Frith WJ (2008) Soft Matter 4:1133
254. Bonn D, Tanaka H, Coussot P, Meunier J (2004) J Phys Condens Matter 16:S4987
255. Mallamace F, Broccio M, Tartaglia P, Chen W-R, Faraone A, Chen SH (2003) Physica A 330:206
256. Crassous JJ, Regisser R, Ballauff M, Willenbacher N (2005) J Rheol 49:851
257. Pham KN, Petekidis G, Vlassopoulos D, Egelhaaf SU, Poon WCK, Pusey PN (2008) J Rheol 52:649
258. Swallen SF, Kearns KL, Mapes MK, Kim YS, McMallon RJ, Ediger MD, Wu T, Yu L, Satija S (2007) Science 315:353
259. Fuchs M, Cates ME (2003) Faraday Discuss 123:267
260. Kobelev V, Schweizer KS (2005) Phys Rev E 71:021401
261. Coussot P (2006) Lect Notes Phys 688:69
262. Moller PCF, Mewis J, Bonn D (2006) Soft Matter 2:274
263. Shah SA, Chen YL, Ramakrishnan S, Schweizer KS, Zukoski CF (2003) J Phys Condens Matter 15:4751–4778
264. Poon WCK (1998) Curr Opin Colloid Interface Sci 3:593
265. Stiakakis E, Vlassopoulos D, Likos CN, Roovers J, Meier G (2002) Phys Rev Lett 89:208302
266. Mayer C, Stiakakis E, Zaccarelli E, Likos CN, Sciortino F, Tartaglia P, Löwen H, Vlassopoulos D (2007) Rheol Acta 46:611
267. Mayer C, Zaccarelli E, Stiakakis E, Likos CN, Sciortino F, Munam A, Gauthier M, Hadjichristidis N, Iatrou H, Tartaglia P, Löwen H, Vlassopoulos D (2008) Nat Mater 7:780
268. Imhof A, Dhont JKG (1995) Phys Rev Lett 75:1662
269. Rogers S, Vlassopoulos D, Callaghan PT (2008) Phys Rev Lett 100:128304
270. Cloitre M, Borrega R, Leibler L (2000) Phys Rev Lett 95:4819
271. Mewis J, Wagner NJ (2009) Adv Colloid Interface Sci 147–148:214
272. Dullaert K, Mewis J (2005) Rheol Acta 45:23

273. Dullaert K, Mewis J (2005) J Rheol 49:1213
274. Giannelis EP, Krishnammorti R, Manias E (1999) Adv Polym Sci 138:107
275. Usiki A, Kojima M, Okada A, Fukushima Y, Kamigaito O (1993) J Mater Res 8:1179
276. Mackay ME, Dao TT, Tuteja A, Ho DL, Horn BV, Kim H-C, Hawker CJ (2003) Nat Mater 2:762
277. Balazs AC, Emrick T, Russell TP (2006) Science 314:1107
278. Solomon MJ, Almusallam AS, Seefeldt KF, Somwangthanaroj A, Varadan P (2001) Macromolecules 34:1864
279. Varadan P, Solomon MJ (2001) Langmuir 17:2918
280. Sepehr M, Carreau PJ, Moan M, Ausias G (2004) J Rheol 48:1023
281. Rajabian M, Dubois C, Grmela M, Carreau PJ (2008) Rheol Acta 47:701
282. Vermant J, Ceccia S, Dolgovskij MK, Maffettone PL, Macoscko CW (2007) J Rheol 51:429
283. Ciprari D (2002) Macromolecules 43:2981
284. Gersappe D (2002) Phys Rev Lett 89:058301
285. Wu CI (2005) Compos Sci Technol 65:635
286. Lin Y, Böker A, He J, Sill K, Xiang H, Abetz C, Li X, Wang X, Emrick T, Long S, Wang Q, Balazs A, Russell TP (2005) Nature 434:55
287. Jiang W, Kim BS, Rutka JT, Chan WCW (2008) Nat Nanotechnol 3:145
288. Mackay ME, Tuteja A, Duxbury PM, Hawker CJ, Van Horn B, Guan Z, Chen G, Krishnan RS (2006) Science 311:1740
289. Tyagi S, Lee JY, Buxton G, Balasz AC (2004) Macromolecules 37:9160
290. Hooper JB, Schweizer KS (2006) Macromolecules 39:5133
291. Nakatami AI, Chen W, Schmidt RG, Gordon GW, Han CC (2001) Polymer 42:3713
292. Sen S, Xie Y, Kumar SK, Yang H, Bansel A, Ho DL, Hall L, Hooper JB, Schweizer KS (2007) Phys Rev Lett 98:128302
293. Tuteja A, Duxbury PM, Mackay ME (2008) Phys Rev Lett 100:077801
294. Erguney FM, Lin H, Mattice WL (2005) Polymer 46:6154
295. Binder K, Lai PY, Wittmer J (1994) Faraday Discuss 98:97
296. Botuchov I, Leibler L (2000) Phys Rev E 62:R41
297. Klos J, Pakula T (2004) Macromolecules 37:8145
298. Kim JU, O'Shaughnessy B (2006) Macromolecules 39:413
299. Kim JU, Matsen MW (2008) Macromolecules 41:246
300. Yezek L, Schärtl W, Chen Y, Gohr K, Schmidt M (2003) Macromolecules 36:4226
301. Hasagawa R, Aoki Y, Doi M (1996) Macromolecules 29:6656
302. Bockstaller MR, Thomas EL (2004) Phys Rev Lett 93:166106
303. Kim BJ, Chiu JJ, Yi GR, Pine DJ, Kramer EJ (2005) Adv Mater 17:2618
304. Thompson XRT, Ginzburg VV, Matsen MW, Balasz AC (2001) Science 292:2469
305. Kim BJ, Fredickson GH, Kramer EJ (2008) Macromolecules 41:436
306. Pryamitsyn V, Ganesan V (2006) Macromolecules 39:8499
307. Sides SW, Kim BJ, Kramer EJ, Fredrickson GH (2006) Phys Rev Lett 96:250601
308. Bockstaller MR, Lapetnikov Y, Margel S, Thomas EL (2003) J Am Chem Soc 125:5276
309. Filippidi E, Michailidou V, Loppinet B, Rühe J, Fytas G (2007) Langmuir 23:5139
310. Lee J-Y, Thomson RB, Jasnow D, Balazs AC (2002) Phys Rev Lett 89:155503
311. Warren S, Disalvo FJ, Wiesner U (2007) Nat Mater 6:156
312. Kim BJ, Bang J, Hawker CJ, Kramer EJ (2006) Mcromolecules 39:4108
313. Stratford K, Adhikari R, Pagonabarraga I, Desplat J-C, Cates ME (2005) Science 309:2198
314. Lee JY, Shou Z, Balazs AC (2003) Phys Rev Lett 91:136103
315. Kempe MD, Kornfield JA (2003) Phys Rev Lett 90:115501
316. Araki T, Tanaka H (2008) J Phys Condens Matter 20:072101
317. Ginzburg VV (2005) Macromolecules 38:2362
318. Gupta S, Zhang Q, Emrick T, Balazs AC, Russell TP (2006) Nat Mater 5:229
319. Buxton GA, Lee JY, Balazs AC (2003) Macromolecules 36:9631
320. Yoon J, Lee W, Caruge JM, Thomas EL, Koi S, Prasad P (2006) Appl Phys Lett 88:091202
321. Cheng W, Gorishnyy T, Krikorian V, Fytas G, Thomas EL (2006) Macromolecules 39:9614
322. Thomson RB, Rasmussen KO, Lookman T (2004) Nano Lett 4:2455

323. Starr FW, Schroeder TB, Glotzer SC (2002) Macromolecules 35:4481
324. Bansal A, Yang H, Li C, Cho K, Benicewicz BC, Kumar SK, Schadler LS (2005) Nat Mater 4:693
325. Lee JJ, Su KE, Chan EP, Zhang Q, Emrick T, Crosby AJ (2007) Macromolecules 40:7755
326. Kropka JM, Sakai VG, Geen PF (2008) Nano Lett 8:1061
327. Cheng W, Gomopoulos N, Fytas G, Gorishnyy T, Walish J, Thomas EL, Hiltner A, Baer E (2008) Nano Lett 8:1423
328. Courty A, Mermet A, Albouy PA, Duval E, Pileni PM (2005) Nat Mater 4:395
329. Cheng W, Wang J, Jonas U, Fytas G, Stefanou N (2006) Nat Mater 5:830
330. Mazurenko DA, Shan X, Stiefelhagen JCP, Graf CM, van Blaaderen A, Dijkhuis JI (2007) Phys Rev B 75:161102R
331. Patterson GD, Latham JP (1980) J Polym Sci Macromol Rev 15:1
332. Still T, Retsch M, Sainidou R, Jonas U, Spahn P, Hellmann G, Fytas G (2008) Nano Lett 8:3194
333. Penciu RS, Fytas G, Economou EN, Steffen W, Yannopoulos SN (2000) Phys Rev Lett 85:4622
334. Vermant J, Vanderbril S, Dewitte C, Moldenaers P (2008) Rheol Acta 47:835
335. Vermant J, Cioccolo G, Golapan Nair K, Moldenaers P (2004) Rheol Acta 43:529
336. Binks BP, Horozov TS (2006) Colloidal particles at liquid interfaces. Cambridge University Press, Cambridge
337. Stratford K, Adhikari R, Pagonabarraga I, Desplat J-C, Cates ME (2005) Science 309:2198
338. Cates ME, Clegg PS (2008) Soft Matter 4:2132
339. Herzig EM, White KA, Schofield AB, Poon WCK, Clegg PS (2007) Nat Mater 6:966
340. Callaghan PT (2008) Rheol Acta 47:243
341. Vermant J, Solomon MJ (2005) J Phys Condens Matter 17:R187
342. Dhont JKG, Briels WJ (2008) Rheol Acta 47:257
343. Olmsted PD (2008) Rheol Acta 47:283
344. Cates ME, Fielding SM (2006) Adv Phys 55:799
345. Manneville S (2008) Rheol Acta 47:301
346. Lee YS, Wagner NJ (2006) Ind Eng Chem Res 45:7015
347. Krishnamurthy L, Wagner NJ, Mewis J (2006) J Rheol 50:1347
348. Osuji CO, Kim C, Weitz DA (2008) Phys Rev E 77:060402(R)
349. Scirocco R, Vermant J, Mewis J (2005) J Rheol 49:551
350. Royall CP, Vermolen ECM, van Blaaderen A (2008) J Phys Condens Matter 20:404225
351. Leunissen ME, Sullivan MT, Chaikin PM, van Blaaderen A (2008) J Chem Phys 128:164508
352. Li Q, Jonas U, Zhao XS, Kappl M (2008) Asia Pac J Chem Eng 3:255
353. Nelson EK, Braun PV (2007) Science 318:924
354. Hales K, Chen Z, Wooley KL, Pochan DJ (2008) Nano Lett 8:2023
355. Chen T, Zhang Z, Glotzer SC (2007) PNAS 104:717
356. Glotzer SC (2004) Science 306:519
357. van Blaaderen A (2006) Nature 439:545
358. Glotzer SC, Solomon MJ (2007) Nat Mater 6:557
359. Srinivas G, Pitera JW (2008) Nano Lett 6:611
360. Bianchi E, Largo J, Tartaglia P, Zaccarelli E, Sciortino F (2006) Phys Rev Lett 97:168301
361. Kegel WK, Breed D, Elsesser M, Pine DJ (2006) Langmuir 22:7135
362. Manoharan VN, Elsesser ME, Pine DJ (2003) Science 301:483
363. Zerrouki D, Rotenberg B, Abramson S, Baudry J, Goubault C, Leal-Calderon F, Pine DJ, Bibette J (2006) Langmuir 22:57
364. Kim J-W, Lee D, Shum HC, Weitz DA (2008) Adv Mater 20:3239
365. Hermans TM, Broeren MAC, Gomopoulos N, Schoot PAM, Van Genderen MHP, Sommerdijk NAM, Fytas G, Meijer EW (2009) Nature Nanotechnology (DOI:10.1038/nnano.2009.232)
366. Mirkin CA, Letsinger RL, Mucic RC, Storhoff JJ (1996) Nature 382:607
367. Proupin-Pérez M, Cosstick R, Liz-Marzan LM, Salgueirino-Maceira V, Brust M (2005) Nucleotides Nucleic Acids 24:1075

368. Park SY, Lytton-Jean AKR, Lee B, Weigand S, Schatz G, Mirkin CA (2008) Nature 451:553
369. Largo J, Tartaglia P, Sciortino F (2007) Phys Rev E 76:011402
370. Largo J, Starr FW, Sciortino F (2007) Langmuir 23:5896
371. Schmatko T, Bozorgui B, Geerts N, Frenkel D, Eiser E, Poon WCK (2007) Soft Matter 3:703
372. Askin A, Dziedzic JM, Smith PW (1982) Opt Lett 7:276
373. Bosanac L, Aabo T, Bendix PM, Odderrshede LB (2008) Nano Lett 5:1486
374. Stegeman GI, Segev M (1999) Science 286:1518
375. Peccianti M, Conti C, Assanto C, De Luca A, Umeton C (2004) Nature 432:733
376. Reece PJ, Wright EM, Dholakia K (2007) Phys Rev Lett 98:203902
377. El-Ganainy R, Christodoulides R, Musslimani DN, Ziad H, Rotschild C, Segev M (2007) Opt Lett 32:3185
378. Mitsui T, Wakayama Y, Onodera T, Takaya Y, Oikawa H (2008) Nano Lett 8:853
379. Brader JM, Voigtmann Th, Cates ME, Fuchs M (2007) Phys Rev Lett 98:058301
380. Sollich P, Lequeux F, Hebraud H, Cates ME (1997) Phys Rev Lett 78:2020
381. Fielding SM, Sollich P, Cates ME (2000) J Rheol 44:323
382. Fielding SM, Cates ME, Sollich P (2009) Soft Matter 5:2378
383. van den Noort A, Otter WK, Briels WJ (2007) Europhys Lett 80:28003
384. van den Noort A, Briels WJ (2007) Macromol Theory Simul 16:742
385. Kindt P, Briels WJ (2007) J Chem Phys 127:134901
386. van den Noort A, Briels WJ (2008) J Non-Newtonian Fluid Mech 152:148
387. Malevanets A, Kapral R (1999) J Chem Phys 110:8605
388. Tao YG, Götze IO, Gompper G (2008) J Chem Phys 128:144902
389. Padding JT, Louis AA (2004) Phys Rev Lett 93:220601
390. Phung TN, Brady JF, Bossis G (1996) J Fluid Mech 313:181
391. Ripoll M, Winkler RG, Gompper G (2006) Phys Rev Lett 96:188302
392. Dhont JKG, Lettinga MP, Dogic Z, Lenstra TAJ, Wang H, Rathgeber S, Carletto P, Willner L, Frielinghaus H, Lindner P (2003) Faraday Discuss 123:157
393. Beris AN, Stiakakis E, Vlassopoulos D (2008) J Non-Newtonian Fluid Mech 152:76
394. Dullaert K, Mewis J (2006) J Non-Newtonian Fluid Mech 139:21
395. Roussel N, Le Roy R, Coussot P (2004) J Non-Newtonian Fluid Mech 117:85
396. Dullaert K, Mewis J (2005) J Colloid Interface Sci 287:542
397. Seth JR, Cloitre M, Bonnecaze R (2008) J Rheol 52:1241
398. Ballesta P, Besseling R, Isa L, Petekidis G, Poon WCK (2009) Phys Rev Lett 101:258301
399. Besseling R, Weeks ER, Schofield AB, Poon WCK (2007) Phys Rev Lett 99:028301
400. Degre G, Joseph P, Tabeling P, Lerouge S, Cloitre M (2006) Appl Phys Lett 89:024104
401. Reynaert S, Moldenaers P, Vermant J (2007) Phys Chem Chem Phys 9:6463
402. Reynaert S, Brooks CF, Moldenaers P, Vermant J, Fuller GG (2008) J Rheol 52:261
403. Brooks CF, Fuller GG, Frank CW, Robertson CR (1999) Langmuir 15:2450
404. Monteux C, Jung E, Fuller GG (2007) Langmuir 23:3975
405. Basavaraj MG, Fuller GG, Fransaer J, Vermant J (2006) Langmuir 22:6605

Nonlinear Rheological Properties of Dense Colloidal Dispersions Close to a Glass Transition Under Steady Shear

Matthias Fuchs

Abstract The nonlinear rheological properties of dense colloidal suspensions under steady shear are discussed within a first principles approach. It starts from the Smoluchowski equation of interacting Brownian particles in a given shear flow, derives generalized Green–Kubo relations, which contain the transients dynamics formally exactly, and closes the equations using mode coupling approximations. Shear thinning of colloidal fluids and dynamical yielding of colloidal glasses arise from competition between a slowing down of structural relaxation because of particle interactions, and enhanced decorrelation of fluctuations caused by the shear advection of density fluctuations. The integration through transients (ITT) approach takes account of the dynamic competition, translational invariance enters the concept of wavevector advection, and the mode coupling approximation enables one to explore quantitatively the shear-induced suppression of particle caging and the resulting speed-up of the structural relaxation. Extended comparisons with shear stress data in the linear response and in the nonlinear regime measured in model thermo-sensitive core-shell latices are discussed. Additionally, the single particle motion under shear observed by confocal microscopy and in computer simulations is reviewed and analysed theoretically.

Keywords Colloidal dispersion · Flow curve · Glass transition · Integration through transients approach · Linear viscoelasticity · Mode coupling theory · Nonlinear rheology · Non-equilibrium stationary state · Shear modulus · Steady shear

Contents

List of Abbreviations and Symbols 56
1 Introduction 57

M. Fuchs (✉)
Fachbereich Physik, Universität Konstanz, 78457 Konstanz, Germany
e-mail: matthias.fuchs@uni-konstanz.de

2	Microscopic Approach	60
	2.1 Interacting Brownian Particles	60
	2.2 Integration Through Transients Approach	62
	2.3 A Microscopic Model: Brownian Hard Spheres	72
	2.4 Accounting for Hydrodynamic Interactions	72
	2.5 Comparison with Other MCT Inspired Approaches to Sheared Fluids	73
3	Microscopic Results in Linear Response Regime	74
	3.1 Shear Moduli Close to the Glass Transition	76
	3.2 Distorted Structure Factor	81
4	Universal Aspects of the Glass Transition in Steady Shear	83
5	Simplified Models	89
	5.1 Isotropically Sheared Hard Sphere Model	89
	5.2 Schematic $F_{12}^{(\dot\gamma)}$-model	95
6	Comparison of Theory and Experiment	104
	6.1 ISHSM and Single Particle Motion Under Steady Shear	104
	6.2 $F_{12}^{(\dot\gamma)}$-Model and Shear Stresses in Equilibrium and Under Flow in a Polydisperse Dispersion	106
	6.3 $F_{12}^{(\dot\gamma)}$-Model and Flow Curves of a Simulated Supercooled Binary Liquid	110
7	Summary and Outlook	113
References		113

List of Abbreviations and Symbols

G_∞	Shear modulus of a solid (transverse elastic constant or Lame-coefficient)
η_0	Newtonian viscosity of a fluid
σ	Shear stress
$\dot\gamma$	Shear rate
$g(t)$	Time dependent shear modulus; in the linear response regime denoted as $g^{lr}(t)$ of the quiescent system; generalized one if including dependence on shear rate
τ	Maxwell (final or α-relaxation) time of structural relaxation
$G'(\omega)$	Storage modulus in linear response
$G''(\omega)$	Loss modulus in linear response
η	Shear viscosity; defined via $\eta = \sigma(\dot\gamma)/\dot\gamma$
S_q	Equilibrium structure factor
R_H	Hydrodynamic radius of a colloidal particle (radius $a = R_H$ assumed)
d	Colloid diameter ($d = 2a = 2R_H$ assumed throughout)
$k_B T$	Thermal energy
η_s	Solvent viscosity
D_0	Stokes Einstein Sutherland diffusion coefficient at infinite dilution
Pe_0	Bare Peclet number
Pe	Dressed Peclet or Weissenberg number
ϕ	Packing fraction $\phi = \frac{4\pi}{3} R_H^3 n$ of spheres of radius R_H at number density n
ε	Separation parameter in MCT giving the relative distance in a thermodynamic control parameter to its value at the glass transition
λ	MCT exponent parameter

G'_∞	Instantaneous isothermal shear modulus
η_∞	High frequency viscosity
σ^+	Dynamic yield stress of a shear molten glass
HI	Hydrodynamic/solvent induced interactions
MCT	Mode coupling theory
ITT	Integrations through transients approach
SO	Smoluchowski operator Ω
PY	Percus–Yevick theory giving the approximate PY S_q of a hard sphere fluid
ISHSM	Isotropically sheared hard spheres model
$F_{12}^{(\dot\gamma)}$	Schematic model without spatial resolution considering a single correlator

1 Introduction

Rheological and elastic properties under flow and deformations are highly characteristic for many soft materials like complex fluids, pastes, sands, and gels, viz. soft (often metastable) solids of dissolved macromolecular constituents [1]. Shear deformations, which conserve volume but stretch material elements, often provide the simplest experimental route to investigate the materials. Moreover, solids and fluids respond in a characteristically different way to shear, the former elastically, the latter by flow. The former are characterized by a shear modulus G_∞, corresponding to a Hookian spring constant, the latter by a Newtonian viscosity η_0, which quantifies the dissipation.

Viscoelastic materials exhibit both elastic and dissipative phenomena depending on external control parameters like temperature and/or density, and depending on frequency or time-scale of experimental observation. Viscoelastic fluids differ from pastes and sands in the importance of thermal fluctuations causing Brownian motion, which enables them to explore their phase space without external drive like shaking that would be required to fluidize granular systems. The change between fluid and solid like behavior in viscoelastic materials can have diverse origins, including phase transitions of various kinds, like freezing and micro-phase separation, and/or molecular mechanisms like entanglement formation in polymer melts. One mechanism existing quite universally in dense particulate systems is the glass transition, that structural rearrangements of particles become progressively slower [2] because of interactions/collisions, and that the structural relaxation time grows dramatically.

Maxwell was the first to describe the viscoelastic response at the fluid-to-glass transition phenomenologically. He introduced a time-dependent shear modulus $g(t)$ describing the response of a viscoelastic fluid to a time-dependent shear deformation,

$$\sigma(t) = \int_0^t dt'\, g(t-t')\, \dot\gamma(t') . \tag{1}$$

Here σ is the (transverse) stress, the thermodynamic average of an off-diagonal element of the microscopic stress tensor, and $\dot{\gamma}(t)$ is the time-dependent shear rate impressed on the material starting at time $t = 0$. Maxwell chose the Ansatz $g(t) = G_\infty \exp\{-(t/\tau)\}$, which interpolates between elastic behavior $\sigma(t \to 0) \approx G_\infty \gamma(t)$ for short times $t \ll \tau$ and dissipative behavior, $\sigma(t) \approx \eta_0 \dot{\gamma}(t)$ for long times $t \gg \tau$; the strain $\gamma(t)$ is obtained from integrating up the strain rate, $\gamma(t) = \int_0^t dt' \dot{\gamma}(t')$. Maxwell found the relation $\eta_0 = G_\infty \tau$ which connects the structural relaxation time and the glass modulus G_∞ to the Newtonian viscosity. He thus explained the increase of the viscosity at the glass transition by the slowing down of the structural dynamics (viz. the increase of τ), and provided a definition of an idealized glass state where $\tau = \infty$. It responds purely elastically.

Above relation (1) between σ and $\dot{\gamma}$ is exact in linear response, where nonlinear contributions in $\dot{\gamma}$ are neglected in the stress. The linear response modulus (to be denoted as $g^{lr}(t)$) itself is defined in the quiescent system and describes the small shear-stress fluctuations always present in thermal equilibrium [1, 3]. Often, oscillatory deformations at fixed frequency ω are applied and the frequency dependent storage- ($G'(\omega)$) and loss- ($G''(\omega)$) shear moduli are measured in or out of phase, respectively. The former captures elastic while the latter captures dissipative contributions. Both moduli result from Fourier-transformations of the linear response shear modulus $g^{lr}(t)$, and are thus connected via Kramers–Kronig relations.

The stationary, nonlinear rheological behavior under steady shearing provides additional insight into the physics of dense colloidal dispersions [1, 3]. Here, the shear rate is constant, $\dot{\gamma}(t) \equiv \dot{\gamma}$, and the stress $\sigma(\dot{\gamma})$ in the stationary state achieved after waiting sufficiently long (taking $t \to \infty$ in (1)) is of interest. Equation (1) may be interpreted under flow to state that the non-linearity in the stress vs shear rate curve (the relation $\sigma(\dot{\gamma})$ is termed "flow curve") results from the dependence of the (generalized) time-dependent shear modulus $g(t, \dot{\gamma})$ on shear rate. The (often) very strong decrease of the viscosity, defined via $\eta(\dot{\gamma}) = \sigma(\dot{\gamma})/\dot{\gamma}$, with increasing flow rate is called "shear thinning," and indicates that the particle system is strongly affected by the solvent flow. One may thus wonder whether the particles' non-affine, random motion relative to the solvent differs qualitatively from the Brownian motion in the quiescent solution. Taylor showed that this is the case for dilute solutions. A single colloidal particle moves super-diffusively for long times along the direction of the flow. Its mean squared non-affine displacement grows with the third power of time, much faster than the linear in time growth familiar from diffusion in the quiescent system[1]. A priori it is thus not clear, whether the mechanisms relevant during glass formation in the quiescent system also dominate the nonlinear rheology. Solvent mediated interactions (hydrodynamic interactions, HI), which do not affect the equilibrium phase diagram, may become crucially important. Also, shear may cause ordering or layering of the particles leading to heterogeneities of various kinds [4].

---

[1] This effect that flow speeds up the irreversible mixing is one mechanism active when stirring a solution. The non-affine motion even in laminar flow prevents the effect that stirring backwards would reverse the motion of the dissolved constituents.

Within a number of theoretical approaches a connection between steady state rheology and the glass transition has been suggested. Brady worked out a scaling description of the rheology based on the concept that the structural relaxation arrests at random close packing [5]. In the soft glassy rheology model, the trap model of glassy relaxation by Bouchaud was generalized by Cates and Sollich and coworkers to describe mechanical deformations and ageing [6–8]. The mean field approach to spin glasses was generalized to systems with broken detailed balance in order to model flow curves of glasses under shear [9, 10]. The application of these novel approaches to colloidal dispersions has led to numerous insights, but has been hindered by the use of unknown parameters in the approaches.

Dispersions consisting of colloidal, slightly polydisperse (near) hard spheres arguably constitute one of the most simple viscoelastic systems, where a glass transition has been identified. It has been studied in detail by dynamic light scattering measurements [11–19], confocal microscopy [20], linear [21, 22], and non-linear rheology [23–33]. Computer simulations are also available [34–36]. Mode coupling theory (MCT) has provided a semi-quantitative explanation of the observed glass transition phenomena, albeit neglecting ageing effects [37] and decay processes at ultra-long times that may cause (any) colloidal glass to flow ultimately [2, 38, 39]. It has thus provided a microscopic approach recovering Maxwell's phenomenological picture of the glass transition; G_∞ and τ could be calculated starting from the particle interactions as functions of the thermodynamic control parameters. MCT was also generalized to include effects of shear on time dependent fluctuations [40–42], and, within the *integrations through transients* (ITT) approach, to describe quantitatively all aspects of stationary states under steady shearing [43–45].

The MCT-ITT approach thus provides a microscopic route to calculate the generalized shear modulus $g(t,\dot{\gamma})$ and other quantities characteristic of the quiescent and the stationary state under shear flow. While MCT has been reviewed thoroughly, see, e.g., [2, 38, 39], the MCT-ITT approach shall be reviewed here, including its recent tests by experiments in model colloidal dispersions and by computer simulations. The recent developments of microscopy techniques to study the motion of individual particles under flow and the improvements in rheometry and preparation of model systems, provide detailed information to scrutinize the theoretical description, and to discover the molecular origins of viscoelasticity in dense colloidal dispersions even far away from thermal equilibrium.

The outline of the review is as follows. First the microscopic starting points, the formally exact manipulations, and the central approximations of MCT-ITT are described in detail. Section 3 summarizes the predictions for the viscoelasticity in the linear response regime and their recent experimental tests. These tests are the quantitatively most stringent ones, because the theory can be evaluated without technical approximations in the linear limit; important parameters are also introduced here. Section 4 is central to the review, as it discusses the universal scenario of a glass transition under shear. The shear melting of colloidal glasses and the key physical mechanisms behind the structural relaxation in flow are described. Section 5 builds on the insights in the universal aspects and formulates successively simpler models which are amenable to complete quantitative analysis. In the next section, those models are compared to experimental data on the microscopic particle motion obtained by confocal microscopy, to data on the macroscopic stresses in dispersions of novel model core-shell particles close to equilibrium and under steady flow, and to simulations providing the same information for binary supercooled mixtures. In the last section, recent generalizations and open questions are addressed.

2 Microscopic Approach

MCT considers interacting Brownian particles, predicts a purely kinetic glass transition, and describes it using only equilibrium structural input, namely the equilibrium structure factor S_q [3, 46] measuring thermal density fluctuations. MCT-ITT extends this statistical mechanics, particle based many-body approach to dispersions in steady flow assuming a linear solvent velocity profile, but neglecting the solvent otherwise.

2.1 Interacting Brownian Particles

N spherical particles with radius R_H are considered, which are dispersed in a volume V of solvent (viscosity η_s). Homogeneous shear is imposed corresponding to a constant linear solvent velocity profile. The flow velocity points along the x-axis and its gradient along the y-axis. The motion of the particles (with positions $\mathbf{r}_i(t)$ for $i = 1,\ldots,N$) is described by N coupled Langevin equations [46]

$$\zeta \left(\frac{d\mathbf{r}_i}{dt} - \mathbf{v}^{\text{solv}}(\mathbf{r}_i) \right) = \mathbf{F}_i + \mathbf{f}_i \,. \tag{2}$$

Solvent friction is measured by the Stokes friction coefficient $\zeta = 6\pi \eta_s R_H$. The interparticle forces $\mathbf{F}_i = -\partial/\partial \mathbf{r}_i U(\{\mathbf{r}_j\})$ derive from potential interactions of particle i with all other colloidal particles; U is the total potential energy. The solvent shear-flow is given by $\mathbf{v}^{\text{solv}}(\mathbf{r}) = \dot{\gamma} y \hat{\mathbf{x}}$, and the Gaussian white noise force satisfies (with α, β denoting directions)

$$\langle f_i^\alpha(t) f_j^\beta(t') \rangle = 2\zeta k_B T \, \delta_{\alpha\beta} \, \delta_{ij} \, \delta(t-t') \,,$$

where $k_B T$ is the thermal energy. Each particle experiences interparticle forces, solvent friction, and random kicks from the solvent. Interaction and friction forces on each particle balance on average, so that the particles are at rest in the solvent on average, giving for their affine motion $\langle \mathbf{r}_i(t) \rangle = \mathbf{r}_i(0) + \dot{\gamma} t \, y_i(0) \hat{\mathbf{x}}$. The Stokesian friction is proportional to the particle's motion *relative to* the solvent flow at its position; the latter varies linearly along the y-direction. The random force on the level of each particle satisfies the fluctuation dissipation relation.

Even though (2) has been obtained under the assumption that solvent fluctuations are close to equilibrium, the Brownian particle system described by it may reach macroscopic states far from thermal equilibrium. Moreover, under (finite) shear, this holds generally because the friction force from the solvent in (2) cannot be derived from a conservative force field. It has non-vanishing curl, and thus the stationary distribution function Ψ describing the probability of the particle positions $\mathbf{r}_i$ cannot be of Boltzmann–Gibbs type [47].

The microscopic starting (2) of MCT-ITT already carries two important approximations. The first is the neglect of hydrodynamic interactions (HI), which would arise from the proper treatment of the solvent flow around moving particles [3, 46]. As vitrification is observed in molecular systems without HI, MCT-ITT assumes that HI are not central to the glass formation of colloidal dispersions. The interparticle forces are assumed to dominate and to hinder and/or prevent structural rearrangements close to arrest into an amorphous, metastable solid. MCT-ITT assumes that pushing the solvent away only provides some additional instantaneous friction, and thus lets short-time transport properties (like single and collective short-time diffusion coefficients, high frequency viscoelastic response, etc.) depend on HI. The second important approximation in (2) is the assumption of an homogeneous shear rate $\dot{\gamma}$. This assumption may be considered as a first step, before heterogeneities and confinement effects are taken into account. The interesting phenomena of shear localization and shear banding and shear driven clustering [48–52] are therefore not addressed. All difficulties in (2) are thus connected to the many-body interactions given by the forces $\mathbf{F}_i$, which couple the N Langevin equations. In the absence of interactions, $\mathbf{F}_i \equiv 0$, (2) immediately leads to the super-diffusive particle motion mentioned in the introduction, which often is termed "Taylor dispersion" [46].

As is well known, the considered microscopic Langevin equations, are equivalent to the reformulation of (2) as a Smoluchowski equation; it is a variant of a Fokker–Planck equation [47]. It describes the temporal evolution of the distribution function $\Psi(\{\mathbf{r}_i\},t)$ of the particle positions

$$\partial_t \Psi(\{\mathbf{r}_i\},t) = \Omega\, \Psi(\{\mathbf{r}_i\},t)\,, \tag{3a}$$

employing the Smoluchowski operator (SO) [3, 46],

$$\Omega = \sum_{j=1}^{N} \left[D_0 \frac{\partial}{\partial \mathbf{r}_j} \cdot \left(\frac{\partial}{\partial \mathbf{r}_j} - \frac{1}{k_B T} \mathbf{F}_j \right) - \dot{\gamma} \frac{\partial}{\partial x_j} y_j \right]\,, \tag{3b}$$

built with the Stokes–Einstein–Sutherland diffusion coefficient $D_0 = k_B T/\zeta$ of a single particle. Averages performed with the distribution function Ψ agree with those obtained from the explicit Lagevin equations.

The Smoluchowski equation is a conservation law for the probability distribution in coordinate space,

$$\partial_t \Psi + \nabla \cdot \mathbf{j} = \partial_t \Psi + \sum_{i=1}^{N} \frac{\partial}{\partial \mathbf{r}_i} \cdot \mathbf{j}_i = 0\,,$$

formed with probability current **j**. Stationary distributions, which clearly obey $\partial_t \Psi_s = 0$, which are not of equilibrium type, are characterised by a non-vanishing probability flux $\mathbf{j}_i^s \neq 0$, where

$$\mathbf{j}_i^s = D_0 \left[-\frac{\partial}{\partial \mathbf{r}_i} + \frac{1}{k_B T} \mathbf{F}_i + \dot{\gamma} y_i \hat{\mathbf{x}} \right] \Psi_s \, .$$

Under shear, $\mathbf{j}_s$ cannot vanish, as this would require the gradient term to balance the term proportional to $\dot{\gamma}$ which, however, has a non-vanishing curl; the "potential conditions" for an equilibrium stationary state are violated under shear [47].

The ITT approach formally exactly solves the Smoluchowski equation, following the transients dynamics into the stationary state. In this way the kinetic competition between Brownian motion and shearing, which arises from the stationary flux, is taken into account in the stationary distribution function. To compute it explicitly, but approximatively, using ideas based on MCT, MCT-ITT approximates the obtained averages by following the transient structural changes encoded in the transient density correlator.

2.2 Integration Through Transients Approach

2.2.1 Generalized Green–Kubo Relations

Formally, the H-theorem valid for general Fokker–Planck equations states that the solution of (3) becomes unique at long times [47]. Yet, because colloidal particles have a non-penetrable core and exhibit excluded volume interactions, corresponding to regions where the potential is infinite, and the proof of the H-theorem requires fluctuations to overcome all barriers, the formal H-theorem may not hold for non-dilute colloidal dispersions. Nevertheless, we assume that the system relaxes into a unique stationary state at long times, so that $\Psi(t \to \infty) = \Psi_s$ holds. This assumption is self-consistent, because later on MCT-ITT finds that under shear all systems are "ergodic" and relax into the stationary state. In cases where phase space decomposes into disjoint pockets ("nonmixing dynamics"), the distribution function calculated in (4) averages over all compartments, and thus cannot be used.

As alread stated, homogeneous, amorphous systems are assumed so that the stationary distribution function Ψ_s is translationally invariant but anisotropic. The formal solution of the Smoluchowski equation for the time-dependent distribution function

$$\Psi(t) = e^{\Omega t} \Psi_e \tag{4a}$$

can, by taking a derivative and integrating it up to $t = \infty$, be brought into the form [43, 44]

$$\Psi_s = \Psi_e + \frac{\dot{\gamma}}{k_B T} \int_0^\infty dt \, \Psi_e \, \sigma_{xy} \, e^{\Omega^\dagger t} \, , \tag{4b}$$

where the adjoint Smoluchowski $\Omega^\dagger$ operator arises from partial integrations over the particle positions (anticipating that averages built later on with Ψ are done by integrating out the particle positions). It acts on the quantities to be averaged with Ψ_s. The assumption of spatial homogeneity rules out the considerations of thermodynamic states where the equilibrium system would, e.g., be crystalline. The equilibrium state is described by Ψ_e, which denotes the equilibrium canonical distribution function, $\Psi_e \propto e^{-U/(k_B T)}$, which is the time-independent solution of (3a) for $\dot{\gamma} = 0$; in (4b) it gives the initial distribution at the start of shearing (at $t = 0$). The potential part of the stress tensor $\sigma_{xy} = -\sum_{i=1}^{N} F_i^x y_i$ entered via $\Omega \Psi_e = \dot{\gamma} \sigma_{xy} \Psi_e$. The simple, exact result (4b) is central to the ITT approach as it connects steady state properties to time integrals formed with the shear-dependent dynamics. Advantageously, the problem to perform steady state averages, denoted by $\langle \ldots \rangle^{(\dot{\gamma})}$, has been simplified to performing equilibrium averages, which will be denoted as $\langle \ldots \rangle$ in the following, and contain the familiar Ψ_e. The transient dynamics integrated up in the second term of (4b) contains slow intrinsic particle motion, whose handling is central to the MCT-ITT approach. Generalized Green–Kubo relations, formally valid for arbitrary $\dot{\gamma}$, can be derived from (4b).

The adjoint Smoluchowski operator was obtained using in the partial integrations over the particle positions the incompressibility condition, Trace$\{\kappa\} = 0$, which should always holds for the solvents of interest in this review. It takes the explicit form (where boundary contributions are neglected throughout, simplifying the partial integrations):

$$\Omega^\dagger = \sum_i (\partial_i + \mathbf{F}_i + \mathbf{r}_i \cdot \kappa^T) \cdot \partial_i .$$

This formula already uses a handy notation[2] employing the shear rate tensor $\kappa = \dot{\gamma} \hat{\mathbf{x}} \hat{\mathbf{y}}$ (that is, $\kappa_{\alpha\beta} = \dot{\gamma} \delta_{\alpha x} \delta_{\beta y}$), and dimensionless quantities. They are introduced by using the particle diameter d as unit of length (throughout we convert $d = 2R_H$), the combination d^2/D_0 as unit of time, and $k_B T$ as unit of energy, whereupon the shear rate turns into the bare Peclet number $\text{Pe}_0 = \dot{\gamma} d^2/D_0$. It measures the effect of affine motion with the shear flow compared to the time it takes a single Brownian particle to diffuse its diameter d. One of the central questions of the nonlinear rheology of dense dispersions concerns the origin of very strong shear-dependences in the viscoelasticity for (vanishingly) small bare Pe_0 numbers. Thus we will simplify by assuming $\text{Pe}_0 \ll 1$, and search for another dimensionless number characterizing the effect of shear.

The formally exact general result for Ψ_s in (4b) can be applied to compute the thermodynamic transverse stress, $\sigma(\dot{\gamma}) = \langle \sigma_{xy} \rangle / V$. Equation (4b) leads to an exact non-linear Green–Kubo relation:

$$\sigma(\dot{\gamma}) = \dot{\gamma} \int_0^\infty dt\, g(t, \dot{\gamma}) , \tag{5a}$$

---

[2] The simplified notation with dimensionless quantities is used in the sections containing formal mainpulations, and in a number of original publications.

where the generalized shear modulus $g(t,\dot{\gamma})$ depends on shear rate via the Smoluchowski operator from (3b):

$$g(t,\dot{\gamma}) = \frac{1}{k_B T V} \langle \sigma_{xy} e^{\Omega^{\dagger} t} \sigma_{xy} \rangle^{(\dot{\gamma}=0)}. \tag{5b}$$

This relation is nonlinear in shear rate, because $\dot{\gamma}$ appears in the time evolution operator $\Omega^{\dagger}$, the adjoint of (3b). In MCT-ITT, the slow stress fluctuations in $g(t,\dot{\gamma})$ will be approximated by following the slow structural rearrangements, encoded in the transient density correlators.

However, before discussing approximations, it's worth pointing out that formally exact explicit expressions for arbitrary steady-state averages can be obtained from (4b). Using the definition $f(\dot{\gamma}) \equiv \langle f_{\mathbf{q}=\mathbf{0}} \rangle^{(\dot{\gamma})}/V$, where $\mathbf{q}$ is the wavevector, and $f_{\mathbf{q}=\mathbf{0}} = \int d\mathbf{r} f(\mathbf{r})$ denotes the integral over an arbitrary density $f(\mathbf{r})$, one finds the general generalized Green–Kubo relation:

$$f(\dot{\gamma}) = \langle f_{\mathbf{q}=\mathbf{0}} \rangle/V + \frac{\dot{\gamma}}{V} \int_0^{\infty} dt \, \langle \sigma_{xy} e^{\Omega^{\dagger} t} \Delta f_{\mathbf{q}=\mathbf{0}} \rangle, \tag{5c}$$

where the symbol ΔX for the fluctuation in X was introduced, $\Delta X = X - \langle X \rangle$, because all mean values (which are constants for these purposes) drop out of the ITT integrals, leaving only the fluctuating parts to contribute. Generalizations of (5c) valid for structure functions (see, e.g., (6b)) and stationary correlation functions (see (8)) are presented in [44]. Note that all the averages, denoted $\langle \ldots \rangle$, are evaluated within the (Boltzmann) equilibrium distribution Ψ_e. Why only $\mathbf{q} = 0$ appears in (5c) is discussed in Sect. 2.2.2.

It is these generalized Green–Kubo relations (5c) which are formally exact even for arbitrary strong flows, and which form the basis for approximations in the MCT-ITT approach. These approximations are guided by the evident aspect that slow dynamics strongly affects the time integral in (5c). Therefore, in MCT-ITT approximations are employed that aim at capturing the slow structural dynamics close to a glass transition. It would be interesting also to employ the generalized Green–Kubo relations in other contexts, where, e.g., entanglements lead to slow dynamics in polymer melts.

2.2.2 Aspects of Translational Invariance

The generalized Green–Kubo relations contain quantities integrated/averaged over the whole sample volume. Thus, the aspect of translational invariance/homogeneity does not become an issue in (5) yet. A system is translational invariant if the correlation between two points $\mathbf{r}$ and $\mathbf{r}'$ depends on the distance $\mathbf{r} - \mathbf{r}'$ between the two points only. The correlation must not change if both points are shifted by the same amount. (Additionally, any quantity depending on one space point $\mathbf{r}$ only, must be constant.) A system would be isotropic if, additionally, the correlation only depended on the length of the distance vector, $|\mathbf{r} - \mathbf{r}'|$; but this obviously cannot be expected, because shear flow breaks rotational symmetry of the SO in (3b). Shear flow also breaks translational symmetry in the SO of (3b), and therefore it is a priori surprising, that translational invariance holds under shear. Moreover, discussion of translational invariance introduces the concept of an advected wavevector, which will become important later on.

The time-dependent distribution function $\Psi(t)$ from (4a) can be used to show that a translationally invariant equilibrium distribution function Ψ_e leads to a translationally invariant steady state distribution Ψ_s, even though the SO in (3b) is not translationally invariant itself. To show this, a point in coordinate space $(\mathbf{r}_1, \ldots, \mathbf{r}_N)$ shall be denoted by Γ, and shall be shifted, $\Gamma \to \Gamma'$, with $\mathbf{r}'_i = \mathbf{r}_i + \mathbf{a}$ for all i; $\mathbf{a}$ is an arbitrary constant vector. This gives

$$\Omega^\dagger(\Gamma) = \Omega^\dagger(\Gamma') - \mathbf{a}\,\kappa^T\,\mathbf{P}, \quad \text{with } \mathbf{P} = \sum_i \partial_i,$$

explicitly stating that the SO is not translationally invariant. From (4a) it follows that

$$\Psi(\Gamma',t) = e^{\Omega(\Gamma)t - \mathbf{P}\kappa\mathbf{a}t}\,\Psi_e(\Gamma),$$

where $\Psi_e(\Gamma') = \Psi_e(\Gamma)$ was used. The SO Ω and the operator $\mathbf{P}\kappa\mathbf{a}$ commute, because the shear rate tensor satisfies $\kappa \cdot \kappa = 0$, and because the sum of all internal forces vanishes due to Newton's third law:

$(\mathbf{P}\kappa\mathbf{a})\,\Omega - \Omega\,(\mathbf{P}\kappa\mathbf{a})$

$$= \sum_{ij}\left\{\left[\partial_i\left(\partial_j \cdot \frac{\partial U}{\partial \mathbf{r}_j}\right) - \left(\partial_j \cdot \frac{\partial U}{\partial \mathbf{r}_j}\right)\partial_i\right]\kappa\mathbf{a} - \left[(\mathbf{a}\kappa^T\partial_i)(\mathbf{r}_j\kappa^T\partial_j) - (\partial_j\kappa\mathbf{r}_j)(\partial_i\kappa\mathbf{a})\right]\right\}$$

$$= \sum_j\left\{\left[\partial_j\cdot\left(\frac{\partial}{\partial \mathbf{r}_j}\left(\sum_i \frac{\partial U}{\partial \mathbf{r}_i}\kappa\mathbf{a}\right)\right)\right] - \left[(\mathbf{a}\kappa^T\cdot\kappa^T\partial_j)\right]\right\} = 0.$$

Therefore, the Baker–Hausdorff theorem [53] simplifies (4a) to

$$\Psi(\Gamma',t) = e^{\Omega(\Gamma)t}\,e^{-\mathbf{P}\kappa\mathbf{a}t}\,\Psi_e(\Gamma) = e^{\Omega(\Gamma)t}\,e^{-(\Sigma_i \mathbf{F}_i)\kappa\mathbf{a}t}\,\Psi_e(\Gamma) = e^{\Omega(\Gamma)t}\,\Psi_e(\Gamma),$$

where the last equality again holds because the sum of all internal forces vanishes. Therefore,

$$\Psi(\Gamma',t) = \Psi(\Gamma,t)$$

holds, proving that the time-dependent and consequently the stationary distribution function $\Psi_s(\Gamma) = \lim_{t\to\infty}\Psi(\Gamma,t)$ are translationally invariant. This applies at least in cases without spontaneous symmetry breaking. Formally, the role of such symmetry breaking is to discard some parts of the steady state distribution function and keep others (with the choice dependent on initial conditions). The distributions developed here discard nothing, and would therefore average over the disjoint symmetry-related states of a symmetry-broken system.

Appreciable simplifications follow from translational invariance for steady-state quantities of wavevector-dependent fluctuations:

$$f_{\mathbf{q}}(\Gamma, t) = e^{\Omega^{\dagger} t} \sum_{i} X_i^f(\Gamma) \, e^{i\mathbf{q}\cdot\mathbf{r}_i} \,,$$

where, e.g., $X_i^\rho = 1$ describes density fluctuations $\rho_{\mathbf{q}}(t)$, while $X_i^\sigma = \delta_{\alpha\beta} + (1/2)\sum_j'(r_i^\alpha - r_j^\alpha) du(|\mathbf{r}_i - \mathbf{r}_j|)/dr_i^\beta$ gives the stress tensor element $\sigma_{\alpha\beta}(\mathbf{q})$ for interactions described by the pair-potential u. Translational invariance in an infinite sheared system dictates that averages are independent of identical shifts of all particle positions. As the integral over phase space must agree for either integration variables Γ or Γ', steady-state averages can be non-vanishing for zero wavevector only:

$$\frac{1}{V} \langle f_{\mathbf{q}}(t) \rangle^{(\dot\gamma)} = f_0(\dot\gamma) \, \delta_{\mathbf{q},0} \,.$$

The average density $n = N/V$ and the shear stress $\sigma(\dot\gamma) = \langle \sigma_{xy} \rangle^{(\dot\gamma)}/V$ are important examples. Wavevector-dependent steady-state structure functions under shear become anisotropic but remain translationally invariant, so that introduction of a single wavevector suffices. The structure factor built with density fluctuations shall be abbreviated by

$$S_{\mathbf{q}}(\dot\gamma) = \frac{1}{N} \langle \delta\rho_{\mathbf{q}}^* \, \delta\rho_{\mathbf{q}} \rangle^{(\dot\gamma)} \,. \tag{6a}$$

It needs to be kept apart from the equilibrium structure factor, denoted by

$$S_q = \frac{1}{N} \langle \delta\rho_{\mathbf{q}}^* \, \delta\rho_{\mathbf{q}} \rangle \,, \tag{6b}$$

which is obtained by averaging over the particle positions using the equilibrium distribution function Ψ_e. It will be one of the hallmarks of a shear molten, yielding glass state, that even in the limit of vanishing shear rate both structure factors do not agree: $S_{\mathbf{q}}(\dot\gamma \to 0) \neq S_q$ in a shear molten glassy state.

Translational invariance of sheared systems takes a special form for two-time correlation functions, because a shift of the point in coordinate space from Γ to Γ' gives

$$\langle \delta f_{\mathbf{q}}^* \, e^{\Omega^{\dagger} t} \, \delta g_{\mathbf{k}} \rangle^{(\dot\gamma)} = e^{-i(\mathbf{k}\cdot\kappa t + \mathbf{k} - \mathbf{q})\cdot\mathbf{a}} \, \langle \delta f_{\mathbf{q}}^* \, e^{\Omega^{\dagger} t} \, \delta g_{\mathbf{k}} \rangle^{(\dot\gamma)} \,,$$

while obviously both averages need to agree. Therefore, a fluctuation with wavevector $\mathbf{q}$ is correlated with a fluctuation of $\mathbf{k} = \mathbf{q}(t)$ with the *advected* wavevector

$$\mathbf{q}(t) = \mathbf{q} - \mathbf{q} \cdot \kappa t \tag{7}$$

at the later time t; only then does the exponential in the last equation become unity; fluctuations with other wavevector combinations are decorrelated. The advected wavevector's y-component decreases with time as $q_y(t) = q_y - \dot\gamma t \, q_x$, corresponding to an (asymptotically) decreasing wavelength, which the shear-advected fluctuation exhibits along the y-direction; see Fig. 1. Taking into account this time-dependence

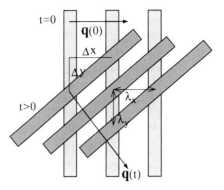

Fig. 1 Shear advection of a fluctuation with initial wavevector in x-direction, $\mathbf{q}(t{=}0) = q(1,0,0)^T$, and advected wavevector at later time $\mathbf{q}(t>0) = q(1,-\dot\gamma t,0)^T$; from [45]. While λ_x is the wavelength in x-direction at $t = 0$, at later time t, the corresponding wavelength λ_y in (negative) y-direction obeys $\lambda_x/\lambda_y = \Delta x/\Delta y = \dot\gamma t$. At all times, $\mathbf{q}(t)$ is perpendicular to the planes of constant fluctuation amplitude. Note that the magnitude $q(t) = q\sqrt{1+(\dot\gamma t)^2}$ increases with time. Brownian motion, neglected in this sketch, would smear out the fluctuation

of the wavelength of fluctuations, a stationary time-dependent correlation function characterized by a single wavevector can be defined:

$$C_{fg;\mathbf{q}}(t) = \frac{1}{N} \langle \delta f_{\mathbf{q}}^* \, e^{\Omega^\dagger t} \, \delta g_{\mathbf{q}(t)} \rangle^{(\dot\gamma)}. \quad (8)$$

Application of (4, 5) is potentially obstructed by the existence of conservation laws, which may cause a zero eigenvalue of the (adjoint) SO, $\Omega^\dagger$. The time integration in (4b, 5) would then not converge at long times. This possible obstacle when performing memory function integrals, and how to overcome it, is familiar from equilibrium Green–Kubo relations [54]. For Brownian particles, only the density ρ is conserved. Yet density fluctuations do not couple in linear order to the shear-induced change of the distribution function [44]. The (equilibrium) average

$$\langle \sigma_{xy} \, e^{\Omega^\dagger t} \, \rho_{\mathbf{q}} \rangle = 0$$

vanishes for all $\mathbf{q}$, at finite $\mathbf{q}$ because of translational invariance, and at $\mathbf{q}=0$ because of inversion symmetry. Thus, the projector Q can be introduced:

$$Q = 1 - P, \quad \text{with } P = \sum_{\mathbf{q}} \delta\rho_{\mathbf{q}} \rangle \frac{1}{NS_q} \langle \delta\rho_{\mathbf{q}}^*. \quad (9)$$

It projects any variable into the space perpendicular to linear density fluctuations. Introducing it into (5c) is straightforward because couplings to linear density can not arise in it anyway. One obtains

$$f(\dot\gamma) = \langle f_{\mathbf{q}=\mathbf{0}} \rangle / V + \frac{\dot\gamma}{V} \int_0^\infty dt \, \langle \sigma_{xy} \, Q \, e^{Q\Omega^\dagger Q t} \, Q \Delta f_{\mathbf{q}=\mathbf{0}} \rangle, \quad (5d)$$

The projection step is exact, and also formally redundant at this stage; but it will prove useful later on, when approximations are performed.

2.2.3 Coupling to Structural Relaxation

The generalized Green–Kubo relations leave us with the problem of how to approximate time-dependent correlation functions in (5). Their physical meaning is that, at time zero, an equilibrium stress fluctuation arises; the system then evolves under internal and shear-driven motion until time t, when its correlation with a fluctuation $\Delta f_{\mathbf{q}=\mathbf{0}}$ is determined. Integrating up these contributions for all times since the start of shearing gives the difference of the shear-dependent quantities to the equilibrium ones. During the considered time evolution, the projector Q prevents linear couplings to the conserved particle density.

The time dependence and magnitudes of the correlations in (5) shall now be approximated by using the overlaps of both the stress and $\Delta f_{\mathbf{q}=\mathbf{0}}$ fluctuations with appropriately chosen *"relevant slow fluctuations."* For the dense colloidal dispersions of interest, the relevant structural rearrangements are assumed to be *density fluctuations*. Because of the projector Q in (5d), the lowest nonzero order in fluctuation amplitudes, which we presume dominant, must then involve pair-products of density fluctuations, $\rho_{\mathbf{k}} \rho_{\mathbf{p}}$ (in simplified notation dropping δ).

The mode coupling approximation may be summarized as a rule that applies to all fluctuation products that exhibit slow structural relaxations but whose variables cannot couple linearly to the density. Their time-dependence is approximated as:

$$Q\, e^{Q \Omega^\dagger Q t}\, Q \approx \sum_{\mathbf{k}>\mathbf{p}} Q \rho_{\mathbf{k}(-t)} \rho_{\mathbf{p}(-t)} \rangle \frac{\Phi_{\mathbf{k}(-t)}(t)\, \Phi_{\mathbf{p}(-t)}(t)}{N^2 S_k S_p} \langle \rho_{\mathbf{k}}^* \rho_{\mathbf{p}}^* Q \tag{10a}$$

The fluctuating variables are thereby projected onto pair-density fluctuations, whose time-dependence follows from that of the transient density correlators $\Phi_{\mathbf{q}(t)}(t)$, defined in (12). These describe the relaxation (caused by shear, interactions and Brownian motion) of density fluctuations with equilibrium amplitudes. Higher order density averages are factorized into products of these correlators, and the reduced dynamics containing the projector Q is replaced by the full dynamics. The entire procedure is written in terms of *equilibrium* averages, which can then be used to compute nonequilibrium steady states via the ITT procedure. The normalization in (10a) is given by the equilibrium structure factors such that the pair density correlator with reduced dynamics, which does not couple linearly to density fluctuations, becomes approximated to:

$$\langle \rho_{\mathbf{k}}^* \rho_{\mathbf{p}}^* Q\, e^{Q \Omega^\dagger Q t}\, Q \rho_{\mathbf{k}(t)} \rho_{\mathbf{p}(t)} \rangle \approx N^2 S_k S_p\, \Phi_{\mathbf{k}}(t)\, \Phi_{\mathbf{p}}(t). \tag{10b}$$

This equation can be considered as central approximation of the MCT [38] and MCT-ITT approach. While the projection onto density pairs, which is also contained/implied in (10a), may be improved upon systematically by including higher order density or other fluctuations, see [55, 56] for examples, no systematic way to improve upon the breaking of averages in (10b) has been discovered up to now, to the knowledge of the author.

The mode coupling approximations introduced above can now be applied to the exact generalized Green–Kubo relations (5d). Steady state expectation values are approximated by projection onto pair density modes, giving

$$f(\dot{\gamma}) \approx \langle f_0 \rangle / V + \frac{\dot{\gamma}}{2V} \int_0^\infty dt \sum_{\mathbf{k}} \frac{k_x k_y(-t) S'_{k(-t)}}{k(-t) S_k^2} V_{\mathbf{k}}^f \, \Phi_{\mathbf{k}(-t)}^2(t) \,, \qquad (11a)$$

with t the time since switch-on of shear. To derive this, the property $\Phi_{\mathbf{k}}^* = \Phi_{-\mathbf{k}} = \Phi_{\mathbf{k}}$ was used; also the restriction $\mathbf{k} > \mathbf{p}$ when summing over wavevectors was dropped, and a factor $\frac{1}{2}$ introduced in order to have unrestricted sums over $\mathbf{k}$. Within (11a) we have already substituted the following explicit result for the equal-time correlator of the shear stress with density products:

$$\langle \sigma_{xy} \, Q \, \rho_{\mathbf{k}(-t)} \, \rho_{\mathbf{p}(-t)} \rangle = N \frac{k_x k_y(-t)}{k(-t)} S'_{k(-t)} \, \delta_{\mathbf{k}(-t), -\mathbf{p}(-t)} = \frac{N}{\dot{\gamma}} \partial_t S_{q(-t)} \, \delta_{\mathbf{k}, -\mathbf{p}} \,. \qquad (11b)$$

It's an exact equality using the equilibrium distribution function and (6)

$$\langle \sigma_{xy} \, Q \, \rho_{\mathbf{k}} \, \rho_{\mathbf{k}}^* \rangle = \langle \sigma_{xy} \, \rho_{\mathbf{k}} \, \rho_{\mathbf{k}}^* \rangle = \int d\Gamma \Psi_e \left(-\sum_i F_i^x y_i \right) \rho_{\mathbf{k}} \, \rho_{\mathbf{k}}^*$$

$$= \int d\Gamma \Psi_e \left(\sum_i \partial_i^x y_i \right) \rho_{\mathbf{k}} \, \rho_{\mathbf{k}}^* = i k_x \sum_{ij} \langle y_i (e^{i\mathbf{k}(\mathbf{r}_i - \mathbf{r}_j)} - e^{-i\mathbf{k}(\mathbf{r}_i - \mathbf{r}_j)}) \rangle$$

Equation (11a), as derived via the mode-coupling rule detailed above, contains a "vertex function" $V_{\mathbf{k}}^f$, describing the coupling of the desired variable f to density pairs. This denotes the following quantity, computed using familiar thermodynamic equalities:

$$V_{\mathbf{k}}^f \equiv \langle \rho_{\mathbf{k}}^* \rho_{\mathbf{k}} \, Q \, \Delta f_0 \rangle / N = \langle \rho_{\mathbf{k}}^* \rho_{\mathbf{k}} \, \Delta f_0 \rangle / N - S_0 \left(S_k + n \frac{\partial S_k}{\partial n} \right) \frac{\partial \langle f_0 \rangle / V}{\partial n} \bigg)_T \,. \qquad (11c)$$

In ITT, the slow stress fluctuations in $g(t, \dot{\gamma})$ are approximated by following the slow structural rearrangements, encoded in the transient density correlators. The generalized modulus becomes, using the approximation (10a) and the vertex (11b):

$$g(t, \dot{\gamma}) = \frac{k_B T}{2} \int \frac{d^3 k}{(2\pi)^3} \frac{k_x^2 k_y k_y(-t)}{k k(-t)} \frac{S'_k S'_{k(-t)}}{S_k^2} \Phi_{\mathbf{k}(-t)}^2(t) \,, \qquad (11d)$$

Summation over wavevectors has been turned into integration in (11d) considering an infinite system.

The familiar shear modulus of linear response theory describes thermodynamic stress fluctuations in equilibrium, and is obtained from (5b, 11d) by setting $\dot{\gamma}=0$ [1, 3, 57]. While (5b) then gives the exact Green–Kubo relation, the approximation (11d) turns into the well-studied MCT formula (see (17)). For finite shear rates, (11d) describes how affine particle motion causes stress fluctuations to explore shorter and shorter length scales. There the effective forces, as measured by the gradient of the direct correlation function, $S'_k/S_k^2 = nc'_k = n\partial c_k/\partial k$, become smaller, and vanish asympotically, $c'_{k\to\infty} \to 0$; the direct correlation function c_k is connected to the structure factor via the Ornstein–Zernicke equation $S_k = 1/(1-nc_k)$, where $n = N/V$ is the particle density. Note that the equilibrium structure suffices to quantify the effective interactions, while shear just pushes the fluctuations around on the "equilibrium energy landscape".

While in the linear response regime, modulus and density correlator are measurable quantities; outside the linear regime, both quantities serve as tools in the ITT approach only. The transient correlator and shear modulus provide a route to the stationary averages because they describe the decay of equilibrium fluctuations under external shear and their time integral provides an approximation for the stationary distribution function. Determination of the frequency dependent moduli under large amplitude oscillatory shear has only recently become possible [58], and requires an extension of the present approach to time dependent shear rates in (3) [59].

2.2.4 Transient Density Correlator

In ITT, the evolution towards the stationary distribution at infinite times is approximated by following the slow structural rearrangements, encoded in the transient density correlator $\Phi_\mathbf{q}(t)$. It is defined by [43, 44]

$$\Phi_\mathbf{q}(t) = \frac{1}{NS_q} \langle \delta\rho_\mathbf{q}^* e^{\Omega^\dagger t} \delta\rho_{\mathbf{q}(t)} \rangle^{(\dot{\gamma}=0)} . \qquad (12)$$

It describes the fate of an equilibrium density fluctuation with wavevector $\mathbf{q}$, where $\rho_\mathbf{q} = \sum_{j=1}^N e^{i\mathbf{q}\cdot\mathbf{r}_j}$, under the combined effect of internal forces, Brownian motion, and shearing. Note that because of the appearance of Ψ_e in (4), the average in (12) can be evaluated with the equilibrium canonical distribution function, while the dynamical evolution contains Brownian motion and shear advection. The normalization is given by S_q, the equilibrium structure factor [3, 46] for wavevector modulus $q = |\mathbf{q}|$. The *advected* wavevector from (7) enters in (12). The time-dependence in $\mathbf{q}(t)$ results from the affine particle motion with the shear flow of the solvent. Again, irrespective of the use of Ψ_e in (12), or Ψ_s in (8), in both cases translational invariance under shear dictates that at a time t later, the density fluctuation $\delta\rho_\mathbf{q}^*$ has a nonvanishing overlap only with the advected fluctuation $\delta\rho_{\mathbf{q}(t)}$. Figure 1 again applies, where a non-decorrelating fluctuation is sketched under shear. In the case of vanishing Brownian motion, viz. $D_0 = 0$ in (3b), we find $\Phi_\mathbf{q}(t) \equiv 1$, because the advected wavevector takes account of simple affine particle motion. The relaxation of $\Phi_\mathbf{q}(t)$ thus heralds decay of structural correlations by Brownian motion, affected by shear.

2.2.5 Zwanzig–Mori Equations of Motion

Structural rearrangements of the dispersion affected by Brownian motion is encoded in the transient density correlator. Shear induced affine motion, viz. the case $D_0 = 0$, is not sufficient to cause $\Phi_\mathbf{k}(t)$ to decay. Brownian motion of the quiescent correlator $\Phi_k^{(\dot{\gamma}=0)}(t)$ leads at high densities to a slow structural process which arrests at long times in (metastable) glass states. Thus the combination of structural relaxation and shear is interesting. The interplay between intrinsic structural motion and shearing in $\Phi_\mathbf{k}(t)$ is captured by (1) first a formally exact Zwanzig–Mori type equation of motion, and (2) second a mode coupling factorisation in the memory function built with longitudinal stress fluctuations [43–45]. The equation of motion for the transient density correlators is

$$\partial_t \Phi_\mathbf{q}(t) + \Gamma_\mathbf{q}(t) \left\{ \Phi_\mathbf{q}(t) + \int_0^t dt'\, m_\mathbf{q}(t,t')\, \partial_{t'} \Phi_\mathbf{q}(t') \right\} = 0, \qquad (13)$$

where the initial decay rate $\Gamma_\mathbf{q}(t) = D_0 q^2(t)/S_{q(t)}$ generalizes the familiar result from linear response theory to advected wavevectors; it contains Taylor dispersion mentioned in the introduction, and describes the short time behavior, $\Phi_\mathbf{q}(t \to 0) \to 1 - \Gamma_\mathbf{q}(0) t + \dots$.

2.2.6 Mode-Coupling Closure

The memory equation contains fluctuating stresses and, like $g(t, \dot{\gamma})$ in (11d), is calculated in mode coupling approximation using (10a) giving:

$$m_\mathbf{q}(t,t') = \frac{1}{2N} \sum_\mathbf{k} V_{\mathbf{qkp}}(t,t')\, \Phi_{\mathbf{k}(t')}(t-t')\, \Phi_{\mathbf{p}(t')}(t-t'), \qquad (14a)$$

where we abbreviated $\mathbf{p} = \mathbf{q} - \mathbf{k}$. The vertex generalizes the expression in the quiescent case, see (18c) below, and depends on twice capturing that shearing decorrelates stress fluctuations [43–45]:

$$V_{\mathbf{qkp}}(t,t') = \frac{S_{\mathbf{q}(t)} S_{\mathbf{k}(t')} S_{\mathbf{p}(t')}}{q^2(t) q^2(t')} \mathcal{V}_{\mathbf{qkp}}(t) \mathcal{V}_{\mathbf{qkp}}(t'),$$
$$\mathcal{V}_{\mathbf{qkp}}(t) = \mathbf{q}(t) \cdot \left(\mathbf{k}(t)\, nc_{\mathbf{k}(t)} + \mathbf{p}(t)\, nc_{\mathbf{p}(t)} \right). \qquad (14b)$$

With shear, wavevectors in (14b) are advected according to (7).

The summarized MCT-ITT equations form a closed set of equations determining rheological properties of a sheared dispersion from equilibrium structural input [43–45]. Only the static structure factor S_q is required to predict (1) the time dependent shear modulus within linear response, $g^{lr}(t) = g(t, \dot{\gamma} = 0)$, and (2) the stationary stress $\sigma(\dot{\gamma})$ from (5a).

2.3 A Microscopic Model: Brownian Hard Spheres

In the microscopic ITT approach, the rheology is determined from the equilibrium structure factor S_q alone. This holds at low enough frequencies and shear rates, and excludes a single time scale, to be denoted by the parameter t_0 in (22b), which needs to be found by matching to the short time dynamics. This prediction has as a consequence that the moduli and flow curves should be a function only of the thermodynamic parameters characterizing the present system, viz. its structure factor. Because the structure factor for simple fluids, far from demixing and other phase separation regions, can be mapped onto the one of hard spheres, the system of hard spheres plays a special role in the MCT-ITT approach. It provides the most simple microscopic model where slow structural dynamics can be studied. Moreover, other experimental systems can be mapped onto it by chosing an effective packing fraction $\phi_{\text{eff}} = (4\pi/3)nR_H^3$ and particle radius so that the structure factors agree.

The claim that the rheology follows from S_q is supported if the rheological properties of a dispersion only depend on the effective packing fraction, if particle size is taken account of properly. Obviously, appropriate scales for frequency, shear rate and stress magnitudes need to be chosen to observe this; see Sect. 6.2. The dependence of the rheology (via the vertices) on S_q suggests that $k_B T$ sets the energy scale as long as repulsive interactions dominate the local packing. The length scale is set by the average particle separation, which can be taken to scale with R_H. The time scale of the glassy rheology within ITT is given by t_0, which should scale with the measured dilute diffusion coefficient D_0. Thus the rescaling of the rheological data can be done with measured parameters alone.

Because the hard sphere system thus provides the most simple system to test and explore MCT-ITT, numerical calculations only for this model will be reviewed in the present overview. Input for the structure factor is required, which, for simplicity, will be taken from the analytical Percus–Yervick (PY) approximation [2, 3]. Straightforward discretization of the wavevector integrals will be performed as discussed below, and in detail in the quoted original papers.

2.4 Accounting for Hydrodynamic Interactions

Solvent-particle interactions (viz. the HI) act instantaneously if the particle microstructure differs from the equilibrium one, but do not themselves determine the equilibrium structure [3, 46]. If one assumes that glassy arrest is connected with the ability of the system to explore its configuration space and to approach its equilibrium structure, then it appears natural to assume that the solvent particle interactions are characterized by a finite time scale τ_{HI} and that they do not shift the glass transition nor affect the frozen glassy structure. HI would thus only lead to an increase of the high frequency viscosity above the solvent value; this value shall be denoted as η_∞:

$$g(t,\dot{\gamma}) \to g(t,\dot{\gamma}) + \eta_\infty \delta(t - 0+). \tag{15a}$$

The parameter η_∞ would thus characterize a short-time, high frequency viscosity and model viscous processes which require no structural relaxation. It can be measured from the high frequency dissipation

$$G''(\omega \to \infty) = \eta_\infty \, \omega \,. \tag{15b}$$

For identical reasoning, the short time diffusion in the collective (and single particle) motion will also be affected by HI. The simplest approximation is to adjust the initial decay rate

$$\Gamma_\mathbf{q}(t) = D_s q^2(t)/S_{q(t)} \,, \tag{15c}$$

where the collective short time diffusion coefficient D_s accounts for HI and other (almost) instantaneous effects which affect the short time motion, and which are not explicitly included in the MCT-ITT approach.

The naive picture sketched here is not correct for a number of reasons. It is well known that for hard spheres without HI the quiescent shear modulus diverges for short times, $g^{\text{lr,HSnoHI}}(t \to 0) \sim t^{-1/2}$. Lubrication forces, which keep the particles apart, eliminate this divergence and render $g^{\text{lr,HI}}(t \to 0)$ finite [60]. Thus, the simple separation of the modulus into HI and potential part is not possible for short times, at least for particles with a hard core. Moreover, comparison of simulations without and with HI has shown that the increase of $(\eta_0 - \eta_\infty)/\eta_\infty$ depends somewhat on HI, and thus not just on the potential interactions as implied.

Nevertheless the sketched picture provides the most basic view of a glass transition in colloidal suspensions, connecting it with the increase of the structural relaxation time τ. Increased density or interactions cause a slowing down of particle rearrangements which leave the HI relatively unaffected, as these solvent mediated forces act on all time scales. Potential forces dominate the slowest particle rearrangements because vitrification corresponds to the limit where they actually prevent the final relaxation of the microstructure. The structural relaxation time τ diverges at the glass transition, while τ^{HI} stays finite. Thus close to arrest a time scale separation is possible, $\tau \gg \tau^{\text{HI}}$.

2.5 Comparison with Other MCT Inspired Approaches to Sheared Fluids

The MCT-ITT approach aims at describing the steady state properties of a concentrated dispersion under shear. Stationary averages are its major output, obtained via the integration through transients (ITT) procedure from (approximate) transient fluctuation functions, whose strength is the equilibrium one, and whose dynamics originates from the competition between Brownian motion and shear induced decorrelation. In this respect, the MCT-ITT approach differs from the interesting recent generalization of MCT to sheared systems by Miyazaki, Reichman and coworkers [40, 41]. These authors considered the stationary but time-dependent fluctuations around the steady state, whose amplitude is the stationary correlation function, e.g., in the case of density fluctuations it is the distorted structure factor $S_\mathbf{q}(\dot{\gamma})$ (6a). In the approach by Miyazaki et al. this structure factor is an input-quantity required to calculate the dynamics, while it is an output quantity, calculated in MCT-ITT from the equilibrium S_q of (6b). Likewise, the stationary stress as function of shear rate, viz. the flow curve $\sigma(\dot{\gamma})$, is a quantity calculated in MCT-ITT, albeit using mode coupling approximations, while in the approach of [40, 41] additional ad-hoc approximations beyond the mode coupling approximation are required to access $\sigma(\dot{\gamma})$. Thus, while the scenario of a non-equilibrium transition between a shear-thinning fluid and a shear molten glass, characterized by universal aspects in, e.g., $\sigma(\dot{\gamma})$ – see the discussion in Sect. 4 – forms the core of the MCT-ITT results, this scenario cannot be directly addressed based on [40, 41].

Because the recent experiments and simulations reviewed here concentrated on the universal aspects of the novel non-equilibrium transition, focus will be laid on the MCT-ITT approach. Reassuringly, however, many similarities between the MCT-ITT equations and the results by Miyazaki and Reichman exist, even though these authors used a different, field theoretic approach to derive their results. This supports the robustness of the mechanism of shear-advection in (7) entering the MCT vertices in (11d, 14), which were derived independently in [40, 41] and [43–45] from quite different theoretical routes. This mechanism had been known from earlier work on the dynamics of critical fluctuations in sheared systems close to phase transition points [61], on current fluctuations in simple liquids [62], and on incoherent density fluctuations in dilute solutions [63]. Different possibilities also exist to include shear into MCT-inspired approaches, especially the one worked out by Schweizer and coworkers including strain into an effective free energy [42]. This approach does not recover the (idealized) MCT results reviewed below but starts from the extended MCT where no true glass transition exists and describes a crossover scenario without, e.g., a true dynamic yield stress as discussed below.

3 Microscopic Results in Linear Response Regime

Before turning to the properties of the stationary non-equilibrium states under shear, it is useful to investigate the quiescent dispersion close to vitrification. Consensus on the ultimate mechanism causing glassy arrest may yet be absent; however the so-called "cage effect" has led to a number of fruitful insights into glass formation in dense colloidal dispersions. For example, it was extended to particles with a short ranged attraction leading to at first surprising predictions [64–70].

Figure 2 shows a section of the cell of a Monte Carlo simulation of hard disks moving in two dimensions (for simplicity of visualization). The density is just below freezing and the sample was carefully equilibrated. Only 100 disks were simulated, so that finite size effects cannot be ruled out. Picking out a disk, it is surrounded

Fig. 2 Positions of hard disks in two dimensions from a Monte Carlo simulation at a density close to freezing; courtesy of Th. Franosch. A particle and its shell of neighbours is highlighted by different colors/shadings

by a shell of on average 6 neighbours (in 2 dimension, of 12 neighbours in 3 dimension), which hinder its free motion. In order for the central particle to diffuse at long times, it needs to escape the shell of neighbours. In order for a gap in this shell to open at higher concentrations, the neighbours have to be able to move somewhat themselves. Yet each neighour is hindered by its own shell of neighbours, to which the originally picked particle belongs. Thus one can expect a cooperative feedback mechanism that with increasing density or particle interactions particle rearrangements take more and more time. It appears natural that, consequently, stress fluctuations also slow down and the system becomes viscoelastic.

MCT appears to capture the cage-effect in supercooled liquids and predicts that it dominates the slow relaxation of structural correlations close to the glass transition [2, 38]. Density fluctuations play an important role because they are well suited to describe the structure of the particle system and its relaxation. Moreover, stresses that decay slowly because of the slow particle rearrangements, MCT argues, can also be approximated by density fluctuations using effective potentials. Density fluctuations are not great at large wavelengths; it is wavelengths corresponding to the average particle distance that turn out to be the dominant ones. In agreement with the picture of the caging of particles by structural correlations, the MCT glass transition is independent of whether the particles move ballistically in between interactions with their neighbors (say collisions for hard spheres) or by diffusion. Structural arrest happens whenever the static density correlations for wavelengths around the average particle distance are strong enough. The arrest of structural correlations entails an increase in the viscosity of the dispersion connected to the existence of a slow Maxwell-process in the shear moduli. While the MCT solutions for density fluctuations have been thoroughly reviewed, the viscoelastic spectra have not been presented in such detail.

3.1 Shear Moduli Close to the Glass Transition

3.1.1 MCT Equations and Results for Hard Spheres

The loss and storage moduli of small amplitude oscillatory shear measurements [1, 3] follow from (5b) in the linear response case at $\dot{\gamma} = 0$:

$$G'(\omega) + iG''(\omega) = i\omega \int_0^\infty dt\, e^{-i\omega t}\, g^{lr}(t) \,. \tag{16a}$$

Here, the shear modulus in the linear response regime is, again like the transient one in (5b), obtained from equilibrium averaging:

$$g^{lr}(t) = \frac{1}{k_B T V} \langle \sigma_{xy}\, e^{\Omega_e^\dagger t}\, \sigma_{xy} \rangle^{(\dot{\gamma}=0)} \,, \tag{16b}$$

yet, different from the transient one, the equilibrium one contains the equilibrium SO Ω_e, which characterizes the quiescent system:

$$\Omega_e = \sum_{j=1}^N D_0 \frac{\partial}{\partial \mathbf{r}_j} \cdot \left(\frac{\partial}{\partial \mathbf{r}_j} - \frac{1}{k_B T} \mathbf{F}_j \right) \,, \tag{16c}$$

The linear response modulus thus quantifies the small stress fluctuations, which are excited thermally, and relax because of Brownian motion.

Predictions of the (idealized) MCT equations for the potential part of the equilibrium, time-dependent shear modulus $g^{lr}(t)$ of hard spheres for various packing fractions ϕ are shown in Fig. 3 and calculated from the limit of (11d) for vanishing shear rate:

$$g^{lr}(t) \approx \frac{k_B T}{60\pi^2} \int_0^\infty dk\, k^4 \left(\frac{\partial \ln S_k}{\partial k} \right)^2 \Phi_k^2(t) \,, \tag{17}$$

The normalized density fluctuation functions are calculated self-consistently within MCT from (13, 14) at vanishing shear rate [2, 38], which turn into quiescent MCT equations:

$$\partial_t \Phi_q(t) + \Gamma_q \left\{ \Phi_q(t) + \int_0^t dt'\, m_q(t-t')\, \partial_{t'} \Phi_q(t') \right\} = 0 \,, \tag{18a}$$

where the initial decay rate $\Gamma_q = D_s q^2 / S_q$ describes diffusion with a short-time diffusion coefficient D_s differing from the D_0 because of HI; $D_s = D_0$ will be taken for exemplary calculations, while $D_s \neq D_0$ is required for analyzing experimental data. The memory kernel becomes (again with abbreviation $\mathbf{p} = \mathbf{q} - \mathbf{k}$)

$$m_q(t) = \frac{1}{2N} \sum_{\mathbf{k}} V_{qkp}\, \Phi_k(t)\, \Phi_p(t) \,, \tag{18b}$$

Nonlinear Rheology

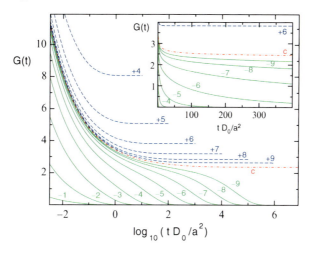

Fig. 3 Equilibrium potential or linear response shear modulus $G(t) = g^{lr}(t)$ (in units of $k_B T/R_H^3$) for Brownian hard spheres with radius $a = R_H$ for packing fractions close to vitrification at ϕ_c; from [71]. Densities are measured by the separation parameter $\varepsilon = (\phi - \phi_c)/\phi_c = \pm 10^{-|n|/3}$, and labels denote the value n. Positive values belong to glass ($\varepsilon > 0$), negative to fluid states ($\varepsilon < 0$); the label c gives the transition. The *inset* shows a subset of the curves on a linear time axis; the increase of $g^{lr}(t)$ for short times cannot be resolved

$$V_{qkp} = \frac{S_q S_k S_p}{q^4} \left(\mathbf{q} \cdot [\mathbf{k}\, nc_k + \mathbf{p}\, nc_p] \right)^2 . \tag{18c}$$

Packing fractions are conveniently measured in relative separations $\varepsilon = (\phi - \phi_c)/\phi_c$ to the glass transition point, which for this model of hard spheres lies at $\phi_c = 0.516$ [38, 72]. Note that this result depends on the static structure factor $S(q)$, which is taken from Percus–Yevick theory, and that the experimentally determined value $\phi_c^{expt.} = 0.58$ lies somewhat higher [13, 14]. The wavevector integrals were discretized using $M = 100$ wavevectors chosen from $k_{min} = 0.1/R_H$ up to $k_{max} = 19.9/R_H$ with separation $\Delta k = 0.2/R_H$ for Figs. 3–5, or using $M = 600$ wavevectors chosen from $k_{min} = 0.05/R_H$ up to $k_{max} = 59.95/R_H$ with separation $\Delta k = 0.1/R_H$ for Figs. 6 and 7, and in Sect. 6.2 in Fig. 23. Time was discretized with initial step-width $dt = 2\,10^{-7} R_H^2/D_s$, which was doubled each time after 400 steps. Slightly different discretizations in time and wavevector of the MCT equations were used in Sects. 3.2.1 and 4, causing only small quantitative differences whose discussion goes beyond the present review. The quiescent density correlators $\Phi_q(t)$ corresponding to the following linear response moduli have thoroughly been discussed in [72].

For low packing fractions, or large negative separations, the modulus decays quickly on a time-scale set by the short-time diffusion of well separated particles. The strength of the modulus increases strongly at these low densities, and its behavior at short times presumably depends sensitively on the details of hydrodynamic and potential interactions; thus Fig. 3 is not continued to small times,

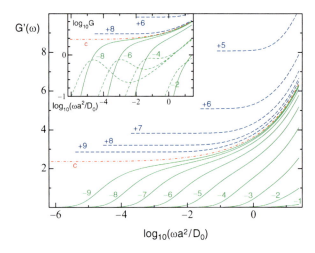

Fig. 4 Storage part of the shear modulus $G'(\omega)$ corresponding to Fig. 3. The *inset* shows storage and loss moduli (only for fluid states) for a number of densities

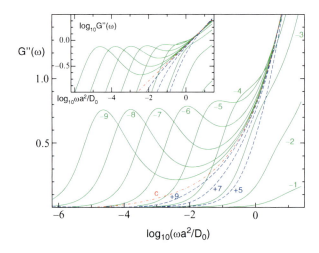

Fig. 5 Loss part of the shear modulus $G''(\omega)$ corresponding to Fig. 3. The *inset* shows the same data in a double logarithmic representation

where the employed model (taken from [72, 73]) is too crude[3]. Approaching the glass transition from below, $\varepsilon \nearrow 0$, little changes in $g^{lr}(t)$ at short times, because the absolute change in density becomes small. Yet, at long times a process in $g^{lr}(t)$

---

[3] The MCT shear modulus at short times depends sensitively on the large cut-off k_{max} for hard spheres [57], $g(t, \dot\gamma = 0) = (n^2 k_B T / 60\pi^2) \int_{k_{min}}^{k_{max}} dk\, k^4 (c'_k)^2 S_k^2 \Phi_k^2(t)$ gives the qualitatively correct [60, 75] short time $g^{lr}(t \to 0) \sim t^{-1/2}$, or high frequency divergence $G'(\omega \gg D_0/R_H^2) \sim \sqrt{\omega}$ only for $k_{max} \to \infty$.

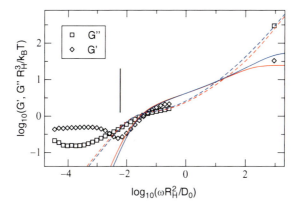

Fig. 6 The reduced storage (*diamonds and solid lines*) and loss (*squares and broken lines*) modulus for a fluid state at effective packing fraction $\phi_{\text{eff}} = 0.540$; from [32]. The *vertical bars* mark the minimal rescaled frequency above which the influence of crystallisation can be neglected. Parameters in the MCT calculation given as *blue lines*: $\varepsilon = -0.01$, $\frac{D_S}{D_0} = 0.15$, and $\eta_\infty = 0.3 k_B T/(D_0 R_H)$; moduli scale factor $c_y = 1.4$. For the *other lines* see [32]

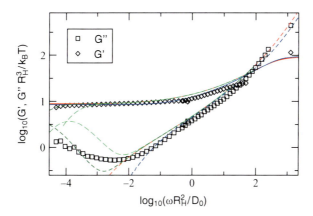

Fig. 7 The moduli for a glass state at effective packing fraction $\phi_{\text{eff}} = 0.622$; storage (*diamonds and solid lines*) and loss (*squares and broken lines*) moduli from [32]. MCT fits are shown as *blue lines* with parameters: $\varepsilon = 0.03$, $\frac{D_S}{D_0} = 0.08$, and $\eta_\infty = 0.3 k_B T/(D_0 R_H)$; moduli scale factor $c_y = 1.4$. For the *other lines* see [32]

becomes progressively slower upon taking ε to zero. It can be considered the MCT analog of the phenomenological Maxwell-process. MCT finds that it depends on the equilibrium structural correlations only, while HI and other short time effects only shift its overall time scale. Importantly, this overall time scale applies to the slow process in coherent and incoherent density fluctuations as well as in the stress fluctuations [74]. This holds even though, e.g., HI are known to affect short time diffusion coefficients and high frequency viscosities differently. Upon crossing the glass transition, a part of the relaxation freezes out and the amplitude G_∞ of the Maxwell-process does not decay; the modulus for long times approaches the elastic constant of the glass $g^{lr}(t \to \infty) \to G_\infty > 0$. Entering deeper into the glassy phase the elastic constant increases quickly with packing fraction.

The corresponding storage $G'(\omega)$ and loss $G''(\omega)$ moduli are shown as functions of frequency in Figs. 4 and 5, respectively. The slow Maxwell-process appears as a shoulder in G' which extends down to lower and lower frequencies when approaching glassy arrest, and reaches zero frequency in the glass, $G'(\omega = 0) = G_\infty$. The slow process shows up as a peak in G'' which in parallel motion (see inset of Fig. 4) shifts to lower frequencies when $\varepsilon \nearrow 0$. Including hydrodynamic interactions into the calculation by adjusting η_∞ would affect the frequency dependent moduli at higher frequencies only. For the range of smaller frequencies which is of interest here, only a small correction would arise.

3.1.2 Comparison with Experiments

Recently, it has been demonstrated that suspensions of thermosensitive particles present excellent model systems for measuring the viscoelasticity of dense concentrations. The particles consist of a solid core of polystyrene onto which a thermosensitive network of poly(N-isopropylacrylamide) (PNIPAM) is attached [31, 32]. The PNiPAM shell of these particles swells when immersed in cold water (10–15°C). Water gets expelled at higher temperatures leading to a considerable shrinking. Thus, for a given number density the effective volume fraction ϕ_eff can be adjusted within wide limits by adjusting the temperature. Senff et al. (1999) were the first to demonstrate the use of these particles as model system for studying the dynamics in concentrated suspensions [23, 24]. The advantage of these systems over the classical hard sphere systems are that dense suspensions can be generated in situ without shear and mechanical deformation. The previous history of the sample can be erased by raising the temperature and thus lowering the volume fraction to the fluid regime.

Frequency dependent moduli were measured spanning a wide density and frequency range by combining different techniques. The moduli exhibit a qualitative change when increasing the effective packing fraction from around 50% to above 60%. For lower densities (see Fig. 6), the spectra $G''(\omega)$ exhibit a broad peak or shoulder, which corresponds to the final or α-relaxation. Its peak position (or alternatively the crossing of the moduli, $G' = G''$) is roughly given by $\omega\tau = 1$. These properties characterize a viscoelastic fluid. For higher density, see Fig. 7, the storage modulus exhibits an elastic plateau at low frequencies. The loss modulus drops far below the elastic one. These observations characterize a soft solid[4].

The linear response moduli are affected by the presence of small crystallites. At low frequencies, $G'(\omega)$ and $G''(\omega)$ increase above the behavior expected for a solution ($G'(\omega \to 0) \to \eta_0 \omega$ and $G''(\omega \to 0) \to c\omega^2$) even at low density, and exhibit

---

[4] The loss modulus rises again at very low frequencies, which may indicate that the colloidal glass at this density is metastable and may have a finite lifetime (an ultra-slow process is discussed in [32]).

elastic contributions (apparent from $G'(\omega) > G''(\omega)$); see Fig. 6. This effect tracks the crystallisation of the system during the measurement after a strong preshearing at $\dot{\gamma} = 100\,s^{-1}$. Only data can be considered which were collected before the crystallisation time; they lie to the right of the vertical bar in Fig. 6. While this experimental restriction limits more detailed studies of the shapes of the spectra close to the glass transition, the use of a system with a rather narrow size distribution provides the quantitatively closest comparison with MCT calculations for a monodisperse hard sphere system. The magnitude of the stresses and the effective densities can be investigated quantitatively.

Included in Figs. 6 and 7 are calculations using the microscopic MCT given by (16–18) evaluated for hard spheres in PY approximation. The only a priori unknown, adjustable parameter is a frequency or time scale, which was adjusted by varying the short time diffusion coefficient D_s appearing in the initial decay rate in (18a). Values for D_s/D_0 are reported in the captions. The viscous contribution to the stress is mimicked by including η_∞ as in (15); it can be directly measured at the highest frequencies. Gratifyingly, the stress values computed from the microscopic approach are close to the measured ones; they are too small by only 40%, which may arise from the approximate structure factors entering the MCT calculation; the Percus–Yevick approximation was used here [3]. In order to compare the shapes of the moduli the MCT calculations were scaled up by a factor $c_y = 1.4$ in Figs. 6 and 7. Microscopic MCT also does not hit the correct value for the glass transition point [2, 38]. It finds $\phi_c^{MCT} = 0.516$, while experiments give $\phi_c^{exp} \approx 0.58$. Thus, when comparing, the relative separation from the respective transition point needs to be adjusted as, obviously, the spectra depend sensitively on the distance to the glass transition; the fitted values of the separation parameter ε are included in the captions.

Overall, the semi-quantitative agreement between the linear viscoelastic spectra and first-principles MCT calculations is very promising. Yet, crystallization effects in the data prevent a closer look, which will be given in Sect. 6.2, where data from a more polydisperse sample are discussed.

3.2 Distorted Structure Factor

3.2.1 Linear Order in $\dot{\gamma}$

The stationary structural correlations of a dense fluid of spherical particles undergoing Brownian motion, neglecting hydrodynamic interactions, change with shear rate $\dot{\gamma}$ in response to a steady shear flow. In linear order, the structure is distorted only in the plane of the flow, while already in second order in $\dot{\gamma}$, the structure factor also changes under shear for wavevectors lying in the plane perpendicular to the flow. Consistent with previous theories, MCT-ITT finds regular expansion coefficients in linear and quadratic order in $\dot{\gamma}$ for fluid (ergodic) suspensions [76]. For the steady state structure factor $S_{\mathbf{q}}(\dot{\gamma})$ of density fluctuations under shearing in plain Couette flow defined in (6a), the change from the equilibrium one in linear order in shear rate is given by the following ITT-approximation:

$$S_q(\dot\gamma) = S_q + \dot\gamma \left\{ \frac{q_x q_y}{q} S_q' \int_0^\infty dt\, \Phi_q^2(t) \right\} + \mathcal{O}(\text{Pe}^2). \tag{19}$$

This relation follows from (11a) in the limit of small $\dot\gamma$, where the quiescent density correlators can be taken from quiescent MCT in (18).

3.2.2 Comparison with Simulations

Figure 8 shows the contribution $\delta S_{\mathbf{q}}^{(\dot\gamma)}$ to the distorted structure factor in leading linear order in $\dot\gamma$ for packing fractions $\phi = 0.36, 0.44$, and 0.46. Data taken from Brownian dynamics simulations at $\phi = 0.43$ and 0.5 from [77] are also included. In both cases the data were divided by a factor $\dot\gamma q_x q_y/q^2$ which is the origin of the trivial anisotropy in the leading linear order. The distortion $\delta S_{\mathbf{q}}^{(\dot\gamma)}$ of the microstructure

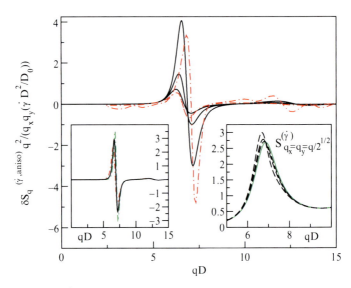

Fig. 8 Contribution $\delta S_{\mathbf{q}}^{\dot\gamma}$ linear in shear rate to the distorted structure factor $S_{\mathbf{q}}(\dot\gamma)$ normalized to $\dot\gamma q_x q_y/q^2$ (*black solid lines*); from [76]. Here D denotes the hard sphere diameter. Decreasing the relative separations from the critical point as $\varepsilon = -0.2, -0.15, -0.1$, the magnitude of $\delta S_{\mathbf{q}}^{(\dot\gamma)}$ increases. The *dashed-dotted red* curves are Brownian dynamics simulation data from [77] using the same normalization at $\varepsilon = -0.259, -0.138$. The *right inset* shows the unnormalized $S_{\mathbf{q}}^{(\dot\gamma)}$ along the extensional axis $q_x = q_y = q/\sqrt{2}$ at $\varepsilon = -0.1$ and, from bottom to top, Pe $= \dot\gamma\tau = 0$ (*green-solid*), $1/8$, $1/4$ and $1/2$ (all *black-dashed*), where $\Phi_{q_p}(t = \tau) = 0.1$ defines τ. Pe/Pe$_0$ = 1.66 holds at this ε. The *left inset* shows the data of the main figure rescaled with the dressed Peclet number, $\delta S_{\mathbf{q}}^{(\dot\gamma,\text{aniso})}/(\text{Pe}\, q_x q_y/q^2)$; for $\varepsilon = -0.1, -0.05$, and -0.01 (with increasing peak height) the values Pe/Pe$_0$ = 1.66, Pe/Pe$_0$ = 8.06, and Pe/Pe$_0$ = 419 are used

grows strongly with ϕ because of the approach to the glass transition. The $\delta S_{\mathbf{q}}^{(\dot{\gamma})}$ is proportional to the α-relaxation time τ, as proven in the left inset of Fig. 8. Here, τ is estimated from $\Phi_{q_p}(t = \tau) = 0.1$, where q_p denotes the position of the primary peak in S_q. Rescaling the data with Pe collapses the curves at different distances to the glass transition. The strongest shear-dependence occurs for the direction of the extensional component of the flow, $q_x = q_y$. Here, the mesoscale order of the dispersion grows; the peak in $\delta S_{\mathbf{q}}^{(\dot{\gamma})}$ increases and sharpens. The ITT results qualitatively agree with the simulations in these aspects [77].

The most important finding of Fig. 8 concerns the magnitude of the distortion of the microstructure, and the dimensionless parameter measuring the effect of shear relative to the intrinsic particle motion. This topic can be discussed using the linear order result, and is not affected by considerations of hydrodynamic interactions, as can be glanced from comparing Brownian dynamics simulations [77] and experiments on dissolved particles [78]. In previous theories, shear rate effects enter when the bare Peclet number Pe$_0$ becomes non-negligible. In the present ITT approach the dressed Peclet/Weissenberg number Pe $= \dot{\gamma}\tau$ governs shear effects; here, τ is the (final) structural relaxation time. Shear flow competes with structural rearrangements that become arbitrarily slow compared to diffusion of dilute particles when approaching the glass transition. The distorted microstructure results from the competition between shear flow and cooperative structural rearrangements. It is thus no surprise that previous theories using Pe$_0$, which is characteristic for dilute fluids or strong flows, had severely underestimated the magnitude of shear distortions in hard sphere suspensions for higher packing fractions; [77, 79] report an underestimate by roughly a decade at $\phi = 0.50$. The ITT approach actually predicts a divergence of $\lim_{\dot{\gamma} \to 0}(S_{\mathbf{q}}^{(\dot{\gamma})} - S_q)/\dot{\gamma}$ for density approaching the glass transition at $\phi_c \approx 0.58$, and for (idealized) glass states, where $\tau = \infty$ holds in MCT following Maxwell's phenomenology, the stationary structure factor becomes non-analytic, and differs from the equilibrium one even for $\dot{\gamma} \to 0$. The distortion $\delta S_{\mathbf{q}}^{(\dot{\gamma})}$ thus qualitatively behaves like the stress, which goes to zero linear in $\dot{\gamma}$ in the fluid, but approaches a yield stress σ^+ in the glass for $\dot{\gamma} \to 0$.

The reassuring agreement of ITT results on $S_{\mathbf{q}}^{(\dot{\gamma})}$ with the data from simulations and experiments shows that in the ITT approach the correct expansion parameter Pe has been identified. This can be taken as support for the ITT-strategy to connect the non-linear rheology of dense dispersions with the structural relaxation studied at the glass transition.

4 Universal Aspects of the Glass Transition in Steady Shear

The summarized microscopic MCT-ITT equations contain a non-equilibrium transition between a shear thinning fluid and a shear-molten glassy state; it is the central novel transition found in MCT-ITT [43]. Close to the transition, (rather) universal predictions can be made about the non-linear dispersion rheology and the steady state properties. Following [32, 80], the central predictions are introduced in this section and summarized in the overview Fig. 9; the following results sections contain more examples. Figure 9 is obtained from the schematic model which is also often used to analyse data, and which is introduced in Sect. 5.2.

A dimensionless separation parameter ε measures the distance to the transition which lies at $\varepsilon = 0$. A fluid state ($\varepsilon < 0$) possesses a (Newtonian) viscosity, $\eta_0(\varepsilon < 0) = \lim_{\dot{\gamma} \to 0} \sigma(\dot{\gamma})/\dot{\gamma}$, and shows shear-thinning upon increasing $\dot{\gamma}$. Via the relation $\eta_0 = \lim_{\omega \to 0} G''(\omega)/\omega$, the Newtonian viscosity can also be taken from the linear response loss modulus at low frequencies, where $G''(\omega)$ dominates over the storage modulus. The latter varies like $G'(\omega \to 0) \sim \omega^2$. A glass ($\varepsilon \geq 0$), in the absence of flow, possesses an elastic constant G_∞, which can be measured in the elastic shear modulus $G'(\omega)$ in the limit of low frequencies, $G'(\omega \to 0, \varepsilon \geq 0) \to G_\infty(\varepsilon)$. Here the storage modulus dominates over the loss one, which drops like $G''(\omega \to 0) \sim \omega$. The high frequency modulus $G'_\infty = G'(\omega \to \infty)$ is characteristic of the particle interactions, see Footnote 4, and exists, except for the case of hard sphere interactions without HI, in fluid and solid states. The dissipation at high frequencies $G''(\omega \to \infty) \to \eta_\infty \omega$ also shows no anomaly at the glass transition and depends strongly on HI and solvent friction.

Enforcing steady shear flow melts the glass. The stationary stress of the shear-molten glass always exceeds a (dynamic) yield stress. For decreasing shear rate, the viscosity increases like $1/\dot{\gamma}$, and the stress levels off onto the yield-stress plateau, $\sigma(\dot{\gamma} \to 0, \varepsilon \geq 0) \to \sigma^+(\varepsilon)$.

Close to the transition, the zero-shear viscosity η_0, the elastic constant G_∞, and the yield stress σ^+ show universal anomalies as functions of the distance to the transition.

The described results follow from the stability analysis of (13, 14) around an arrested, glassy structure f_q of the transient correlator [43, 80]. Considering the time window where $\Phi_\mathbf{q}(t)$ is close to arrest at f_q, and taking all control parameters like density, temperature, etc., to be close to the values at the transition, the stability analysis yields the "factorization" between spatial and temporal dependences

$$\Phi_\mathbf{q}(t) = f_q^c + h_q \, \mathcal{G}(t/t_0, \varepsilon, \dot{\gamma} t_0) + \ldots, \qquad (20)$$

where the (isotropic) glass form factor f_q^c and critical amplitude h_q describe the spatial properties of the metastable glassy state. The critical glass form factor f_q^c gives the long-lived component of density fluctuations right at the transition $\varepsilon = 0$, while $\Phi_\mathbf{q}(t \to \infty, \varepsilon \geq 0, \dot{\gamma} = 0) = f_q > 0$ characterizes states even deep in the glass with f_q obeying [38]

$$\frac{f_q}{1 - f_q} = \frac{1}{2N} \sum_\mathbf{k} V_{qkp} f_k f_p, \qquad (21)$$

with V_{qkp} from (18c) as follows from (13, 14) asymptotically in the limit of vanishing shear rate [80]. Figure 10 shows dynamic light scattering data for the glass form factors at a number of densities in PMMA hard sphere colloids, comparing them to solutions of (21) evaluated for hard spheres using the PY structure factor;

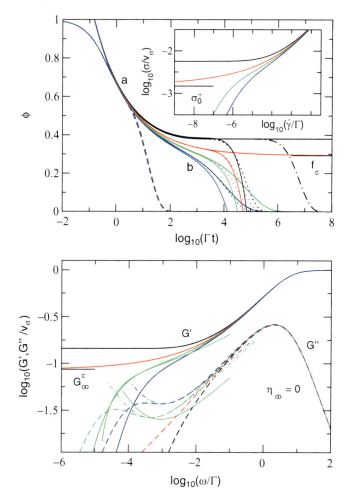

Fig. 9 Overview of the properties of the $F_{12}^{(\dot\gamma)}$-model characteristic for the transition between fluid and yielding glass; from [32]. The *upper panel* shows numerically obtained transient correlators $\Phi(t)$ for $\varepsilon = 0.01$ (*black curves*), $\varepsilon = 0$ (*red*), $\varepsilon = -0.005$ (*green*), and $\varepsilon = -0.01$ (*blue*). The shear rates are $|\dot\gamma/\Gamma| = 0$ (*thick solid lines*), $|\dot\gamma/\Gamma| = 10^{-6}$ (*dotted lines*), and $|\dot\gamma/\Gamma| = 10^{-2}$ (*dashed lines*). For the glass state at $\varepsilon = 0.01$ (*black*), $|\dot\gamma/\Gamma| = 10^{-8}$ (*dashed-dotted-line*) is also included. All curves were calculated with $\gamma_c = 0.1$ and $\eta_\infty = 0$. The *thin solid lines* give the factorization result (20) with scaling functions $\mathcal{G}$ for $|\dot\gamma/\Gamma| = 10^{-6}$; label *a* marks the critical law (22b), and label *b* marks the von Schweidler-law (23). The critical glass form factor f_c is indicated. The *inset* shows the flow curves for the same values for ε. The *thin black bar* shows the yield stress σ_c^+ for $\varepsilon = 0$. The *lower panel* shows the viscoelastic storage (*solid line*) and loss (*broken line*) modulus for the same values of ε. The *thin green lines* are the Fourier-transformed factorization result (20) with scaling function $\mathcal{G}$ taken from the *upper panel* for $\varepsilon = -0.005$. The *dashed-dotted lines* show the fit formula (24) for the spectrum in the minimum-region with $G_{\min}/v_\sigma = 0.0262$, $\omega_{\min}/\Gamma = 0.000457$ at $\varepsilon = -0.005$ (*green*) and $G_{\min}/v_\sigma = 0.0370$, $\omega_{\min}/\Gamma = 0.00105$ at $\varepsilon = -0.01$ (*blue*). The elastic constant at the transition G_∞^c is also marked, while the high frequency asymptote $G_\infty' = G'(\omega \to \infty)$ is not labeled explicitly

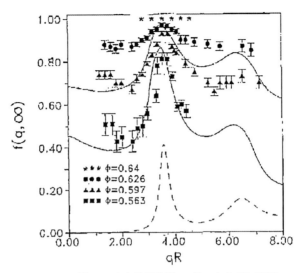

van Megen et al. (1991) Phys. Rev. Lett. 67, 1586

Fig. 10 Glass form factors f_q as function of wavevector q in a colloidal glass of hard spheres for packing fractions ϕ as labeled. Data obtained by van Megen and coworkers by dynamic light scattering are qualitatively compared to MCT computations using the PY-S_q at ϕ values chosen ad hoc to match the experimental data; from [12]. The PY structure factor at the glass transition density $\phi = 0.58$ is shown as *broken line*, rescaled by a factor $1/10$

it is included in Fig. 10 for the packing fraction $\phi_c = 0.58$ of the experimental glass transition. The glass form factor varies with the average particle separation and in phase with the structure factor. Both f_q and h_q thus describe local correlations, the so-called "cage-effect", and can be taken as constants independent on shear rate and density, as they are evaluated from the vertices in (14) at the transition point ($\varepsilon = 0, \dot{\gamma} = 0$).

All time-dependence and (sensitive) dependence on the external control parameters is contained in the function $\mathcal{G}$, which is often called "β-correlator" and obeys the non-linear stability equation [43, 45, 80]

$$\tilde{\varepsilon} - c^{(\dot{\gamma})} (\dot{\gamma} t)^2 + \lambda \, \mathcal{G}^2(t) = \frac{d}{dt} \int_0^t dt' \, \mathcal{G}(t-t') \, \mathcal{G}(t') , \qquad (22a)$$

with initial condition

$$\mathcal{G}(t \to 0) \to (t/t_0)^{-a} . \qquad (22b)$$

The two parameters λ and $c^{(\dot{\gamma})}$ in (22a) are determined by the static structure factor at the transition point, and take values around $\lambda \approx 0.73$ and $c^{(\dot{\gamma})} \approx 0.7$ for the PY S_q for hard spheres. The transition point then lies at packing fraction $\phi_c = \frac{4\pi}{3} n_c R_H^3 \approx 0.52$ (index 'c' for critical), and the separation parameter measures the relative distance, $\tilde{\varepsilon} = C\varepsilon$ with $\varepsilon = (\phi - \phi_c)/\phi_c$ and $C \approx 1.3$. The "critical" exponent a is given by the exponent parameter λ via $\lambda = \Gamma(1-a)^2/\Gamma(1-2a)$, as had been found in quiescent MCT [2, 38].

The time scale t_0 in (22b) provides the means to match the function $\mathcal{G}(t)$ to the microscopic, short-time dynamics. Equations (13, 14) contain a simplified description of the short time dynamics in colloidal dispersions via the initial decay rate $\Gamma_{\mathbf{q}}(t)$. From this model for the short-time dynamics, the time scale $t_0 \approx 1.6\,10^{-2} R_H^2/D_0$ is obtained. Solvent mediated effects on the short time dynamics are well known and are neglected in $\Gamma_{\mathbf{q}}(t)$ in (13). The most simple minded approximation to account for HI is given in (15). It only shifts the value of t_0. Within the ITT approach, this finding holds more generally. Even if HI lead to more substantial changes of (13), all of the mentioned universal predictions would remain true, as long as HI do not affect the mode coupling vertex in (14). As in the quiescent MCT [74], in MCT-ITT hydrodynamic interactions can thus be incorporated into the theory of the glass transition under shear, and amount to a rescaling of the matching time t_0 only.

The parameters ε, λ, and $c^{(\dot{\gamma})}$ in (22a) can be determined from the equilibrium structure factor S_q at or close to the transition and, together with t_0 and the shear rate $\dot{\gamma}$, they capture the essence of the rheological anomalies in dense dispersions. A divergent viscosity follows from the prediction of a strongly increasing final relaxation time in $\mathcal{G}$ in the quiescent fluid phase [2, 38]:

$$\mathcal{G}(t \to \infty, \varepsilon < 0, \dot{\gamma} = 0) \to -(t/\tau)^b, \quad \text{with} \quad \frac{t_0}{\tau} \propto (-\varepsilon)^\gamma. \tag{23}$$

The entailed temporal power law, termed von Schweidler law, initiates the final decay of the correlators, which has a density and temperature independent shape $\tilde{\Phi}_q(\tilde{t})$. In MCT, the (full) correlator thus takes the characteristic form of a two-step relaxation. The final decay, often termed α-relaxation, depends on ε only via the time scale $\tau(\varepsilon)$ which rescales the time, $\tilde{t} = t/\tau$. Equation (22) establishes the crucial time scale separation between t_0 and τ, the divergence of τ, and the stretching (non-exponentiality) of the final decay; it also gives the values of the exponents via $\lambda = \Gamma(1+b)^2/\Gamma(1+2b)$, and $\gamma = (1/a + 1/b)/2$. Using (17), the MCT-prediction for the divergence of the Newtonian viscosity follows [2, 38]. During the final decay the quiescent shear modulus also becomes a function of rescaled time, $\tilde{t} = t/\tau$, leading to $\eta_0 \propto \tau(\varepsilon)$; its initial value is given by the elastic constant at the transition, G_∞^c.

The two asymptotic temporal power-laws of MCT also affect the frequency dependence of G'' in the minimum region. The scaling function $\mathcal{G}$ describes the minimum as crossover between two power laws in frequency. The approximation for the modulus around the minimum in the quiescent fluid becomes [38]

$$G''(\omega) \approx \frac{G_{\min}}{a+b} \left[b\left(\frac{\omega}{\omega_{\min}}\right)^a + a\left(\frac{\omega_{\min}}{\omega}\right)^b \right]. \tag{24}$$

The parameters in this approximation follow from (22, 23) which give $G_{\min} \propto \sqrt{-\varepsilon}$ and $\omega_{\min} \propto (-\varepsilon)^{1/2a}$. Observation of this handy expression requires that the relaxation time τ is (very) large, viz. that time scale separation holds (extremely well) for (very) small $-\varepsilon$; even in the exemplary Fig. 9, the chosen distances to the glass transition are too large in order for the (24) to agree with the true β-correlator $\mathcal{G}$, which is also included in Fig. 9. The reason for this difficulty is the aspect that the expansion in (20) is an expansion in $\sqrt{\varepsilon}$, which requires very small separation parameters ε for corrections to be negligible. For packing fractions too far below the glass transition, the final relaxation process is not clearly separated from the high frequency relaxation. This holds in the experimental data shown in Fig. 6, where the final structural decay process only forms a shoulder.

On the glassy side of the transition, $\varepsilon \geq 0$, the transient density fluctuations stay close to a plateau value for intermediate times which increases when going deeper into the glass:

$$\mathcal{G}(t_0 \ll t \ll 1/|\dot{\gamma}|, \varepsilon \geq 0) \to \sqrt{\frac{\tilde{\varepsilon}}{1-\lambda}} + \mathcal{O}(\varepsilon). \tag{25}$$

Entered into (17), the square-root dependence of the plateau value translates into the square-root anomaly of the elastic constant G_∞, and causes the increase of the yield stress close to the glass transition.

For vanishing shear rate, $\dot{\gamma} = 0$, an ideal glass state exists in the ITT approach for steady shearing. All density correlators arrest at a long time limit, which from (25) close to the transition is given by $\Phi_q(t \to \infty, \varepsilon \geq 0, \dot{\gamma} = 0) = f_q = f_q^c + h_q \sqrt{\tilde{\varepsilon}/(1-\lambda)} + \mathcal{O}(\varepsilon)$. Consequently the modulus remains elastic at long times, $g(t \to \infty, \varepsilon \geq 0, \dot{\gamma} = 0) = G_\infty > 0$. Any (infinitesimal) shear rate, however, melts the glass and causes a final decay of the transient correlators. The function $\mathcal{G}$ initiates the decay around the critical plateau of the transient correlators and sets the common time scale for the final decay under shear:

$$\mathcal{G}(t \to \infty) \to -\sqrt{\frac{c^{(\dot{\gamma})}}{\lambda - \frac{1}{2}}} |\dot{\gamma} t| \equiv -\frac{t}{\tau_{\dot{\gamma}}}, \quad \text{with} \quad \tau_{\dot{\gamma}} = \sqrt{\frac{\lambda - \frac{1}{2}}{c^{(\dot{\gamma})}}} \frac{1}{|\dot{\gamma}|}. \tag{26}$$

Under shear all correlators decay from the plateau as function of $|\dot{\gamma} t|$; see, e.g., Figs. 11, 12, 21, and 22. Steady shearing thus prevents non-ergodic arrest and restores ergodicity. Shear melts a glass and produces a unique steady state at long times. This conclusion is restricted by the already discussed assumption to neglect aging of glassy states. It could remain because of non-ergodicity in the initial quiescent state, which needs to be shear-molten before ITT holds. Ergodicity of the sheared state, however, suggests aging to be unimportant under shear, and that it should be possible to melt initial non-ergodic contributions [8, 9]. The experiments in model colloidal dispersions reported in Sects. 3.1.2 and 6.2 support this notion, as history independent steady states could be achieved at all densities[5].

---

[5] An ultra-slow process causing the metastability of glassy states even without shear may have contributed to restore ergodicity in [32, 33].

The described universal scenario of shear-molten glass and shear-thinnig fluid makes up the core of the MCT-ITT predictions derived from (11–14). Their consequences for the nonlinear rheology will be discussed in more detail in the following sections, while the MCT results for the linear viscoelasticity were reviewed in Sect. 3. Yet, the anisotropy of the equations has up to now prevented more complete solutions of the MCT-ITT equations of Sect. 2. Therefore, simplified MCT-ITT equations become important, which can be analysed in more detail and recover the central stability equations (20, 22). The two most important ones will be reviewed next, before the theoretical picture is tested in comparison with experimental and simulations' data.

5 Simplified Models

Two progressively more simplified models provide insights into the generic scenario of non-Newtonian flow, shear melting and solid yielding which emerge from the ITT approach.

5.1 Isotropically Sheared Hard Sphere Model

On the fully microscopic level of description of a sheared colloidal suspension, affine motion of the particles with the solvent leads to anisotropic dynamics. Yet recent simulation data of steady state structure factors indicate a rather isotropic distortion of the structure for $Pe_0 \ll 1$, even though the Weissenberg number Pe is already large [10, 81]. Confocal microscopy data on concentrated solutions support this observation [30]. The shift of the advected wavevector in (7) with time to higher values, intially is anisotropic, but becomes isotropic at longer times, when the magnitude of $q(t)$ increases along all directions. As the effective potentials felt by density fluctuations evolve with increasing wavevector, this leads to a decrease of friction functions, speed-up of structural rearrangements, and shear-fluidization. Therefore, one may hope that an "isotropically sheared hard spheres model" (ISHSM), which for $\dot{\gamma} = 0$ exhibits the nonlinear coupling of density correlators with wavelength equal to the average particle distance (viz. the "cage-effect"), and which, for $\dot{\gamma} \neq 0$, incorporates shear-advection, captures some spatial aspects of shear driven decorrelation.

5.1.1 Definition of the ISHSM

Thus, in the ISHSM, the equation of motion for the density fluctuations at time t after starting the shear is approximated by that of the quiescent system, namely (18a) (with $\Gamma_q = q^2 D_s/S_q$). The memory function is also taken as isotropic and modeled close to the unsheared situation [43, 80]

$$m_q(t) \approx \frac{1}{2N} \sum_{\mathbf{k}} V_{qkp}^{(\dot{\gamma})}(t) \, \Phi_k(t) \, \Phi_p(t) \, , \qquad (27a)$$

with

$$V_{qkp}^{(\dot{\gamma})}(t) = \frac{n^2 S_q S_k S_p}{q^4} \left[\mathbf{q} \cdot \mathbf{k} \, c_{\bar{k}(t)} + \mathbf{q} \cdot \mathbf{p} \, c_{\bar{p}(t)} \right] \left[\mathbf{q} \cdot \mathbf{k} \, c_k + \mathbf{q} \cdot \mathbf{p} \, c_p \right] \qquad (27b)$$

where $\mathbf{p} = \mathbf{q} - \mathbf{k}$, and the length of the advected wavevectors is approximated by $\bar{k}(t) = k(1 + (t\dot{\gamma}/\gamma_c)^2)^{1/2}$ and equivalently for $\bar{p}(t)$. Note, that the memory function thus only depends on one time, and that shear advection leads to a dephasing of the two terms in the vertex (27b), which form a perfect square in the quiescent vertex of (18c) without shear. This (presumably) is also the dominant effect of shear in the full microscopic MCT-ITT memory kernel in (14). The fudge factor γ_c is introduced in order to correct for the underestimate of the effect of shearing in the ISHSM[6].

The expression for the potential part of the transverse stress may be simplified to

$$\sigma = \dot{\gamma} \int_0^\infty dt \, g(t, \dot{\gamma}), \text{ with } g(t, \dot{\gamma}) \approx \frac{k_B T}{60\pi^2} \int dk \, \frac{k^5 \, c_k' \, S_{\check{k}(t)}'}{\check{k}(t) S_k^2} \, \Phi_{\check{k}(t)}^2(t) \, , \qquad (27c)$$

where, in the last equation (27c), the advected wavevector is chosen as $\check{k}(t) = k(1 + (t\dot{\gamma})^2/3)^{1/2}$, as follows from straight forward isotropic averaging of $\mathbf{k}(t)$. For the numerical solution of the ISHSM for hard spheres using S_q in PY approximation, the wavevector integrals were discretized as discussed in Sect. 3.1.1 and following [72], using $M = 100$ wavevectors from $k_{min} = 0.1/R$ up to $k_{max} = 19.9/R$ with separation $\Delta k = 0.2/R$. Again, time was discretized with initial step-width $dt = 2\,10^{-7} R^2/D_s$, which was doubled each time after 400 steps [82]. The model's glass transition lies at $\phi_c = 0.51591$, with exponent parameter $\lambda = 0.735$ and $c^{(\dot{\gamma})} \approx 0.45/\gamma_c^2$; note that these values still change somewhat if the discretization is made finer. The separation parameter $\varepsilon = (\phi - \phi_c)/\phi_c$, and $\dot{\gamma}$ are the two relevant control parameters determining the rheology.

5.1.2 Transient Correlators

The shapes of the transient density fluctuation functions can be studied with spatial resolution in the ISHSM. Figure 11 displays density correlators at two densities, just below (panel (a)) and just above (panels (b,c)) the transition, for varying shear rates. Panel (b) and (c) compare correlators at different wavevectors to exemplify

---

[6] Except for the introduction of the parameter γ_c, further quantitatively small, but qualitatively irrelevant, differences exist between the ISHSM defined here and used in Sect. 6.1 according to [45], and that originally defined in [43] and shown in Sect. 5.1; see [45] for discussion.

Nonlinear Rheology

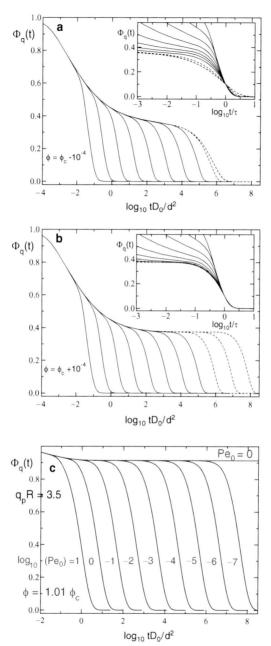

Fig. 11 Normalized transient density correlators $\Phi_q(t)$ of the ISHSM at wavevector $q = 3.4/d$ below (panel (**a**) at $\phi = \phi_c - 10^{-4}$) and above (panel (**b**) at $\phi = \phi_c + 10^{-4}$) the transition for increasing shear rates $\mathrm{Pe}_0 = 9^n * 10^{-8}$ with $n = 0, \ldots, 10$ from *right* to *left*; the distances correspond to $\varepsilon = \pm 10^{-3.53}$. Curves for $n = 9, 10$ carry *short dashes* and for $n = 8$ *long dashes*; note the collapse of the two *short dashed* curves in (a). The *insets* show the data rescaled so as to coincide at $\Phi(t = \tau) = 0.1$. Panel (**c**) shows glass correlators at another wavevector $q = 7/d$ for parameters as labeled; from [80]

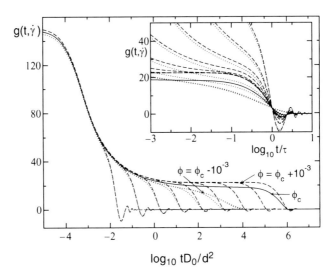

Fig. 12 Transient non-Newtonian shear modulus $g(t, \dot\gamma)$ of the ISHSM in units of $k_B T/d^3$ for the packing fractions $\phi = \phi_c \pm 10^{-3}$ ($\varepsilon = \pm 10^{-2.53}$; *dashed/dotted lines*, respectively) for increasing shear rates $\mathrm{Pe}_0 = 9^n \times 10^{-6}$ with $n = 0, \ldots, 8$ from *right* to *left*; note the collapse of fluid lines for the smallest Pe_0; from [80]. The *solid line* gives $g(t, \dot\gamma)$ for $\phi = \phi_c$ and $\mathrm{Pe}_0 = 10^{-6}$. The *inset* shows the data rescaled so as to coincide at $g(t = \tau, \dot\gamma) = 5$; note the collapse of the $n = 6, 7$ & 8 curves for both $\phi > \phi_c$ and $\phi < \phi_c$

the spatial variation. In almost all cases the shear rate is so small that the bare Peclet number Pe_0 is negligibly small and the short-time dynamics is not affected.

In the fluid case, the final or α-relaxation is also not affected for the two smallest dressed Peclet Pe values, but for larger Pe it becomes faster and less stretched; see the inset of Fig. 11(a).

Above the transition, the quiescent system forms an (idealized) glass [2, 38], whose density correlators arrest at the glass form factors f_q from Fig. 10, and which exhibits a finite elastic constant G_∞ that describes the (zero-frequency) Hookian response of the amorphous solid to a small applied shear strain γ: $\sigma = G_\infty \gamma$ for $\gamma \to 0$; the plateau G_∞ can be seen in Fig. 3 and for intermediate times in Fig. 12. If steady flow is imposed on the system, however, the glass melts for any arbitrarily small shear rate. Particles are freed from their cages and diffusion perpendicular to the shear plane also becomes possible. Any finite shear rate, however small, sets a finite longest relaxation time, beyond which ergodicity is restored; see Figs. 11(b,c) and 12.

The glassy curves at $\varepsilon > 0$, panels (b,c), exhibit a shift of the final relaxation with $\tau_{\dot\gamma}$ from (26) and asymptotically approach a scaling function $\Phi_q^+(t/\tau_{\dot\gamma})$. The master equation for the "yielding" scaling functions Φ_q^+ in the ISHSM can be obtained from eliminating the short-time dynamics in (18a). After a partial integration, the equation with $\partial_t \Phi_q(t) = 0$ is solved by the scaling functions:

$$\Phi_q^+(\tilde{t}) = m_q^+(\tilde{t}) - \frac{d}{d\tilde{t}} \int_0^{\tilde{t}} d\tilde{t}' \, m_q^+(\tilde{t}-\tilde{t}') \, \Phi_q^+(\tilde{t}'), \tag{28a}$$

where $\tilde{t} = t/\tau_{\dot{\gamma}}$, and the memory kernel is given by

$$m_q^+(\tilde{t}) = \frac{1}{2N} \sum_{\mathbf{k}} V_{q,\mathbf{k}}^{(\tilde{\dot{\gamma}})}(\tilde{t}) \, \Phi_k^+(\tilde{t}) \, \Phi_{|\mathbf{q}-\mathbf{k}|}^+(\tilde{t}). \tag{28b}$$

While the vertex is evaluated at fixed shear rate, $\tilde{\dot{\gamma}} = \sqrt{\frac{\lambda - \frac{1}{2}}{c(\dot{\gamma})}}$, it depends on the equilibrium parameters. The initialization for the correlator is given by

$$\Phi_q^+(\tilde{t} \to 0) = f_q, \tag{28c}$$

with glass form factor taken from (21). The two-step relaxation and the shift of the final relaxation with $\tau_{\dot{\gamma}}$ are quite apparent in Fig. 11.

Figure 12 shows the transient shear modulus $g(t,\dot{\gamma})$ of the ISHSM from (27c), which determines the viscosity via $\eta = \int_0^\infty dt\, g(t,\dot{\gamma})$. It is the time derivative of the shear stress growth function $\eta^+(t,\dot{\gamma})$ (or transient start up viscosity; here, the $^+$ labels the shear history [1, 83]), $g(t,\dot{\gamma}) = \frac{d}{dt}\eta^+(t,\dot{\gamma})$, and in the Newtonian-regime reduces to the time dependent shear modulus, $g^{lr}(t)$. The $g(t,\dot{\gamma})$ shows all the features exhibited by the correlator of the density correlators in the ISHSM, and thus the discussion based on the stability analysis in Sect. 4 and the yielding scaling law carries over to it: $g(t,\dot{\gamma}) = G_\infty^c + h_g \mathcal{G} + \ldots$; the dependence of the final relaxation step on rescaled time $t/\tau_{\dot{\gamma}}$ is apparent in the glass curves. However, in contrast to the density correlators $\Phi_q(t)$, the function $g(t,\dot{\gamma})$ becomes negative (oscillatory) in the final approach towards zero, an effect more marked at high Pe. This behavior originates in the general expression for $g(t,\dot{\gamma})$, (27c), where the vertex reduces to a positive function (complete square) only in the absence of shear advection. A overshoot and oscillatory approach of the start up viscosity to the steady state value, $\eta^+(t\to\infty,\dot{\gamma}) \to \eta(\dot{\gamma})$, therefore are generic features predicted from our approach.

5.1.3 Flow Curves

As discussed in Sect. 4, in the fluid, MCT-ITT finds a linear or Newtonian regime in the limit $\dot{\gamma} \to 0$, where it recovers the standard MCT approximation for Newtonian viscosity η_0 of a viscoelastic fluid [2, 38]. Hence $\sigma \to \dot{\gamma}\eta_0$ holds for Pe $\ll 1$, as shown in Fig. 13, where Pe calculated with the structural relaxation time τ is included. As discussed, the growth of τ (asymptotically) dominates all transport coefficients of the colloidal suspension and causes a proportional increase in the viscosity η. For Pe > 1, the non-linear viscosity shear thins, and σ increases sublinearly with $\dot{\gamma}$. The stress vs strain rate plot in Fig. 13 clearly exhibits a broad crossover between the linear Newtonian and a much weaker (asymptotically) $\dot{\gamma}$-independent variation of the stress. In the fluid, the flow curve takes a S-shape in double logarithmic representation, while in the glass it is bent upward only.

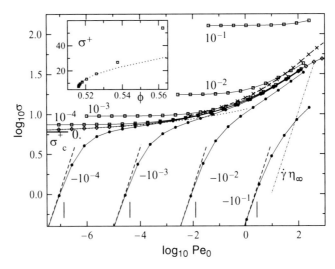

Fig. 13 Steady state shear stress σ in units of $k_B T/d^3$ vs $\mathrm{Pe}_0 = \dot{\gamma} d^2/D_0$, for the ISHSM at various distances from its glass transition, $\phi - \phi_c$ as labeled; *circles* correspond to fluid, *diamonds* to the critical, and *squares* to glassy densities; from [80], where the additional lines are discussed. For the fluid cases, $\phi < \phi_c$, *dashed lines* indicate Newtonian fluid behavior, $\sigma = \eta \dot{\gamma}$, while *vertical bars* mark Pe= $\dot{\gamma}\tau = 1$, with the structural relaxation time taken from $\Phi_{q=7/d}(t=\tau) = 0.1$. The additional stress which would arise from the background solvent viscosity, $\sigma = \dot{\gamma}\eta_\infty$, is marked by a *dot-dashed line*. For the critical density, ϕ_c, the critical yield stress, $\sigma_c^+ = 6.04$, is shown by a *horizontal bar*. The *inset* shows the rise of the dynamical yield stress $\sigma^+ = \sigma(\varepsilon \geq 0, \dot{\gamma} \to 0+)$ in the glass together with a fitted power-law asymptote, $\sigma^+ = \sigma_c^+ + 112\sqrt{\phi - \phi_c}$

Above the transition, the glass melts for any shear rate. Nonetheless, a finite limiting stress (yield stress) must be overcome in order to maintain the flow of the glass:

$$\sigma(\dot{\gamma}, \varepsilon \geq 0) \geq \sigma^+(\varepsilon \geq 0) = \lim_{\dot{\gamma} \to 0} \sigma(\dot{\gamma}, \varepsilon > 0) \ .$$

For $\varepsilon \geq 0$ and $\dot{\gamma} \to 0$, the time $\tau^{(\dot{\gamma})}$ for the final decay, (26), can become arbitrarily slow compared to the time characterizing the decay onto f_q. Inserting the scaling functions Φ^+ from (28) into the expression (27c) for the stress, the long time contribution separates out. Importantly, the integrands containing the Φ^+ functions depend on time only via $\tilde{t} = t/\tau_{\dot{\gamma}} \propto \dot{\gamma} t$, so that nontrivial limits for the stationary stress follow in the limit $\dot{\gamma} \to 0$. In the ISHSM for $\varepsilon \geq 0$, the yield stress is given by

$$\sigma^+ = \frac{k_B T \tilde{\gamma}}{60\pi^2} \int_0^\infty d\tilde{t} \int dk\, k^5 \frac{S_k' S_{k(\tilde{t})}'}{k(\tilde{t}) S_k^2} \left(\Phi_{k(\tilde{t})}^+ (\tilde{t}) \right)^2 , \tag{28d}$$

The existence of a dynamic yield stress in the glass phase is thus seen to arise from the scaling law in (28), which is clearly borne out in Figs. 11 and 12. The yield stress arises from those fluctuations which require the presence of shearing to prevent their arrest. Even though σ^+ requires the solution of dynamical equations, in

MCT-ITT it is completely determined by the equilibrium structure factor S_q. This may suggest a connection of MCT to the potential energy paradigm for glasses, as recently discussed [84, 85]. One might argue that σ^+ arises because the external driving allows the system to overcome energy barriers so that different metastable states can be reached. This interpretation would agree with ideas from spin-glass [9] and soft-glassy rheology [6–8]. MCT-ITT indicates how shear achieves this in the case of colloidal suspensions. It pushes fluctuations to shorter wavelengths where smaller particle rearrangements cause their decorrelation.

The increase of the amplitude of the yielding master functions Φ^+ in (28) originates in the increase of the arrested structure in the unsheared glass (25). In consequence the yield stress should rapidly increase as one moves further into the glass phase, $\sigma^+ - \sigma_c^+ \propto \sqrt{\varepsilon}$ should (approximately) hold; see however [86] for the more complicated rigorous expression. Indeed, the inset of Fig. 12 shows a good fit of this anomalous increase to the numerical data. It is one of the hallmarks of the weakening of the glass upon approaching the glass transition from the low temperature or high density side.

5.2 Schematic $F_{12}^{(\dot{\gamma})}$-model

The universal aspects described in Sect. 4 are contained in any ITT model that contains the central bifurcation scenario and recovers (20, 22). Equation (20) states that spatial and temporal dependences decouple in the intermediate time window. Thus it is possible to investigate ITT models without proper spatial resolution. Because of the technical difficulty to evaluate the anisotropic functionals in (11d, 14), it is useful to restrict the description to few or to a single transient correlator. The best studied version of such a one-correlator model is the $F_{12}^{(\dot{\gamma})}$-model.

5.2.1 Definition and Parameters

In the schematic $F_{12}^{(\dot{\gamma})}$-model [80] a single "typical" density correlator $\Phi(t)$, conveniently normalized according to $\Phi(t \to 0) = 1 - \Gamma t$, obeys a Zwanzig–Mori memory equation which is modeled according to (13)

$$\partial_t \Phi(t) + \Gamma \left\{ \Phi(t) + \int_0^t dt'\, m(t-t')\, \partial_{t'} \Phi(t') \right\} = 0. \tag{29a}$$

The parameter Γ mimics the short time, microscopic dynamics, and depends on structural and hydrodynamic correlations. The memory function describes stress fluctuations which become more sluggish together with density fluctuations, because slow structural rearrangements dominate all quantities. A self consistent approximation closing the equations of motion is made mimicking (14a). In the

$F_{12}^{(\dot{\gamma})}$-model one includes a linear term (absent in (14a)) in order to sweep out the full range of λ values in (21), and in order to retain algebraic simplicity:

$$m(t) = \frac{v_1 \Phi(t) + v_2 \Phi^2(t)}{1 + (\dot{\gamma}t/\gamma_c)^2} \qquad (29b)$$

This model, for the quiescent case $\dot{\gamma} = 0$, was introduced by Götze in 1984 [38, 87] and describes the development of slow structural relaxation upon increasing the coupling vertices $v_i \geq 0$; they mimic the dependence of the vertices in (14b) at $\dot{\gamma} = 0$ on the equilibrium structure given by S_q. Under shear an explicit time dependence of the couplings in $m(t)$ captures the decorrelation by shear in (14b). The parameter γ_c sets a scale that is required in order for the accumulated strain $\dot{\gamma}t$ to matter. Shearing causes the dynamics to decay for long times, because fluctuations are advected to smaller wavelengths where small scale Brownian motion relaxes them. Equations (29a,b) lead, with $\Phi(t) = f^c + (1-f^c)^2 \mathcal{G}(t,\varepsilon,\dot{\gamma})$, and the choice of the vertices $v_2 = v_2^c = 2$, and $v_1 = v_1^c + \varepsilon(1-f^c)/f^c$, where $v_1^c = 0.828$, to the critical glass form factor $f^c = 0.293$ and to the stability equation (22), with parameters

$$\lambda = 0.707, \ c^{(\dot{\gamma})} = 0.586/\gamma_c^2, \text{ and } t_0 = 0.426/\Gamma.$$

The $F_{12}^{(\dot{\gamma})}$-model possesses a line of glass transitions where the long time limit $f = \Phi(t \to \infty)$ jumps discontinuously; it obeys the equivalent equation to (21). The glass transition line is parameterized by $(v_1^c, v_2^c) = ((2\lambda - 1), 1)/\lambda^2$ with $0.5 \leq \lambda < 1$, and $f^c = 1 - \lambda$. The present choice of transition point (v_1^c, v_2^c) is a typical one, which corresponds to the given typical λ-value. The separation parameter ε is the crucial control parameter as it takes the system through the transition.

For simplicity, the quadratic dependence of the generalized shear modulus on density fluctuations is retained from the microscopic (11d). It simplifies because only one density mode is considered, and as, for simplicity, a dependence of the vertex (prefactor) v_σ on shear is neglected:

$$g(t, \dot{\gamma}) = v_\sigma \Phi^2(t) + \eta_\infty \delta(t - 0+). \qquad (29c)$$

The parameter η_∞ characterizes a short-time, high frequency viscosity and models viscous processes which require no structural relaxation, like in the general case (15). Together with Γ, it is the only model parameter affected by HI. Steady state shear stress under constant shearing, and viscosity then follow via integrating up the generalized modulus:

$$\sigma = \eta \, \dot{\gamma} = \dot{\gamma} \int_0^\infty dt \, g(t) = \dot{\gamma} \int_0^\infty dt \, v_\sigma \Phi^2(t) + \dot{\gamma} \, \eta_\infty. \qquad (30)$$

Also, when setting shear rate $\dot{\gamma} = 0$ in (29a, b), so that the schematic correlator belongs to the quiescent, equilibrium system, the frequency dependent moduli are obtained from Fourier transforming:

Nonlinear Rheology

$$G'(\omega) + i G''(\omega) = i\omega \int_0^\infty dt\, e^{-i\omega t}\, v_\sigma\, \Phi^2(t)\big|_{\dot\gamma=0} + i\omega\, \eta_\infty\,. \tag{31}$$

Because of the vanishing of the Fourier-integral in (31) for high frequencies, the parameter η_∞ can be identified as high frequency viscosity:

$$\lim_{\omega\to\infty} G''(\omega)/\omega = \eta_\infty^\omega, \quad \text{with } \eta_\infty^\omega = \eta_\infty\,. \tag{32}$$

At high shear, on the other hand, (29b) leads to a vanishing of $m(t)$, and (29a) gives an exponential decay of the transient correlator, $\Phi(t) \to e^{-\Gamma t}$ for $\dot\gamma \to 0$. The high shear viscosity thus becomes

$$\eta_\infty^{\dot\gamma} = \lim_{\dot\gamma\to\infty} \sigma(\dot\gamma)/\dot\gamma = \eta_\infty + \frac{v_\sigma}{2\Gamma} = \eta_\infty^\omega + \frac{v_\sigma}{2\Gamma}\,. \tag{33}$$

5.2.2 Correlators and Stability Analysis

Representative solutions of the $F_{12}^{(\dot\gamma)}$-model are summarized in Fig. 9; these bring out the discussed universal aspects included in all ITT models. For small separation parameters and shear rates the correlators develop a stretched dynamics located around the critical plateau value f_c, according to (20–22). The discussion of the dynamics around this plateau was a topic of Sect. 4. Figure 14 shows these aspects in the $F_{12}^{(\dot\gamma)}$-model and presents typical correlators and the corresponding β-correlators.

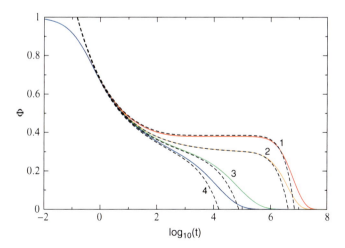

Fig. 14 Numerically obtained transient correlators $\Phi(t)$ (*solid lines*) for $\varepsilon = 0.01$ (*red*, 1), $\varepsilon = 0$ (*orange*, 2), $\varepsilon = -0.005$ (*green*, 3) and $\varepsilon = -0.01$ (*blue*, 4) for the $F_{12}^{(\dot\gamma)}$-model from [86]. All curves were calculated with $\dot\gamma = 10^{-7}$. The *dashed lines* show the corresponding numerically obtained functions $f_c + (1-f_c)^2 \mathcal{G}(t)$

The latter describe how the glassy structure, which is present on intermediate times, is molten either because the density is too low, or the temperature too high, or, alternatively, because of the effect of shearing. For long times $\mathcal{G}(t)$ merges into the linear asymptote $-t/\tau_{\dot\gamma}$ from (26). In the liquid region, for short times $\mathcal{G}(t)$ follows $(t/t_0)^{-a}$ from (22b), and merges into the second power law $-(t/\tau_0)^b$ from (23) for intermediate times with the von MCT Schweidler exponent [38]. In the transition region close to $\varepsilon = 0$, after following $(t/t_0)^{-a}$ the function $\mathcal{G}(t)$ merges directly into the long time asymptote $-t/\tau_{\dot\gamma}$. In the yielding glass region, $\mathcal{G}(t)$ follows $(t/t_0)^{-a}$, arrests on the plateau value $\sqrt{\varepsilon/(1-\lambda)}$ for intermediate times, and merges into the linear asymptote $-t/\tau_{\dot\gamma}$ only for long times. So we can summarize that the short- and long time asymptotes are common for all ε if $\dot\gamma \ne 0$ is common. Figure 15 shows an overview of the properties of $\mathcal{G}(t)$.

The present β-scaling law bears some similarity to that presented by Götze and Sjögren for the description of thermally activated processes in glasses [55, 88]. In both cases, ideal glass states are destroyed by additional decay mechanisms. Yet, the ITT equations and the generalised MCT equations differ qualitatively in the mechanism melting the glass. The similarity between both scaling laws thus underlines the universality of the glass stability analysis, which is determined by quite fundamental principles. In (22), the shear rate can only be a relevant perturbation (at long times) if it appears multiplied by time itself. Symmetry dictates the appearance of $(\dot\gamma t)^2$, because the sign of the shear rate must not matter. The aspect that shear melts the glass determines the negative sign of $(\dot\gamma t)^2$.

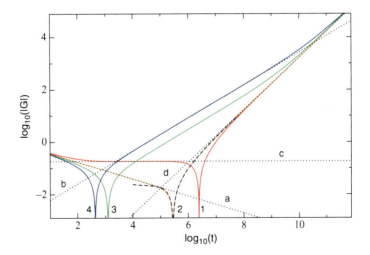

Fig. 15 An overview of the properties of $\mathcal{G}(t)$ (*solid lines*) for the same values for ε and $\dot\gamma$ as in Fig. 14; from [86]. The *dotted lines* show the leading asymptotes for the corresponding time scales: the critical decay $(t/t_0)^{-a}$ (a), the von Schweidler law $-(t/\tau_0)^b$ (b), the arrest on the plateau value $\sqrt{\varepsilon/(1-\lambda)}$ (c) and the shear-induced linear asymptote $-t/\tau_{\dot\gamma}$ (d). The *dashed line* shows a generalization of the latter law evaluated to higher order with a fitted parameter a_1 (at $\varepsilon = 0$)

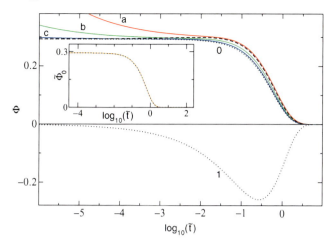

Fig. 16 The numerically determined master function of the yielding-process obeying a "time-shear-superposition principle" (*dotted line*) $\Phi^+(\tilde{t}) = \tilde{\Phi}_0(\tilde{t})$ (label 0); from [86]. The *solid lines* show numerically obtained transient density correlators for $\varepsilon = 0$ and $\dot{\gamma} = 10^{-7}$ (*red*, a), $\dot{\gamma} = 10^{-9}$ (*green*, b) and $\dot{\gamma} = 10^{-12}$ (*blue*, c), plotted as functions of the rescaled time $\tilde{t}$. The plots demonstrate that the rescaled correlators converge to the yield master function $\tilde{\Phi}_0(\tilde{t})$ from the analog of (28) in the $F_{12}^{(\dot{\gamma})}$-model for $\dot{\gamma} \to 0$; the *blue curve* (c) is already quite close to the master curve (0). The *dashed line* shows $\tilde{\Phi}_0(\tilde{t}) + a_1|\dot{\gamma}|^c \tilde{\Phi}_1(\tilde{t})$ for $\dot{\gamma} = 10^{-7}$ and the same numerical value for a_1 as in Fig. 15, using the leading correction $\tilde{\Phi}_1(\tilde{t})$ (1), which is shown in the *lower panel*. This first order expansion already describes quite well the shear-induced decay of the *red curve* (a). The *inset* demonstrates that the master function $\tilde{\Phi}_0(\tilde{t})$ (*dotted line*) can be well approximated by an exponential function (*solid line*); the curves overlap completely

The melting of the glassy structure during the yield process of (28) can be explicitly evaluated in the schematic $F_{12}^{(\dot{\gamma})}$-model at $\varepsilon = 0$. The yield master function does not depend on $\dot{\gamma}$, and while its form is model-dependent, its initial decay follows from the universal stability equation (22). Figure 16 shows numerical results, which can be well approximated by an exponential function.

The qualitative agreement between the transient correlators of the ISHSM and the schematic $F_{12}^{(\dot{\gamma})}$-model support the simplification to disregard the spatial structure. The only cost to be paid, is the fixed plateau value f_c, which can be varied with wavevector in the ISHSM, but not in the schematic model.

5.2.3 Asymptotic Laws of Flow Curves

A major advantage of the simplified $F_{12}^{(\dot{\gamma})}$-model is that it allows for asymptotic expansions that qualitatively capture the flow curves. They can thus be investigated in detail addressing such questions as, e.g., for the existence of power-law shear thinning [1], or the dependence of the yield stress on separation parameter. Figure 17 shows an overview of the numerically obtained flow curves and the corresponding

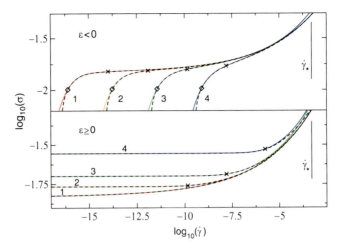

Fig. 17 Overview of the numerically obtained flow curves (*solid lines*) and the asymptotic Λ-formula evaluated numerically (*dashed lines*); from [86]. The liquid curves in the *upper panel* are shown for $\varepsilon = -10^{-7}$ (*red*, 1), $\varepsilon = -10^{-6}$ (*orange*, 2), $\varepsilon = -10^{-5}$ (*green*, 3) and $\varepsilon = -10^{-4}$ (*blue*, 4). The *lower panel* shows the glassy curves for $\varepsilon = 0$ (*red*, 1), $\varepsilon = 10^{-5}$ (*orange*, 2), $\varepsilon = 10^{-4}$ (*green*, 3) and $\varepsilon = 10^{-3}$ (*blue*, 4). *Crosses* mark the points with $|\varepsilon| = \varepsilon_{\dot\gamma} = |\dot\gamma\tau_0|^{\frac{2a}{1+a}}$. The natural upper boundary for the shear rate, $\dot\gamma_*$, where the range of validity of the Λ-formula is *left*, is also indicated. For $\varepsilon < 0$, the natural lower limits for the shear rates, below which the Λ-formula does not describe the flow curves, are marked by *diamonds*

asymptotic results given by the so-called Λ-formula. While the glass flow curves exhibit an upward curvature only, the fluid curves show a characteristic S-shape, where the initial downward curvature changes to an upward one for increasing shear rate. Both behaviors are captured by the asymptotic expansions. For positive separation parameters the range of validity of the Λ-formula is given by $|\varepsilon| \ll 1$ and $|\dot\gamma\tau_0| \ll 1$. These two requirements ensure that $\mathcal{G}(t)$ describes the dynamics of $\Phi(t)$ with a sufficiently high accuracy; see Figs. 14 and 15. For sufficiently small negative separation parameters, the Λ-formula is valid in finite shear rate windows only, as it does not reproduce the linear asymptotes for low shear rates. Precise criteria for the range of its validity are known [86], and Fig. 17 presents an overview of the flow curves and their asymptotic laws.

While the detailed discussion of the flow curves and their asymptotics leads beyond the present review, see [86], the important conclusions from Fig. 17 in the present context are that the universal aspects discussed in Sect. 4 are recovered, that qualitative agreement is obtained with the results of the ISHSM, and that analytical expressions for the flow curves can be obtained. For example, the critical flow curve follows a generalized Herschel–Bulkley law:

$$\sigma(\varepsilon = 0, \dot\gamma) = \sigma_c^+ \sum_{n=0}^{3} c_n |\dot\gamma/\dot\gamma_*|^{mn},$$

where σ_c^+ is the critical dynamic yield stress and $\dot{\gamma}_*$ defines a natural scale for the shear rates in the asymptotic expansion; the upper limit 3 of the summation is discussed in [86]. At the transition, this law describes the flow curve correctly for sufficiently small shear rates; see Fig. 17. This result also implies a generalized power-law weakening of the yield stress $\sigma^+(\varepsilon)$ when approaching the glass transition for $\varepsilon \searrow 0+$, which is shown in the inset of Fig. 13. In fluid states for $\varepsilon < 0$, the flow curves in double logarithmic presentation, viz. $\log_{10}(\sigma)$ as function of $\log_{10}(\dot{\gamma})$, show an inflection point defined by

$$\frac{d^2(\log_{10}(\sigma))}{d(\log_{10}(\dot{\gamma}))^2} = 0.$$

But then in some finite shear rate windows the flow curves can be approximated by the corresponding inflection tangents. The slopes p of the inflection tangents can be interpreted as exponents occurring in some pseudo power laws:

$$\sigma \propto \dot{\gamma}^p, \quad \Leftrightarrow \quad \eta \propto \dot{\gamma}^{p-1}.$$

Figure 18 shows some examples. The asymptotic formula also describes the neighborhood of the inflection point correctly for sufficiently small ε, but does not represent a real power law. In the framework of asymptotic expansions, there is thus no real exponent p. The power-law shear thinning, often reported in the literature, in the ITT-flow curves is thus a trivial artifact of the double logarithmic plot.

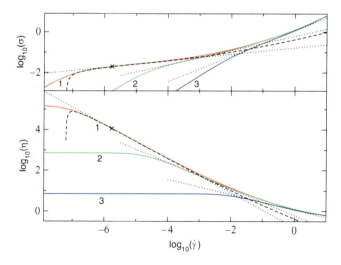

Fig. 18 The *upper panel* shows numerically obtained flow curves (*solid lines*) for $\varepsilon = -10^{-3}$ (*red*, 1), $\varepsilon = -10^{-2}$ (*green*, 2) and $\varepsilon = -10^{-1}$ (*blue*, 3). The *dotted lines* show the corresponding inflection tangents, with exponents $p = 0.16, 0.35,$ and 0.63 from *left* to *right*. The *dashed line* shows the numerically evaluated Λ-formula for $\varepsilon = -10^{-3}$. The shear rate with $\varepsilon = -\varepsilon_{\dot{\gamma}}$ is marked by a *cross*. The *lower panel* shows the corresponding results for the viscosity; from [86]

Rather, the flow curves on the fluid side exhibit a characteristic S-shape. While this shape is rather apparent when plotting stress vs shear rate, plotting the same data as viscosity as function of shear rate hides it, because the vertical axis gets appreciably stretched.

5.2.4 Test of Asymptotics in a Polydisperse Dispersion

While the asymptotic expansions in the previous section provide an understanding of the contents of the MCT-ITT scenario, experimental tests of the asymptotic laws require flow curves over appreciable windows in shear rate.

Figures 19 and 20 show experimental data recently obtained by Siebenbürger et al. [33] on polydisperse dispersions of the thermosensitive core-shell particles introduced in Sect. 3.1.2 [31]. In all cases stationary states were achieved after shearing long enough, proving that ageing could be neglected even for glassy states. Because of the appreciable poyldispersity in particle size (standard deviation 17%) crystallization could efficiently be prevented and flow curves over extremely wide windows could be obtained. Two flow curves from their work can be used to test the asymptotic results.

Figure 19 shows the result for a liquid-like flow curve where the asymptotic Λ-formula holds for approximately four decades. The pseudo power law resulting

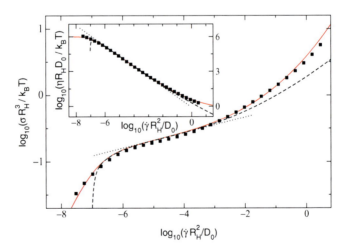

Fig. 19 Reduced flow curves for a core-shell dispersion at an effective volume fraction of $\phi_{\text{eff}} = 0.580$; data from [33], analysis from [86]. Here R_H denotes the hydrodynamic radius and D_0 the self diffusion coefficient of the colloidal particles; $k_B T$ is the thermal energy. The *solid line (red)* shows the result for the fitted $F_{12}^{(\dot{\gamma})}$-model with $v_2^c = 2.0$. The fitted parameters are: $\varepsilon = -0.00042$, $\gamma_c = 0.14$, $v_\sigma = 70 k_B T/R_H^3$, $\Gamma = 80 D_0/R_H^2$, and $\eta_\infty = 0.394 k_B T/R_H D_0$. The *dashed line* shows the corresponding result for the Λ-formula. The *dotted line* shows the inflection tangent of the numerically determined flow curve with a slope of $p = 0.12$. The *inset* shows the corresponding results for the viscosity

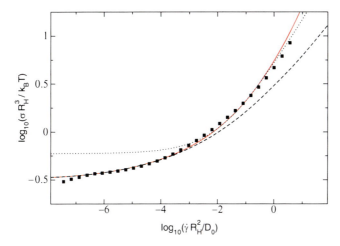

Fig. 20 Reduced flow curves for a core-shell dispersion at an effective volume fraction of $\phi_{\text{eff}} = 0.629$; quantities as defined in the caption of Fig. 19. The *solid line* (*red*) shows the result for the fitted $F_{12}^{(\dot\gamma)}$-model with $v_2^c = 2.0$. The fitted parameters are: $\varepsilon = 0.000021$, $\gamma_c = 0.16$, $v_\sigma = 115 k_B T/R_H^3$, $\Gamma = 120 D_0/R_H^2$, and $\eta_\infty = 0.431 k_B T/R_H D_0$. The *dashed line* shows the corresponding result for the Λ-formula. The *dotted line* shows the fitted Herschel–Bulkley law given by (34) with the analytically calculated exponent $\tilde{m} = 0.489$; data from [33], analysis from [86]

from the inflection tangent of the flow curve holds for approximately two decades within the range of validity of the Λ-formula.

Figure 20 shows the result for a flow curve, where a small positive separation parameter was necessary to fit the flow curve and the linear viscoelastic moduli simultaneously. The data are compatible with the (ideal) concept of a yield stress, but fall below the fit curves for very small shear rates. This indicates the existence of an additional decay mechanism neglected in the present approach [32, 33]. Again, the Λ-formula describes the experimental data correctly for approximately four decades. For higher shear rates, an effective Herschel–Bulkley law

$$\sigma(\dot\gamma_* \ll |\dot\gamma| \ll \Gamma, \varepsilon = 0) = \tilde\sigma_0 + \tilde\sigma_1 |\dot\gamma t_0|^{\tilde m} \tag{34}$$

with constant amplitudes and exponent $\tilde m = 0.49$ can be fitted in a window of approximately two decades. The constant $\tilde\sigma_0$ is not the actual yield stress, σ^+, which is obtained in the limit of vanishing shear rate, $\sigma^+ = \sigma(\dot\gamma \to 0)$, but is larger,

The experimental data of the polydisperse samples, which exhibit structural dynamics over large windows, and their fits with the full schematic model, will be taken up again in Sect. 6.2, where additionally the linear response moduli are considered, as had been done in Sect. 3.1.2 for the less polydisperse samples affected by crystallization.

6 Comparison of Theory and Experiment

As MCT-ITT contains uncontrolled approximations, justification to studying it can be obtained only from its power to rationalize experimental observations. Because the transient density fluctuations are the central quantity in the approach, density correlators shall be considered first. Flow curves have been studied in most detail experimentally and in simulations, and thus are considered next.

6.1 ISHSM and Single Particle Motion Under Steady Shear

Detailed measurements of the stationary dynamics under shear of a colloidal hard sphere glass have recently been obtained by confocal microscopy [30]. Single particle motion was investigated in a shear-molten glass at roughly the wavevector inverse to the average particle separation. Figure 21 shows self-intermediate scattering functions measured for wavevectors along the vorticity direction where neither affine particle motion nor wavevector advection appears. The stationary correlators

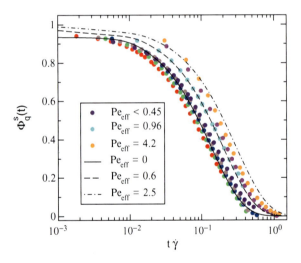

Fig. 21 Steady state incoherent intermediate scattering functions $\Phi_q^s(t)$ as functions of accumulated strain $\dot{\gamma}t$ for various shear rates $\dot{\gamma}$; the data were obtained in a colloidal hard sphere dispersion at packing fraction $\phi = 0.62$ (at $\varepsilon \approx 0.07$) using confocal microscopy [30]; the wavevector points in the vorticity ($\hat{z}$) direction and has $q = 3.8/R$ (at the peak of S_q). The effective Peclet numbers $\text{Pe}_{\text{eff}} = 4R^2\dot{\gamma}/D_s$ are estimated with the short time self diffusion coefficient $D_s \approx D_0/10$ at this concentration [15]. ISHSM calculations with separation parameter $\varepsilon = 0.066$ at $qR = 3.9$ (PY-S_q peaking at $qR = 3.5$), and for strain parameter $\gamma_c = 0.033$, are compared to the data for the Pe_{eff} values labeled. The yielding master function at $\text{Pe}_{\text{eff}} = 0$ lies in the data curves which span $0.055 \leq \text{Pe}_{\text{eff}} \leq 0.45$, but discussion of the apparent systematic trend of the experimental data would require ISHSM to approximate better the shape of the final relaxation process; from [45]

deep in the glass, for shear rates spanning almost two decades, are shown as function of accumulated strain $\dot{\gamma}t$, to test whether a simple scaling $\tau_{\dot{\gamma}} \sim 1/\dot{\gamma}$ as predicted by (28) holds. Small but systematic deviations are apparent which have been interpreted as a power law $\tau_{\dot{\gamma}} \sim \dot{\gamma}^{-0.8}$ [30, 89]. ISHSM computations were performed for a nearby wavevector where S_q is around unity so that coherent and incoherent correlators may be assumed to be similar [90]. Additionally, for the comparison it was assumed that time dependent transient and stationary fluctuation functions agree. The yielding master function from (28) in ISHSM can be fitted to the data measured at small effective Peclet numbers Pe$_{\text{eff}}$, by using for the phenomenological "strain rescaling parameter" $\gamma_c = 0.033$; the smallness of the fitted value, which would be expected to be of order unity, is not yet understood. The effective Peclet number Pe$_{\text{eff}} = 4R^2\dot{\gamma}/D_s$ with $D_s/D_0 = 0.1$ taken from [15] measures the importance of shear relative to the Brownian diffusion time obtained from the short time self diffusion coefficient D_s at the relevant volume fraction. At the larger effective Peclet numbers, Pe$_{\text{eff}} \geq 0.5$, for which the short-time and final (shear-induced) relaxation processes move closer together, the model gives quite a good account of the $\dot{\gamma}$-dependence.

The shape of the final relaxation step in a shear-molten glass can be studied even more closely in recent computer simulations, where a larger separation of short and long time dynamics could be achieved [91]. In these molecular dynamics simulations of an undercooled binary Lenard-Jones mixture, schematic ITT models give a good account of the steady state flow curves, $\sigma(\dot{\gamma})$ [81, 92]; this will be discussed in Sect. 6.3. Figure 22 shows the corresponding stationary self intermediate scattering functions for a wavevector near the peak in S_q, oriented along the vorticity direction, for shear rates spanning more than four decades. Collapse onto a master function when plotted as function of accumulated strain is nicely observed as predicted by (28). At larger shear rates, the correlators raise above the master function; this resembles the behaviour observed in the confocal experiments in Fig. 21, and in the theoretical calculations in Figs. 11 and 16. Assuming again that transient coherent correlators can be fitted to stationary incoherent ones, the shape of the master function can be fitted with the ISHSM, using again an unaccountedly small strain parameter γ_c. After this rescaling, modest but visible differences in the shapes remain: the theoretical master function decays more steeply than that from simulations.

Overall, theory and experiment agree in finding a two step relaxation process, where shear has a strong effect on the final structural relaxation, while the short time diffusion is not much affected. This supports the central MCT-ITT prediction that shearing speeds up the structural rearrangements in a concentrated dispersion close to vitrification. More detailed comparisons await better theoretical calculations where the effect of shear on the stationary density fluctuation functions is taken into account more faithfully than in the ISHSM.

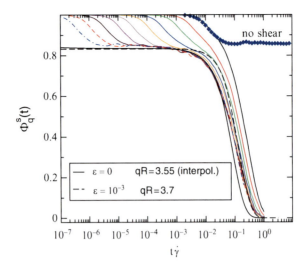

Fig. 22 Steady state incoherent intermediate scattering functions $\Phi_q^s(t)$ measured in the vorticity direction as functions of accumulated strain $\dot\gamma t$ for various shear rates $\dot\gamma$; data from molecular dynamics simulations of a supercooled binary Lenard-Jones mixture below the glass transition are taken from [91]. These collapse onto a yield scaling function at long times. The wavevector is $q = 3.55/R$ (at the peak of S_q). The quiescent curve, shifted to agree with that at the highest $\dot\gamma$, shows ageing dynamics at longer times outside the plotted window. The apparent yielding master function from simulation is compared to those calculated in ISHSM for glassy states at or close to the transition (separation parameters ε as labeled) and at nearby wave vectors (as labeled). ISHSM curves were chosen to match the plateau value f_q, while strain parameters $\gamma_c = 0.083$ at $\varepsilon = 0$ (*solid line*) and $\gamma_c = 0.116$ at $\varepsilon = 10^{-3}$ (*dashed line*) were used; from [45]

6.2 $F_{12}^{(\dot\gamma)}$-Model and Shear Stresses in Equilibrium and Under Flow in a Polydisperse Dispersion

A central result of MCT-ITT concerns the close connection between structural relaxation at the glass transition and the rheological properties far from equilibrium. The ITT approach aims to unify the understanding of these two phenomena, which were introduced wrt. experimental data in Sects. 3.1.1 and 5.2.4, respectively. MCT-ITT requires, as sole input, information on the equilibrium structure (namely S_q), and first gives a formally exact generalization of the shear modulus to finite shear rates, $g(t, \dot\gamma)$, which is then approximated in a consistent way. A novel dense colloidal dispersion serves as experimental model system, whose linear and nonlinear rheology can be determined over very broad windows of control parameters. The generalized modulus $g(t, \dot\gamma)$ can thus be investigated as function of shear rate and time (more precisely frequency), and the MCT-ITT approach can be tested thoroughly. Thermosensitive core-shell particles consisting of a polystyrene core and a crosslinked poly(*N*-isopropylacrylamide) (PNIPAM) shell were synthesized and their slightly polydisperse dispersions (standard deviation 17%) characterized in detail [33]; see Sect. 3.1.1. While their precise structure factor has not been measured yet, the system can well be considered a slightly polydisperse mixture of hard spheres. Because polydispersity prevents crystallization and the lack of attractions prevents demixing and coagulation, this system opens a window on structural relaxation, which can be nicely tuned by changing the effective packing fraction by varying temperature.

Shear stresses measured in non-linear response of the dispersion under strong steady shearing, and frequency dependent shear moduli arising from thermal shear stress fluctuations in the quiescent dispersion were measured and fitted with results from the schematic $F_{12}^{(\dot\gamma)}$-model. Some results from the microscopic MCT for the equilibrium moduli were also included; see Fig. 23. The fits with quiescent MCT for (monodisperse) hard spheres using the PY S_q support the finding of Sect. 3.1.2 that MCT accounts for the magnitude of the stresses at the glass transition semi-quantitatively. Because of the polydispersity of the samples, which is neglected in the calculations performed according to the presentation in Sect. 3.1.2, somewhat larger rescaling factors c_y are required; they are included in Table 1. Also, the critical packing fraction ϕ_c of the glass transition again is reproduced with some small error. Because of polydispersity, the experimental estimate $\phi_{\text{eff}}^c \approx 0.625$ [33] lies somewhat higher than in the (more) monodisperse case [32], which is as expected [93, 94].

Figure 23 gives the comparison of the experimental flow curves and the linear response moduli G' and G'' with theory for five given different effective volume fractions ϕ_{eff} adjusted by varying temperature. On the left-hand side the flow curves $\sigma(\dot\gamma)$ are presented as functions of the bare Peclet number $\text{Pe}_0 = k_B T/(6\pi\eta_s R_H^3)\,\dot\gamma$, on the right-hand side G' and G'' are displayed as functions of the frequency-Peclet or Deborrah number $\text{Pe}_\omega = k_B T/(6\pi\eta_s R_H^3)\,\omega$, calculated with the respective frequency ω. Table 1 gathers the effective volume fractions together with the fit parameters of the $F_{12}^{\dot\gamma}$-model. Note that G' and G'' have been obtained over nearly seven orders of magnitude in frequency, while the flow curves extend over more than eight decades in shear rate.

The generalized shear modulus $g(t,\dot\gamma)$ of the $F_{12}^{\dot\gamma}$-model (cf. 29c) presents the central theoretical quantity used in these fits. Within the schematic model, the vertex prefactor v_σ is kept as a shear-independent quantity. It can easily be obtained from the stress and modulus magnitudes. Hydrodynamic interactions enter through η_∞, which can be obtained from measurements done at high frequencies, and through Γ, which can be obtained via (33) from measurements done at high shear rates. Given these three parameters, both the shapes of the flow curves as well as the shapes of the moduli G' and G'' may be obtained as function of the two parameters ε and $\dot\gamma/\gamma_c$. The former sets the separation to the glass transition and thus (especially) the longest relaxation time, while the latter tunes the effect of the shear flow on the flow curve.

All measured quantities, namely σ, G', and G'' were converted to the respective dimensionless quantities by multiplication with $R_H^3/k_B T$ where R_H is the hydrodynamic radius at the respective temperature. As already discussed above, the experimental control parameters $\dot\gamma$ and ω also were converted by $6\pi\eta_s R_H^3/k_B T$ to the respective Peclet numbers.

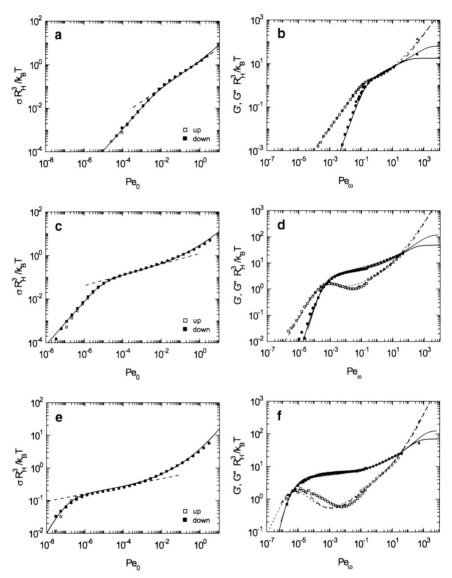

Fig. 23 *Left column*: Reduced flow curves (*filled squares*) for different volume fractions. The *solid lines* are the results of the schematic model, the *dashed line* represent the pseudo power law behaviour; from [33]. *Right column*: Reduced frequency dependent moduli for different volume fractions. *Full symbols/solid lines* represent G′, *hollow symbols/dashed lines* represent G″. *Thick lines* are the results of the schematic model, the *thin lines* the results of the microscopic MCT. Graphs in one row represent the continuous and dynamic measurements at one volume fraction. (**a**) and (**b**) at $\phi_{\text{eff}} = 0.530$, (**c**) and (**d**) at $\phi_{\text{eff}} = 0.595$, (**e**) and (**f**) at $\phi_{\text{eff}} = 0.616$, (**g**) and (**h**) at $\phi_{\text{eff}} = 0.625$, and (**i**) and (**j**) at $\phi_{\text{eff}} = 0.627$

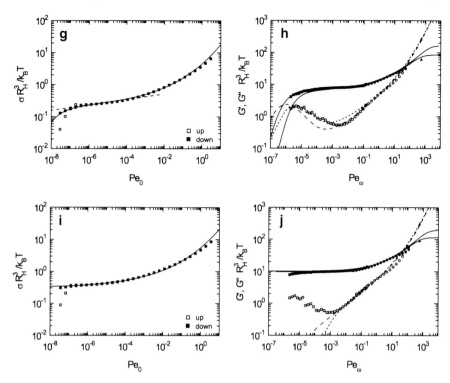

Fig. 23 continued

Table 1 Packing fraction ϕ_{eff} and parameters ν_σ, Γ, γ_c, and η_∞ of the fit using the schematic $F_{12}^{\dot{\gamma}}$-model for the measurements shown in Fig. 23; from [33]. The parameters ε, short time diffusion coefficient D_s/D_0, and rescaling factor c_y from the microscopic linear response calculation using MCT, and the pseudo-power law exponent p are also included

ϕ_{eff}	ε $F_{12}^{(\dot{\gamma})}$-mod.	ν_σ $\left[\frac{k_BT}{R_H^3}\right]$	Γ $\left[\frac{D_0}{R_H^2}\right]$	γ_c	η_∞ $\left[\frac{k_BT}{D_0R_H}\right]$	$\varepsilon^{\text{micro}}$ micro.	D_s/D_0	c_y	p
0.530	−0.072000	18	20	0.0845	0.2250	−0.10	0.3	2.3	0.631
0.595	−0.003500	48	50	0.1195	0.2400	−0.008	0.3	2.3	0.248
0.616	−0.000420	70	80	0.1414	0.3938	−0.001	0.3	2.3	0.117
0.625	−0.000170	85	90	0.1491	0.4250	−0.001	0.3	3.0	0.0852
0.627	0.000021	115	120	0.1622	0.4313	0.002	0.3	3.5	–

Evidently, both the reduced moduli, the Pe number, and the packing fraction depend on the effective particle volume R_H^3. In the polydisperse sample, a distribution of values R_H^3 actually exists, whose variance may be determined by disc centrifugation at low concentration, and which is fixed for one given sample. Close to the glass transition the size distribution is thus (almost) density independent, and $R_H^3(T)$ is the single experimental control parameter, whose small change upon varying temperature T drives the system through the glass transition.

Figure 23 demonstrates that the rheological behavior of a non-crystallizing colloidal dispersion can be modeled in a highly satisfactory manner by five parameters that display only a weak dependence on the effective volume fraction of the particles. Increasing the effective packing fraction drives the system towards the glass transition, viz. ε increases with ϕ_{eff}. Stress magnitudes (measured by v_σ) also increase with ϕ_{eff}, as do high frequency and high shear viscosities; their difference determines Γ. The strain scale γ_c remains around the reasonable value 10%. In spite of the smooth and small changes of the model parameters, the $F_{12}^{(\dot{\gamma})}$-model manages to capture the qualitative change of the linear and non-linear rheology. The measured Newtonian viscosity increases by a factor around 10^5. The elastic modulus G' at low frequencies is utterly negligible at low densities, while it takes a rather constant value around $10 k_B T / R_H^3$ at high densities. An analogous observation holds for the steady state shear stress $\sigma(\dot{\gamma})$, which at high densities takes values around $0.3 k_B T / R_H^3$ when measured at lowest shear rates. For lower densities, shear rates larger by a factor around 10^7 would be required to obtain such high stress values.

At volume fractions around 0.5 the suspension is Newtonian at small Pe_0. Approaching the glass transition leads to a characteristic S-shape of the flow curves and the Newtonian region becomes more and more restricted to the region of smallest Pe_0. Concomitantly, a pronounced minimum in G'' starts to develop, separating the slow structural relaxation process from faster, rather density independent motions, while G' exhibits a more and more pronounced plateau. At the highest density (Fig. 23, panels i and j), the theory would conclude that a yielding glass is formed, which exhibits a finite elastic shear modulus (elastic constant) $G_\infty = G'(\omega \to 0)$, and a finite dynamic yield stress, $\sigma^+ = \sigma(\dot{\gamma} \to 0)$. The experiment shows, however, that small deviations from this glass like response exist at very small frequencies and strain rates. Description of this ultra-slow process requires extensions of the present MCT-ITT which are discussed in [32].

Considering the low frequency spectra in $G'(\omega)$ and $G''(\omega)$, microscopic MCT and schematic model provide completely equivalent descriptions of the measured data. Differences in the fits in Fig. 23 for $Pe_\omega \le 1$ only remain because of slightly different choices of the fit parameters which were not tuned to be close. These differences serve to provide some estimate of uncertainties in the fitting procedures. Main conclusion of the comparisons is the agreement of the moduli from microscopic MCT, schematic ITT model, and from the measurements. This observation strongly supports the universality of the glass transition scenario which is a central line of reasoning in the ITT approach to the non-linear rheology.

6.3 $F_{12}^{(\dot{\gamma})}$-Model and Flow Curves of a Simulated Supercooled Binary Liquid

In large scale molecular dynamics simulations an 80:20 binary mixture of Lennard-Jones (LJ) particles at constant density was supercooled under shear. This model

Nonlinear Rheology   111

has well known equilibrium properties and many aspects that can be understood consistently within MCT [95]. To account for shearing, it was used together with Lees-Edwards boundary conditions and the SLLOD equations of motion to develop a linear velocity profile. Note that in the simulation, solvent effects are obviously lacking, and the simulated flow curves thus provide support for the notion that shear thinning can arise from shear-induced speed up of the structural relaxation; it is evident in Fig. 22. Because the microscopic motion is Newtonian, the theoretical description of this model goes beyond the framework of Sect. 2. Yet the universality of the structural long time dynamics, predicted by MCT [74] and also retained in MCT-ITT, supports the application of MCT-ITT to the simulation data. Moreover, the independence of the glassy dynamics on the employed microscopic motion was explicitly confirmed in simulations of the mixture [96]. This supercooled simple liquid has been characterized quite extensively under shear [10, 50, 81, 91, 92], and thus fits of the flow curves provide challenging tests to the schematic $F_{12}^{(\dot{\gamma})}$-model.

Figure 24 shows the stress-shear rate dependence as flow curves, ranging from supercooled states to the glassy regime; LJ units are used as described in [81]. The solid lines are fits to the simulation data with the $F_{12}^{(\dot{\gamma})}$-model, which reproduce the transition from a shear-thinning fluid to a yielding glass quite well. Coming

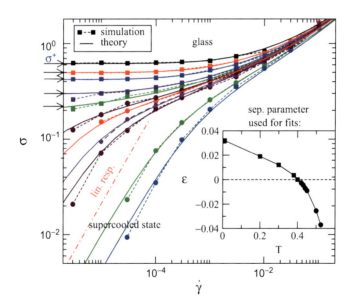

Fig. 24 Flow curves $\sigma(\dot{\gamma})$ reaching from the supercooled to the glassy state of a simulated binary LJ mixture. The data points correspond to the temperatures $T = 0.525, 0.5, 0.45, 0.44, 0.43, 0.42, 0.4, 0.38, 0.3, 0.2$, and 0.01 in LJ-units (from *bottom* to *top*). $F_{12}^{(\dot{\gamma})}$-model curves fitted by eye are included as *lines*. The *inset* shows the relation between the fitted separation parameters and temperature. Units are converted by $\sigma = 1.5\sigma_{\text{theo}}$ and $\dot{\gamma} = 1.3\dot{\gamma}_{\text{theo}}\Gamma$; from [92]. The arrows mark the values of the extrapolated dynamic yield stresses $\sigma^+(\varepsilon)$

from high shear rates, the flow curves of the supercooled state pass to the linear response regime in the lower left corner, indicated by a dashed-dotted line with slope 1. On approaching the transition point, the linear response regime shifts to lower and lower shear rates. Beyond the fluid domain, the existence of a dynamic yield stress $\sigma^+ = \lim_{\dot\gamma \to 0} \sigma > 0$ is supported by the simulation results, which sustain a stress plateau over three decades in shear rate. Best $F_{12}^{(\dot\gamma)}$-model fits are obtained for a $T_c = 0.4$, suggesting a slightly lower transition temperature [97] as determined from the simulations of the quiescent system, where $T_c = 0.435$ was found [95]. The reason may be the ergodicity restoring processes which were also observed in the colloidal experiments shown in Figs. 7, 20, and 23.

The stress plateau is best developed for temperatures deep in the glassy phase extending over about two decades in shear rate. Its onset is shifted toward progressively lower $\dot\gamma$ as the temperature is increased toward T_c. This makes an estimate of the dynamic yield stress, $\sigma^+(T) \equiv \sigma(T; \dot\gamma \to 0)$, a difficult task for temperatures below but close to T_c. Nevertheless, an estimate of $\sigma^+(T)$ is interesting because it highlights the anomalous weakening of the glass when heating to T_c. Testing the MCT predictions below T_c has previously not been possible in simulations because of problems in reaching the equilibrated or steady state at sufficiently low shear rates. Figure 25 significantly supports the notion of a glass transition under shear as it presents the first simulations result exhibiting the predicted anomalous softening in an elastic property of the glass upon approaching the transition from the glass side, viz. upon heating.

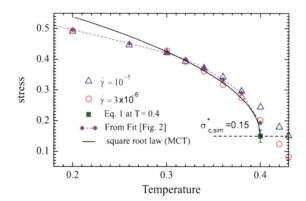

Fig. 25 Dynamic yield stress estimated from the simulations of a supercooled binary LJ mixture under steady shear shown in Fig. 24, and its temperature dependence (in LJ units); from [81]. The estimate uses the stress values for the two lowest simulated shear rates, namely $\dot\gamma = 10^{-5}$ (*triangle*) and $\dot\gamma = 3 \times 10^{-6}$ (*circle*); the extrapolation with the $F_{12}^{(\dot\gamma)}$-model is shown by *diamonds*. At temperatures below $T = 0.38$, (almost) the same shear stress is obtained for both values of $\dot\gamma$ and the extrapolation, indicating the presence of a yield stress plateau

7 Summary and Outlook

The present review explored the connection between the physics of the glass transition and the rheology of dense colloidal dispersions, including in strong steady shear flow. A microscopic theoretical approach for the shear-thinning of concentrated suspensions and the yielding of colloidal glasses was presented, which builds on the MCT of idealized glass transitions. The extension to strongly driven stationary states uses the so-called ITT approach, which leads to a scenario of shear melting a glass, whose universal aspects can be captured in simplified schematic models. Consecutive generalizations of ITT to arbitrary time-dependent states far from equilibrium [59] and to arbitrary flow geometries [98] have yielded a non-Newtonian constitutive equation applicable to concentrated dispersions in arbitrary homogeneous flows (not reviewed here), albeit still under the approximation that hydrodynamic interactions are neglected. Within the theory, this approximation becomes valid close to the glass transition and for weak but nonlinear flows, where the slow structural relaxation dominates the system properties, and where hydrodynamic interactions only affect the overall time scale.

The structural dynamics under flow is predicted to result from a competition between local particle hindrance (termed cage effect) and the compression/stretching (i.e., advection) of the wavelength of fluctuations induced by the affine particle motion with the flow. Measurements of the single particle motion in the stationary state under shear support the theoretical picture that shear speeds up the structural dynamics, while instantaneous structural correlations remain rather unaffected. Model dispersions made of thermo-sensitive core-shell particles allow investigation of the close vicinity of the transition. Measurements of the equilibrium stress fluctuations, viz. linear storage and loss moduli, and measurements of flow curves, viz. nonlinear steady state shear stress vs shear rate, for identical external control parameters verify that the glassy structural relaxation can be driven by shearing and in turn itself dominates the low shear or low frequency rheology.

In the employed theoretical approach, the equilibrium structure factor S_q captures the particle interactions. Theory misses an ultra-slow decay of all glassy states, and neglects (possible) ageing effects.

Acknowledgment It is a great pleasure to thank all my colleagues for the enjoyable and fruitful collaboration on this topic. I especially thank Mike Cates for introducing me to rheology, and Matthias Ballauff for his inspiring studies. Kind hospitality in the group of John Brady, where part of this review was written, is gratefully acknowledged. Financial support is acknowledged by the Deutsche Forschungsgemeinschaft in SFB-TR6, SFB 513, IRTG 667, and via grant Fu 309/3.

References

1. Larson RG (1999) The structure and rheology of complex fluids. Oxford University Press, New York
2. Götze W, Sjögren L (1992) Rep Prog Phys 55:241

3. Russel WB, Saville DA, Schowalter WR (1989) Colloidal dispersions. Cambridge University Press, New York
4. Laun HM, Bung R, Hess S, Loose W, Hess O, Hahn K, Hädicke E, Hingmann R, Schmidt F, Lindner P (1992) J Rheol 36:743
5. Brady JF (1993) J Chem Phys 99:567
6. Sollich P, Lequeux F, Hébraud P, Cates ME (1997) Phys Rev Lett 78:2020
7. Sollich P (1998) Phys Rev E 58:738
8. Fielding S, Sollich P, Cates ME (2000) J Rheol 44:323
9. Berthier L, Barrat J-L, Kurchan J (2000) Phys Rev E 61:5464
10. Berthier L, Barrat J-L (2002) J Chem Phys 116:6228
11. Pusey PN, van Megen W (1987) Phys Rev Lett 59:2083
12. Megen W, Pusey PN (1991) Phys Rev A 43:5429
13. van Megen W, Underwood SM (1993) Phys Rev Lett 70:2766
14. van Megen W, Underwood SM (1994) Phys Rev E 49:4206
15. van Megen W, Mortensen TC, Müller J, Williams SR (1998) Phys Rev E 58:6073
16. Hébraud P, Lequeux F, Munch J, Pine D (1997) Phys Rev Lett 78:4657
17. Beck C, Härtl W, Hempelmann R (1999) J Chem Phys 111:8209
18. Bartsch E, Eckert T, Pies C, Sillescu H (2002) J Non-Cryst Solids 802:307
19. Eckert T, Bartsch E (2003) Faraday Discuss 123:51
20. Weeks ER, Crocker JC, Levitt AC, Schofield A, Weitz DA (2000) Science 287:627
21. Mason TG, Weitz DA (1995) Phys Rev Lett 75:2770
22. Zackrisson M, Stradner A, Schurtenberger P, Bergenholtz J (2006) Phys Rev E 73:011408
23. Senff H, Richtering W (1999) J Chem Phys 111:1705
24. Senff H, Richtering W, Norhausen Ch, Weiss A, Ballauff M (1999) Langmuir 15:102
25. Petekidis G, Vlassopoulos D, Pusey P (1999) Faraday Discuss 123:287
26. Petekidis G, Vlassopoulos D, Pusey PN (2004) J Phys Condens Matter 16:S3955
27. Petekidis G, Moussaid A, Pusey PN (2002) Phys Rev E 66:051402;
28. Petekidis G, Vlassopoulos D, Pusey PN (2003) Faraday Discuss 123:287
29. Pham KN, Petekidis G, Vlassopoulos D, Egelhaaf SU, Pusey PN, Poon WCK (2006) Europhys Lett 75:624
30. Besseling R, Weeks ER, Schofield AB, Poon WC (2007) Phys Rev Lett 99:028301
31. Crassous JJ, Siebenbürger M, Ballauf M, Drechsler M, Henrich O, Fuchs M (2006) J Chem Phys 125:204906
32. Crassous JJ, Siebenbürger M, Ballauf M, Drechsler M, Hajnal D, Henrich O, Fuchs M (2008) J Chem Phys 128:204902
33. Siebenbürger M, Fuchs M, Winter H, Ballauff M (2009) Viscoelasticity and shear flow of concentrated, non-crystallizing colloidal suspensions: Comparison with Mode-Coupling Theory. J Rheol 53:707–726
34. Phung T, Brady J, Bossis G (1996) J Fluid Mech 313:181
35. Strating P (1999) Phys Rev E 59:2175
36. Doliwa B, Heuer A (2000) Phys Rev E 61:6898
37. Purnomo EH, van den Ende D, Mellema J, Mugele F (2006) Europhys Lett 76:74
38. Götze W (1991) In: Hansen JP, Levesque D, Zinn-Justin J (eds) Liquids, freezing and glass transition. Session LI of Les Houches summer schools of theoretical physics, North-Holland, Amsterdam, 287, 1989
39. Götze W (1999) J Phys Condens Matter 11:A1
40. Miyazaki K, Reichman DR (2002) Phys Rev E 66:050501
41. Miyazaki K, Reichman DR, Yamamoto R (2004) Phys Rev E 70:011501
42. Kobelev V, Schweizer KS (2005) Phys Rev E 71:021401
43. Fuchs M, Cates ME (2002) Phys Rev Lett 89:248304
44. Fuchs M, Cates ME (2005) J Phys Condens Matter 17:S1681
45. Fuchs M, Cates ME (2009) A mode coupling theory for Brownian particles in homogeneous steady shear flow. J Rheol 53:957–1000
46. Dhont JKG (1996) An introduction to dynamics of colloids. Elsevier, Amsterdam
47. Risken H (1989) The Fokker–Planck equation. Springer, Berlin

48. Dhont JKG, Briels W (2008) J Rheol Acta 47:257–281
49. Bender J, Wagner NJ (1996) J Rheol 40:899
50. Varnik F, Bocquet L, Barrat JL (2004) J Chem Phys 120:2788
51. Ganapathy R, Sood AK (2006) Phys Rev Lett 96:108301
52. Ballesta P, Besseling R, Isa L, Petekidis G, Poon WCK (2008) Phys Rev Lett 101:258301
53. Van Kampen NG (2007) Stochastic processes in physics and chemistry. North Holland, Amsterdam
54. Forster D (1975) Hydrodynamic fluctuations, broken symmetry, and correlation functions. WA Benjamin, Reading, MA
55. Götze W, Sjögren L (1987) Z Phys B 65:415
56. Schofield J, Oppenheim I (1992) Physica A 187:210
57. Nägele G, Bergenholtz J (1998) J Chem Phys 108:9893
58. Miyazaki K, Wyss HM, Reichman DR, Weitz DA (2006) Europhys Lett 75:915
59. Brader JM, Voigtmann Th, Cates ME, Fuchs M (2007) Phys Rev Lett 98:058301
60. Lionberger RA, Russel WB (1994) J Rheol 38:1885
61. Onuki A, Kawasaki K (1979) Ann Phys (NY) 121:456
62. Kawasaki K, Gunton JD (1973) Phys Rev A 8:2048
63. Indrani AV, Ramaswamy S (1995) Phys Rev E 52:6492
64. Bergenholtz J, Fuchs M (1999) Phys Rev E 59:5706
65. Dawson K, Foffi G, Fuchs M, Gotze W, Sciortino F, Sperl M, Tartaglia P, Voigtmann T, Zaccarelli E (2001) Phys Rev E 63:011401
66. Fabbian L, Götze W, Sciortino F, Tartaglia P, Thiery F (1999) Phys Rev E 59:R1347–R1350
67. Sciortino F (2009) Nonlinear rheological properties of dense colloidal dispersions close to a glass transition under steady shear. Adv Polymer Sci. doi:10.1007/12_2009_30
68. Sciortino F (2003) Nat Mater 1:145–146
69. Pham KN, Puertas AM, Bergenholtz J, Egelhaaf SU, Moussaid A, Pusey PN, Schofield AB, Cates ME, Fuchs M, Poon WCK (2002) Science 296:104–106
70. Poon WCK, Pham KN, Egelhaaf SU, Pusey PN (2003) J Phys Cond Matt 15:S269–S275
71. Fuchs M, Ballauff M (2005) Colloids Surf A 270/271:232
72. Franosch T, Fuchs M, Götze W, Mayr MR, Singh AP (1997) Phys Rev E 55:7153
73. Fuchs M, Mayr MR (1999) Phys Rev E 60:5742
74. Franosch T, Götze W, Mayr MR, Singh AP (1998) J Non-Cryst Solids 235/237:71
75. Verberg R, de Schepper IM, Feigenbaum MJ, Cohen EGD (1997) J Stat Phys 87:1037
76. Henrich O, Pfeifroth O, Fuchs M (2007) J Phys Condens Matter 19:205132
77. Szamel G (2001) J Chem Phys 114:8708
78. Johnson SJ, de Kruif CG, May RP (1988) J Chem Phys 89:5909
79. Lionberger RA, Russel WB (2000) Adv Chem Phys 111:399
80. Fuchs M, Cates ME (2003) Faraday Discuss 123:267
81. Varnik F, Henrich O (2006) Phys Rev B 73:174209
82. Fuchs M, Götze W, Hofacker I, Latz A (1991) J Phys Condens Matter 3:5047–5071
83. Zausch J, Horbach J, Laurati M, Egelhaaf SU, Brader JM, Voigtmann Th, Fuchs M (2008) J Phys Condens Matt 20:404210
84. Angelani L, Leonardo RD, Ruocco G, Scala A, Sciortino F (2000) Phys Rev Lett 85:5356
85. Broderix K, Bhattachrya KK, Cavagna A, Zippelius Z (2000) Phys Rev Lett 85:5360
86. Hajnal D, Fuchs M (2009) Eur Phys J E, in print. doi:10.1140/epje/i2008-10361-0; also at arXiv:0807.1288
87. Götze W (1984) Z Phys B 56:139
88. Fuchs M, Götze W, Hildebrand S, Latz A (1992) J Phys Condens Matter 4:7709
89. Saltzman EJ, Yatsenko G, Schweizer KS (2008) J Phys Condens Matter 20:244129
90. Pusey PN (1978) J Phys A 11:119
91. Varnik F (2006) J Chem Phys 125:164514
92. Henrich O, Varnik F, Fuchs M (2005) J Phys Condens Matter 17:S3625
93. Götze W, Voigtmann Th (2003) Phys Rev E 67:021502
94. Foffi G, Götze W, Sciortino F, Tartaglia P, Voigtmann Th (2003) Phys Rev Lett 91:085701
95. Kob W, Andersen HC (1995) Phys Rev E 51:4626; 52:4134
96. Gleim T, Kob W, Binder K (1998) Phys Rev Lett 81:4404
97. Flenner E, Szamel G (2005) Phys Rev E 72:011205
98. Brader JM, Cates ME, Fuchs M (2008) Phys Rev Lett 101:138301

Micromechanics of Soft Particle Glasses

Roger T. Bonnecaze and Michel Cloitre

Abstract Soft glasses encompass a broad class of materials at the boundaries between polymers, granular dispersions, and colloidal glasses. Although they display a huge diversity of compositions and architectures, soft glasses share a common structure as well as generic static and flow properties. In this chapter, we show that the dense amorphous microstructure of soft glasses, combined with the existence of repulsive elastohydrodynamic interactions mediated by the solvent, lie at the heart of their behavior. These two basic ingredients are incorporated into a micromechanical model and a dynamic molecular-like simulation. Our theory successfully predicts near-equilibrium quantities such as the pair distribution function and shear moduli, the slip properties that are observed when soft glasses are sheared along solid surfaces, as well as the bulk shear rheology. These results, which connect properties at the particle scale to macroscopic behavior, provide predictive tools for the design of materials with a desired rheological response.

Keywords Elastohydrodynamic interactions · Glass transition · Linear viscoelasticity · Nonlinear rheology · Polymer-colloid materials · Shear-thinning · Wall slip

Contents

1 Introduction 118
2 Generic Properties of Soft Particle Suspensions and Glasses 120
   2.1 Composition and Architecture 120
   2.2 Origin of Particle Elasticity 125

R.T. Bonnecaze
Department of Chemical Engineering and Texas Materials Institute,
The University of Texas at Austin, Austin, TX 78712, USA
e-mail: rtb@che.utexas.edu

M. Cloitre (✉)
Matière Molle et Chimie (UMR 7167, ESPCI-CNRS), ESPCI-ParisTech,
10 rue Vauquelin, 75005 Paris, France
e-mail: michel.cloitre@espci.fr

    2.3 Phase Behavior of Soft Particle Dispersions: Suspensions and Glasses 126
    2.4 Interaction Pair Potential in the Dense State Limit 128
3 Near-Equilibrium Properties of Soft Particle Glasses 131
    3.1 Introduction 131
    3.2 Micromechanical Model 132
    3.3 Near-Equilibrium Radial Distribution Function 134
    3.4 Shear Moduli and Osmotic Pressure 136
    3.5 Comparison with Experimental Data 138
4 Wall Slip and Surface Rheology 140
    4.1 Ubiquity of Wall Slip Phenomena in High-Solid Dispersions 140
    4.2 Generic Features of Wall Slip 141
    4.3 Theory of Soft Lubrication 144
5 Shear Rheology of Soft Glasses 148
    5.1 Generic Properties of the Nonlinear Rheology of Soft Glasses 148
    5.2 Model Description 151
    5.3 Simulation Results 153
6 Outlook and Open Questions 155
References 157

1 Introduction

Concentrated suspensions of soft particles, also named soft glasses, form an important class of materials at the frontier between polymer solutions, granular materials, and colloidal glasses. Soft glasses are made from soft and deformable particles dispersed in a solvent at large volume fractions that are well above close-packing. Soft glasses encompass a wide range of materials including concentrated emulsions, colloidal pastes, multilamellar vesicles, star polymers, copolymer micelles, and clay suspensions [1]. For example, Vlassopoulos and Fytas describe how macromolecular and colloidal chemistry can be used to generate well-defined soft particles presenting a rich variety of phase states and materials properties [2]. The structure and the dynamics of biological fluids, tissues, and the intracellular cytoplasm are analogous to a crowded suspension of repulsive soft colloidal particles [3, 4]. In soil mechanics and geology, heterogeneous materials such as mud, slurries and lavas can be described as highly concentrated suspensions of deformable particulate aggregates of various size and composition [5]. All these materials exhibit both solid-like and liquid-like properties, with the solid–liquid transition taking a variety of forms. This behavior is exploited industrially to formulate food or personal care products and to process high performance materials such as films, coatings, solid inks, and ceramics. Therefore, understanding and predicting the deformation and flow of soft glasses in terms of their constituents and structure pose an outstanding challenge for real life applications as well as for statistical and condensed-matter physics.

Soft glasses are named in reference to hard glasses, with which they share some common features such as nonergodicity and caged-dynamics. It is well known that monodisperse, hard-sphere suspensions form glasses when the volume fraction exceeds a value of about 0.58. Colloidal glasses are out-of-equilibrium materials, where particles are kinetically trapped into a metastable, disordered configuration [6]. Each particle is constrained in a cage formed by a small number of neighbors, which restrict and eventually arrest macroscopic motion. At short times, particles move within their cages, a process which constitutes β-relaxation. At much longer times, which are usually experimentally inaccessible, particles eventually escape from their cages via a process termed α-relaxation, which represents the longest relaxation process of the glass. The application of an external stress exceeding the strength of the cages forces the particles to move past one another over large distances and causes macroscopic flow. The flow properties and the rheology of hard-sphere glasses, both in linear and nonlinear regimes, have been studied experimentally at length during the last decade. On the theoretical side, the soft glassy rheology model provides a simple and useful description of glass rheology [7, 8]. More recently, extensions of the mode coupling theory have appeared to be extremely powerful for predicting the linear and nonlinear rheology of colloidal glasses [9–11]. Fuchs focuses on some of these models, which enjoy promising success in capturing important features of the rheology of hard-sphere suspensions near the glass transition [12].

Compared to conventional glasses, soft glasses exhibit some new and interesting features. The particles interact through potentials that are much weaker than in hard-sphere glasses and that can be tuned at will through the composition and the architecture. Consequently, whereas monodisperse, hard-sphere glasses cannot exceed close-packing at an approximate volume fraction of 0.64, soft glasses can be formed at much higher volume fractions. The particles then develop repulsive forces of elastic origin at contact, which control the cage elasticity and other macroscopic properties. Unlike hard-sphere glasses, which become solids at close-packing, soft glasses can still flow at such high volume fractions, albeit slowly, because the particles are able to change their shape by deforming elastically [13]. The solvent that lubricates the contact plays an important role in transmitting the elastic interactions through the glass. In this chapter, we show how this subtle interplay between disorder and solvent-mediated elastic interactions can be incorporated into a micromechanical description that quantitatively accounts for the peculiar static and dynamic properties of these materials.

The outline of the chapter is as follows. In Sect. 2 we review some of the generic properties shared by soft particle glasses. Our objective is to emphasize the underlying universality present in the phase diagram of these materials, although the composition and the architecture of the particles, as well as the interactions between them, can be very diverse. In Sect. 3 we present the essence of our micromechanical model, focusing first on the prediction of the static properties such as osmotic pressure, elastic modulus and pair distribution function. Section 4 is central to the chapter since it introduces solvent-mediated elastic interactions, which constitute one of the specificities of soft glasses. These interactions explain in particular the slip behavior of microgel pastes and concentrated emulsions near rigid surfaces. Section 5 presents a 3D model that incorporates the elastohydrodynamic contact interactions to the dense amorphous structure of soft glasses. Once implemented in a molecular dynamics simulation, the model provides quantitative predictions of the flow properties of soft glasses, which involve the solvent viscosity and the glass elasticity as the two main parameters. The predictions quantitatively agree with the experimental behavior measured for microgel pastes and concentrated emulsions, highlighting the universality of our description.

2 Generic Properties of Soft Particle Suspensions and Glasses

2.1 Composition and Architecture

This section presents an overview of the great variety of soft particles encountered both in fundamental science and in applications. We propose a classification based on composition and architecture, distinguishing colloidal-like particles, network particles, polymer–colloid systems, and surfactant particles, as illustrated in Fig. 1 and discussed below.

2.1.1 Colloidal-Like Particles

The preparation of highly concentrated colloidal dispersions requires an efficient stabilization strategy that keeps the particles apart, preventing aggregation and gelation at high solid content. This is achieved by chemically and/or physically modifying the surface of the particles. These modifications confer some degree of

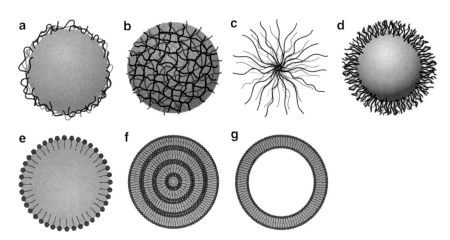

Fig. 1 Various types of soft particles: (**a**) solid particle covered with adsorbed or grafted polymer chains; (**b**) microgel particle; (**c**) star polymer; (**d**) block copolymer micelle; (**e**) emulsion droplet; (**f**) multilamellar vesicle; (**g**) liposome

softness and deformability to initially rigid particles. Although an extensive review of the subject is far beyond the scope of this chapter, it is useful to describe some representative situations.

Electrostatic stabilization has been known for a long time as an efficient way to stabilize particles against aggregation [14]. Ionic groups, either dissociated or adsorbed on the colloidal surfaces, leave their counterions in solution, and these small, mobile ions form electrical double-layers through which the particles interact. At low ionic strength, double-layers expand and interparticle repulsion acts over long distances, leading to the formation of crystals or glasses [15, 16]. Interestingly, the double-layers surrounding the particles are sensitive to compression and/or distort easily under an external force, making crystals and glasses soft and highly deformable [16, 17]. Historically, this problem has been worked out for the case of model spherical particles such as silica particles or polystyrene lattices. More recently, it has been shown that the same description applies to aqueous suspensions of charged disk-like Laponite particles [18–20]. Laponite is a synthetic clay with diameter around 30 nm and thickness of 1 nm and is nearly monodisperse. The structural charge borne by the surface of the platelets is negative, whereas that of the edges depends on the pH. Although the phase behavior of Laponite suspensions is relatively complicated and not fully understood, they exhibit a liquid to soft solid transition driven by repulsive interactions at low ionic strength [20]. This system has appeared during the last few years as one of the most studied examples of soft glassy materials [18, 21, 22].

Steric stabilization is another well-established method of stabilizing colloidal suspensions of submicron to micron size [23]. The particles are coated with a layer of adsorbed or grafted polymer chains that provides a steric repulsion of entropic origin and helps disperse the particles by counterbalancing van der Waals attraction (Fig. 1a). The polymeric nature of the adsorbed or grafted layer softens the interparticle interactions and makes the particles intrinsically deformable. Many polymer chain/particle combinations have been synthesized and studied, and are described in the literature. Several popular colloidal systems consist of silica particles covered with various polymers such as polydimethylsiloxane [24], stearyl alcohol [25], alkyl chains [26], and polyethylene oxide [27]. Polymethylmethacrylate and polystyrene particles grafted with polymer chains have also been used extensively. For a review on the impressive literature on the subject we refer the interested reader to Vlassopoulos and Fytas [2].

For electrostatic and steric stabilization, the particles can be viewed effectively as colloids consisting of a soft and deformable corona surrounding a rigid core. Colloidal particles with bulk elastomeric properties are also available. These particles, which are generally of submicron size, are developed and used as reinforcement additives to improve the impact resistance of various polymer matrices [28–30]. The rubber of choice is often a styrene/butadiene copolymer. The presence of chemical groups at the matrix–filler interface leads to improved adhesion between them. Typically, the addition of about 30% by volume of these elastomeric particles increases the impact strength of a brittle glassy polymer like polystyrene by up to a factor of 10. For some applications, particles with more complex architecture have been developed. For instance, core–shell particles consisting of an elastomeric core of polybutadiene or n-butyl-acrylate and a thin rigid shell of polymethylmethacrylate are currently used to reinforce polymers like poly(vinylchloride) and polycarbonate.

2.1.2 Network Particles

Network particles refer to colloidal objects consisting of crosslinked and/or entangled polymer chains. Microgel particles certainly form the most important class of network particles, both for fundamental science and applications. They consist of an intramolecular crosslinked polymeric network swollen by a solvent (Fig. 1b). Because of this architecture, they are partially impenetrable, just like colloids, but at the same time inherently soft and deformable like polymers. Microgels have become central components of advanced functional colloidal materials [31], with promising applications in the fields of bioencapsulation and controlled targeted drug release [32], metal ion adsorption [33], photonic materials [34], and rheology control [35, 36]. A rich literature describes how the synthesis, composition, and architecture of microgels (monomer composition, crosslink density, particle size, surface charge, and functional groups) can be customized to meet the requirements of demanding applications [37]. The size of individual particles can span several orders of magnitude, from 10 nm to 1 μm or more. They can be made from various polymers or copolymers, leading to an enormous variety of products. Historically, the first systems to be developed operated in organic solvents [38–40]. Today, neutral or polyelectrolyte water-soluble microgels offer the richest opportunities in terms of novel and environmentally safe applications. Neutral water-swellable microgels are based on poly(N-isopropylacrylamide) (PNIPAm) [41] or poly(N-vinylcaprolactam) [42]. A non-exhaustive list of polyelectrolyte microgels include particles made from poly(acrylic acid) [43, 44], poly(methylmethacrylic acid) ([45] and its copolymers with poly(methylmethacrylate) [46] or ethylacrylate [47], and poly(2-vinylpyridine) [48]. Biopolymers, which form physically crosslinked networks, are also used to create biocompatible microgels [49, 50].

Another distinctive feature that makes microgel particles extremely attractive is their capacity to change their volume almost reversibly when the properties of the suspending medium are modified. Like their macroscopic counterparts, microgels swell up to the point where their modulus becomes equal to the difference between the osmotic pressure inside the polymer network and the osmotic pressure of the solution. The swelling of neutral particles, which primarily depends on the solvent quality, can be finely tuned through small variations of temperature [38]. Particles comprising thermosensitive monomers undergo a volume transition with temperature [51, 52]. In the case of polyelectrolyte microgels, the osmotic pressure of counterions is responsible for swelling. This renders ionic microgels highly sensitive to variations of pH or ionic strength [53]. The synthesis and the use of microgels combining complex response to pH and temperature have also been described [54, 55].

At swelling equilibrium the osmotic pressures of the solvent inside and outside microgels balance each other. A direct consequence is that any change of the osmotic pressure of the continuous phase can induce osmotic deswelling. This effect is crucial in complex formulations, which generally comprise many components. For instance an increase of the ionic strength provokes the osmotic deswelling of ionic microgels [56]. Similarly the addition of excluded linear free chains causes the deswelling of neutral and polyelectrolyte microgels [57–60]. Osmotic deswelling is responsible for a reduction in particle size and a decrease of volume fraction. In general these modifications alter the macroscopic behavior of suspensions in a dramatic way and complicate the interpretation of experiments since the volume fraction is not known accurately [61].

2.1.3 Polymer–Colloid Particles

Polymer–colloids refer to particles consisting of a solid core surrounded by a polymeric corona. This broad class of materials encompasses systems as different as star polymers, block copolymer micelles, and grafted particles.

Multiarm star polymers have recently emerged as ideal model polymer–colloids, with properties interpolating between those of polymers and hard spheres [62–64]. They are representatives of a large class of soft colloids encompassing grafted particles and block copolymer micelles. Star polymers consist of f polymer chains attached to a solid core, which plays the role of a topological constraint (Fig. 1c). When the functionality f is large, stars are virtually spherical objects, and for $f = \infty$ the hard sphere limit is recovered. A considerable literature describes the synthesis, structure, and dynamics of star polymers both in melt and in solution (for a review see [2]).

Block copolymers in selective solvents exhibit a remarkable capacity to self-assemble into a great variety of micellar structures. The final morphology depends on the molecular architecture, the block composition, and the affinity of the solvent for the different blocks. The solvophobic blocks constitute the core of the micelles, while the soluble blocks form a soft and deformable corona (Fig. 1d). Because of this architecture, micelles are partially impenetrable, just like colloids, but at the same time inherently soft and deformable like polymers. Most of their properties result from this subtle interplay between colloid-like and polymer-like features. In applications, micelles are used to solubilize in solvents otherwise insoluble compounds, to compatibilize polymer blends, to stabilize colloidal particles, and to control the rheology of complex fluids in various formulations. A rich literature describes the phase behavior, the structure, the dynamics, and the applications of block-copolymer micelles both in aqueous and organic solvents [65–67].

The morphology of a micelle is primarily determined by the composition of the copolymer and the incompatibility between the blocks and the solvent. Symmetric block copolymers produce micelles in which the core and the corona have comparable volume, leading to colloidal particles akin to the sterically stabilized particles described above. By contrast, very asymmetric copolymers form star-like particles with a very small rigid core and a much larger deformable corona. Interestingly, colloidal and star-like micellar solutions have markedly different structure factors and dynamics and, in principle, the interaction potential between micelles can be tuned continuously between hard-sphere-like and star-like behavior by changing the block copolymer composition and architecture [68]. The structure of the concentrated phases is another interesting issue. Very often block copolymer micelles form ordered crystalline phases at high concentration, but disordered glassy phases have also be obtained and used as model systems of micellar glasses. Star-like micelles switch from ordered crystalline phases to disordered glasses when the degree of aggregation increases [69]. Defects coming from some intrinsic polydispersity of the block copolymers also play an important role [70, 71].

2.1.4 Liquid Dispersions: Emulsions, Vesicles, and Liposomes

This section concerns dispersions of two liquid phases dispersed one into the other using surfactants. Emulsions belong to this class of materials. They consist of a mixture of two immiscible fluids, one of which, generally oil, is dispersed as small droplets in the continuous phase of the other fluid, generally water. The interfaces are stabilized by a surfactant, preventing coalescence over a reasonable period of time (Fig. 1e). Emulsions are generally obtained by intensively shearing a mixture of two immiscible liquids in the presence of one or several surfactants. A rich literature describes the properties of emulsions, their phase behavior, and the processes at work during emulsification [72]. Recent works have focused on the preparation of well-defined monodisperse emulsions using specific shearing procedures [73] or microfluidic techniques [74].

In concentrated solutions, amphiphilic molecules are known to self-assemble into a variety of spatially organized structures, which include lyotropic liquid crystals [75]. Although mesophases of different symmetries have been reported, smectic lamellar phases occupy a large portion of the phase diagram of amphiphilic systems [76, 77]. These phases sometimes persist when they are diluted, leading to the formation of fluctuating sheet-like membranes in solution. Due to their anisotropy and to their high flexibility, lamellar phases under steady shear flow can present different orientations, depending on the volume fraction and shear intensity [78]. Although a well-defined orientation is generally obtained at low and high shear rates, the membranes are wrapped around a spherical core to form multilamellar vesicles (Fig. 1f), which are close-packed and fill the space at intermediate shear rates. The size of these objects results from a balance between the viscous stress associated with the shearing motion and the elastic stress of the membranes. An interesting result is that these multilamellar vesicles are quenched when the shear flow is stopped. Their structure is metastable but relaxes extremely slowly over a few days to several months, depending on the lamellar composition.

Multilamellar vesicles belong to the general class of liposomes. Liposomes are small particles made out of membranes filled with active substances. The membranes, which are usually made of amphiphilic molecules such as phospholipids, can be multilayers as described above or bilayers (Fig. 1g). Because liposomes encapsulate one or several aqueous regions inside hydrophobic membranes, dissolved hydrophilic substances cannot readily pass through the lipids. This makes liposomes extremely attractive for drug delivery applications.

2.2 Origin of Particle Elasticity

All the systems presented in the previous section, although very different in composition and structure, can be viewed as dispersions of soft and elastic spherical particles. However, the origin of the particle elasticity can be very different, depending on the architecture and composition.

The elasticity of emulsion droplets comes from the interfacial tension of the oil–water interface. When a strain is exerted on the droplets, their shape is changed and the area of the interfaces increases, storing energy that is released when the droplets recover their initial shape [79]. The energy scale that controls the cost of small deformations is the surface energy ΓR^2, where Γ is the interfacial tension and R the radius of the droplets.

The elasticity of multilamellar vesicles can be discussed in reference to that of emulsion droplets. The crystalline lamellar phase constituting the vesicles is characterized by two elastic moduli, one accounting for the compression of the smectic layers, $\overline{B}$, and the second for the bending of the layers, K [80]. The combination $\sqrt{K\overline{B}}$ has the dimension of a surface tension and plays the role of an effective surface tension when the lamellae undergo small deformations [80]. This result is valid for multilamellar vesicles of arbitrary shapes [81, 82]. Like for emulsion droplets, the quantity $\sqrt{K\overline{B}}\, R^2$ is the energy scale that determines the cost of small deformations.

The elasticity of microgel particles has essentially the same origin as that of their macroscopic counterparts. Simple relationships exist between the elastic modulus and the difference of osmotic pressure between a gel and its solvent bath at swelling equilibrium, both for neutral and charged gels [83]. Any parameter that acts to increase or decrease the swelling will change the elastic modulus in the opposite direction. For neutral gels, the osmotic pressure inside the polymer network results from the solvent–polymer mixing free energy so that the particle elasticity depends essentially on the solvent quality and of the crosslink density [84]. For polyelectrolyte gels, the osmotic pressure is dominated by the counterions associated with the fixed charges borne by the polymer network. The gel elasticity strongly depends on the degree of ionization and on the ionic strength [85, 86]. In core–shell latex consisting of a polystyrene core covered with crosslinked PNIPAm shell, the elasticity results essentially from the outer polymer layer [87]. In hairy particles consisting of linear chains adsorbed or grafted onto a solid core, the elasticity arises from the entropy of the dangling chains [88].

While providing useful guides, this approach neglects the fact that the elastic moduli of polymer networks or polymer–colloid particles are often extremely dependent on the conditions of synthesis [86] and preparation [49, 50]. Moreover, it is not obvious that predictions holding for macroscopic gels can be transposed to microgels of micron size, which may have a heterogeneous structure or where finite size effects play an important role [89]. Recently, advanced micromanipulation techniques have been used to directly characterize the swelling behavior and the elastic properties of micron-sized microgels [90].

2.3 Phase Behavior of Soft Particle Dispersions: Suspensions and Glasses

The phase behavior of concentrated suspensions made from soft and deformable particles can be described in analogy to that of hard-sphere suspensions. The latter form glasses when their volume fraction exceeds a value $\Phi_g \cong 0.58$ [6, 91, 92]. Colloidal glasses are out-of-equilibrium materials in which particles are kinetically trapped into a metastable, disordered configuration [6, 93]. The cage picture has been proven extremely successful in describing the dynamics of colloidal glasses [92, 94]. It considers that each particle in a glass is constrained in a cage formed by a small number of neighbors, which restrict and eventually arrest macroscopic motion. At short times, particles move within their cages, a process which constitutes β-relaxation. At much longer times, which are usually experimentally inaccessible, particles eventually escape from their cages via a process termed α-relaxation, which represents the longest relaxation process of the glass [92, 93, 95].

Above Φ_g, soft particle suspensions interacting through purely repulsive interactions also form glasses. This is well documented in the literature for the case of emulsions [96] and microgels [97–101]. Soft particle glasses share common features with hard-sphere glasses, such as nonergodicity and caged dynamics. However, whereas hard-sphere glasses cannot exceed close-packing at an approximate volume fraction of 0.64, soft particles can be arranged at much higher volume fractions due to their softness and deformability. Above close-packing, particles adapt their shape due to steric constraints by forming flat facets at contact. Many concentrated dispersions share this generic structure irrespective of the origin of elasticity, as depicted in Fig. 2. Particles oppose external deformations by exerting repulsive forces through their contacting facets. These repulsive forces are responsible for the elastic moduli and for the osmotic pressure of the suspensions, which will be studied in Sect. 3. Each particle is surrounded by many neighbors with whom it interacts, just as if it were trapped in a cage. Cages persist until a stress exceeding the cage elasticity is applied, which is at the origin of yielding and flow, as will be shown in Sect. 5. We shall refer to concentrated suspensions made of soft particles at a volume fraction above close-packing as soft particle glasses.

The existence of short-range attractive interactions between particles leads to a much richer phase behavior, as illustrated in Fig. 3. This situation can be achieved by adding a nonadsorbing polymer to the suspensions, which induces an effective depletion attraction between the particles [105]. Such polymer–colloid mixtures can be viewed as model systems of complex fluids and are involved in many practical

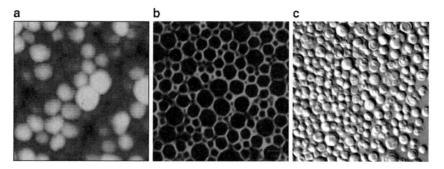

Fig. 2 Generic structure of concentrated dispersions above close-packing: (**a**) polyelectrolyte microgels [47] (particle diameter $d \approx 0.2\,\mu\text{m}$); (**b**) oil in water emulsion [102] ($d \approx 2\,\mu\text{m}$); (**c**) multilamellar vesicles [103] ($d \approx 5\,\mu\text{m}$). (Pictures are reproduced with permission of the authors)

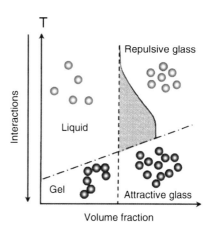

Fig. 3 Phase diagram of concentrated particle dispersions in the presence of short-range attractive interaction. Repulsive and attractive glasses are formed at high and low temperatures, respectively. The *hatched* area represents the reentrant liquid region observed when the temperature is varied at constant volume fraction. (Reproduced from [104] with permission of the author)

applications. The range and the depth of the attraction are tuned independently by varying the molecular weight and concentration of the polymer, respectively [106]. At low polymer concentration, the attraction does not play any significant role and the suspension behaves very much like a hard-sphere suspension. At larger polymer concentrations, short-range attractive interactions come into play and the particles tend to stick together, eventually shrinking the confining cage and causing the melting of the glass [104]. In this regime, the effect of attraction is primarily to stabilize liquid states and to shift the glass transition to higher volume fractions. When the polymer concentration is further increased, interparticle bonds become stronger and have a longer lifetime; the dynamics slows down, and a so-called attractive glass is formed. An immediate consequence is that the suspension undergoes a reentrant solid–liquid–solid transition when the strength of the attraction is increased or lowered. This scenario has been observed experimentally in hard-sphere/polymer and microgel/polymer mixtures [107–110]. Many other concentrated suspensions involving short-range attractive forces follow the same qualitative behavior, e.g., concentrated protein solutions [111], copolymer micelles [112–115], stearyl grafted silica particles [116], and star polymer mixtures [117]. However, many questions remain about the nature of the different phases associated with the reentrant transition, the possibility of inducing transitions between attractive and repulsive glasses, and the connection between the local dynamics and the macroscopic rheology.

2.4 Interaction Pair Potential in the Dense State Limit

In the dense limit where they are highly compressed, particles develop repulsive forces at contact due to their elasticity. The resulting potentials are weaker than observed in hard-sphere suspensions. In this section, we show that their exact form depends on the local architecture and the origin of elasticity but that there exist strong similarities and also subtle differences between materials as different as soft elastomeric particles, microgel particles, emulsions, and star polymers. We shall focus on simple situations where the interaction forces are purely repulsive without attractive components. We begin our discussion with the case of elastomeric particles compressed against one another. Figure 4 represents two such elastic spheres, i and j, with radii R_i and R_j and centered at r_i and r_j, respectively; the particles interact elastically through flat facets. The overlap distance between them is: $h_{ij} = R_i + R_j - r_{ij}$, where $r_{ij} = |r_i - r_j|$ is the center-to-center distance. The contacts are assumed to be frictionless and hence exert only a normal repulsive force at contact.

When the particle deformation is small compared to the size of the undeformed spheres, the contacts obey Hertzian contact mechanics. According to Hertz's theory, the elastic energy associated with a single contact is [118]:

$$U_{ij} = 0 \qquad\qquad h_{ij} < 0$$
$$U_{ij} = \frac{8}{15} E^* \left(\frac{h_{ij}}{R_c} \right)^{5/2} R_c^3 \qquad h_{ij} > 0 \quad, \qquad (1)$$

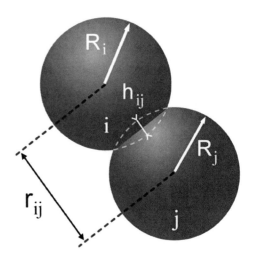

Fig. 4 Two deformable particles (i and j) compressed against each other and interacting elastically through their contacting facet

where E^* is the elastic contact modulus of the particles, which is related to the Young's modulus E and the Poisson's ratio v through: $E^* = E/2(1-v^2)$. E^* accounts for the contact stiffness; the stiffer the particle, the greater is the contact energy for the same deformation. The factor R_c is the reciprocal of the relative curvature defined by $R_c^{-1} = \left(R_i^{-1} + R_j^{-1}\right)^{-1}$. Liu et al. have measured the force required to compress single elastomeric spheres of micron size between two flat plates [119]. They found that estimates from Hertz theory are accurate to within 10% for up to 15% deformation, but the elastic force is higher at larger deformations. For large compression ratios, they derived an accurate expression assuming the Mooney–Rivlin law of nonlinear elasticity. This solution is well represented by the following empirical rule:

$$U_{ij} = 0 \qquad h_{ij} < 0$$
$$U_{ij} = CE^* \left(\frac{h_{ij}}{R_c}\right)^n R_c^3 \qquad h_{ij} > 0 \qquad (2)$$

with $n = 5/2, C = 8/15$ for $h_{ij}/R < 0.1$; $n = 4, C = 32/3$ for $0.1 \leq h_{ij}/R \leq 0.2$; and $n = 6, C = 1580/9$ for $0.2 \leq h_{ij}/R \leq 0.6$. The values of the constant C ensure force continuity over the entire range of compression ratios considered. Equations (1) and (2) provide useful estimates for the interaction potentials between soft elastic particles with bulk elasticity such as elastomeric particles and microgels.

For emulsions in which elasticity has an interfacial origin, Lacasse et al. have shown that the energy of interaction per contact between two compressed droplets can be well approximated at small compression ratios by an anharmonic potential of the form [120]:

$$U_{ij} = 0 \qquad h_{ij} < 0$$
$$U_{ij} = C\Gamma \left(\frac{h_{ij}}{R_c}\right)^\alpha R_c^2 \qquad h_{ij} < 0 \qquad (3)$$

where C is a constant prefactor and Γ is the interfacial tension. Both the constant C and the exponent α depend on the number of interacting neighbours. The exponent α varies from 2.1 at low coordination numbers to 2.6 at large coordination numbers. Note that this form of the interaction energy between compressed droplets is quite similar to the elastic energy for Hertzian contacts given by (1), with the contact elastic modulus E^* being proportional to the Laplace pressure Γ/R. For an emulsion droplet surrounded by 12 neighbors, we have: $E^* = 9.92(\Gamma/R)$ [121]. The validity of expression (3) is limited to situations where the droplets are weakly compressed. Lacasse et al. have also proposed a more general expression which applies over a wider range of compression ratios [120]:

$$U_{ij} = 0 \qquad h_{ij} < 0$$
$$U_{ij} = C\Gamma \left[\left(\frac{R_c}{r_{ij}}\right) - 1\right]^\alpha R_c^2 \qquad h_{ij} > 0 \qquad (4)$$

where again the prefactor C and the exponent α depend on the coordination number. Whereas these expressions describe static interaction between two droplets, dynamic interactions between two moving droplets are much less understood. Recent work suggests that such dynamic interactions involve a combination between interfacial deformation, static surface forces, and hydrodynamic drainage [122].

Star polymers are known to interact through an ultrasoft pair potential that is very different from that of the other soft spheres described above [123]. The energy of interaction between two identical stars with effective diameter σ is of the form:

$$\frac{U_{ij}}{k_B T} = \frac{5}{18} f^{3/2} \left[-\ln(r_{ij}/\sigma) + (1 + f^{1/2}/2)^{-1} \right] \qquad r \leq \sigma$$

$$\frac{U_{ij}}{k_B T} = \frac{5}{18} f^{3/2} (1 + f^{1/2}/2)^{-1} (\sigma/r_{ij}) \exp\left[-f^{1/2}(r_{ij} - \sigma)/2\sigma\right] \qquad r > \sigma$$

(5)

where k_B is the Boltzmann constant. This potential exponentially decays at large distances and crosses over at the corona diameter to a weak logarithmic repulsion.

We compare the variations of the pair potentials for particles with bulk elasticity (1)–(2) and surface elasticity (3)–(4), for star polymers (5), and for hard sphere particles in Fig. 5. The different potentials are normalized in a way that allows direct

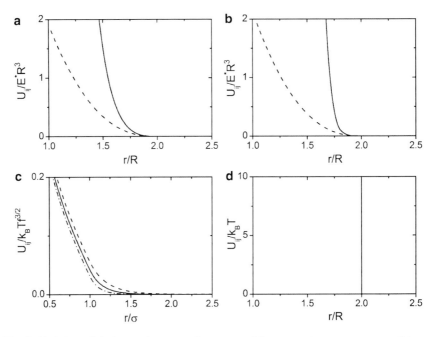

Fig. 5 Variations of the interaction energy between particles versus the center-to-center distance. (a) Hertzian potentials for particles with bulk elasticity: the *dashed line* represents the usual Hertz potential (1), and the *solid line* the generalized Hertzian potential (2). (b) Potentials for emulsions with surface elasticity: the *dashed line* denotes the approximate solution for small compression ratios (3), and the *solid line* the general solution (4). (c) Ultrasoft potentials for star polymers (where $\sigma \approx 1.3 R_G$, with R_G being the radius of gyration of the star, following [123]) (5): *dashed line* $f = 256$; *solid line* $f = 128$; *dashed and dotted line* $f = 64$. (d) Hard-sphere potential

comparison. Although they all are repulsive, there exist significant differences between them. First, we note that the Hertzian pair potentials for particles with bulk elasticity (elastomeric particles, microgels) in Fig. 5a and those for particles with interfacial elasticity (emulsion droplets, multilamellar vesicles) in Fig. 5b are quite similar. We simply note that the Hertzian potentials are slightly softer than their counterparts for liquid droplets. It is also interesting to observe that the approximate forms of the potentials given by relations (1) and (3), which are valid at small compression ratios only, significantly underestimate the energies of interaction at short distances. The potential for stars in Fig. 5c is markedly different from the others, indicating that stars are more easily compressed and deformed than microgels and emulsion droplets. Once the functionality f is large, the dependence of the energy on f is relatively weak, apart from the amplitude factor $f^{3/2}$. Finally, all these potentials differ drastically from the hard-sphere potential shown in Fig. 5d. These results point to the specificities of the pair potentials of soft particles compared to the hard-sphere potential.

3 Near-Equilibrium Properties of Soft Particle Glasses

3.1 Introduction

Concentrated dispersions of soft particles display both solid-like and fluid-like properties. In this section, we discuss the solid-like properties for near-equilibrium dispersions above the volume fraction for random close-packing. Near equilibrium is considered to be a configuration of the dispersion in which the interparticle forces are balanced and there is no net force on each particle. For these jammed systems, entropic effects are negligible and the sum of the pairwise interaction energies determines the free energy of the system. Balancing the forces on each particle ensures a local energy minimum or "near equilibrium" state. This of course is not the absolute free energy minimum, which would be a face-centered cubic lattice structure for monodisperse spheres, for example.

There has been considerable work on the elastic properties of soft particle dispersions composed of compressed emulsions and microgels. The elastic properties of compressed emulsions have been explored experimentally and theoretically [79, 124–129]. Concentrated microgel suspensions [87, 121, 130] and multilamellar vesicles [77, 82, 131, 132] have also been studied extensively. The elasticity of these different systems exhibit interesting analogies that will be analyzed in this section.

The importance of the amorphous glassy microstructure of soft particle dispersions is reflected by the great influence that the particle elastic modulus has on yielding and flow. The yield stresses of colloidal pastes and of emulsions scales like the shear modulus [13, 133, 134]. In Sect. 5, the flow curves of soft particle glasses will be shown to exhibit a remarkable universal behavior in terms of a unique microscopic time scale that involves the shear modulus [13]. In Sect. 4, the slip velocity of microgel pastes and concentrated emulsions will be shown to scale linearly with the shear modulus [102, 135]. In view of this, it is crucial to have predictions of the elastic modulus and of the osmotic pressure of these dispersions in terms of the particle properties and the microstructure.

3.2 Micromechanical Model

To model the elastic properties of dispersions of soft particles, we consider a dispersion of N spheres in a periodic box, as shown in Fig. 6. The particles are either monodisperse with radius R or polydispersed with a Gaussian distribution around a mean radius R. The concentration of particles is above the random close-packed volume fraction of $\phi_c = 0.64$ so that the particles are jammed together and form facets at contact. The contacts are assumed to be purely repulsive and frictionless and hence exert only a normal repulsive force at contact. The total elastic energy stored in the structure is the summation of the pairwise contact energies. Even at the highest volume fraction at near-equilibrium conditions, i.e., without flow, deformation of a particle is no more than 10% of its radius. Thus, the particle deformation is small compared to the size of the undeformed sphere and the contacts obey the Hertzian contact potential given by (1).

The total energy U for a random close-packing of N spheres is obtained as a sum of all the contact energies:

$$U = \sum_{i}^{N} \sum_{j>i}^{N} U_{ij}. \tag{6}$$

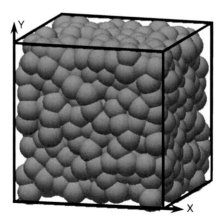

Fig. 6 Periodic box of concentrated dispersion of soft spherical particles. Each pair of particles at contact forms a facet, as shown in Fig. 4, that deforms according to Hertz's theory or similar law

The stress tensor σ is obtained by the Kirkwood formula [136]:

$$\sigma = -\frac{1}{V}\sum_{i}^{N}\sum_{j>i}^{N} r_{ij}\frac{dU_{ij}}{dr_{ij}}, \qquad (7)$$

where V is the system volume. The osmotic pressure π is the mean normal stress:

$$\pi = \frac{1}{3}tr(\sigma). \qquad (8)$$

For an isotropic body with Poisson's ratio $\nu = 0.5$ stretched uniaxially, the elastic energy change is given by [137]:

$$\frac{\Delta U}{V} = \frac{G}{2}\left(\lambda^2 + \frac{2}{\lambda} - 3\right), \qquad (9)$$

where λ is the extension ratio. For small strains, λ is close to unity and can be written as $(1+\varepsilon)$, where ε is a small number. Using this and approximating (9) correctly to $O(\varepsilon)$, the shear modulus is given by:

$$G \cong \frac{2}{3\varepsilon^2}\frac{\Delta U}{V}. \qquad (10)$$

Glassy packings of soft spheres are created first by generating a three-dimensional, periodically replicated random close-packed configurations of hard spheres using a compression algorithm introduced for glasses [138]. The close-packed configuration is compressed by reducing the box size in small steps until the desired density is achieved. After each size decrement, the system is allowed to relax. The relaxation protocol utilizes the conjugate gradient algorithm to minimize the system energy given by (8), which is equivalent to allowing the particles to readjust their positions so that they are in force equilibrium. Polydisperse packings can also be generated by the same protocol.

To compute the shear moduli and the osmotic pressure, we deform the packing in small increments. At each increment, the periodic box is stretched and the particle centers undergo the corresponding affine displacement. Because the initial configuration was relaxed (i.e., at an energy minimum), any slight deformation would be accompanied by an increase in energy, ΔU_∞. From this energy change, the shear modulus calculated with (10) is equivalent to the high-frequency modulus G_∞. In this limit, the relaxation time of the packing particles is slow, and so even though the particles experience unequal forces, they do not have the opportunity to relax. On the other hand, for low-frequency experiments, the deformation time is much greater than the relaxation time. While the packing is being deformed, particles rearrange themselves in order to balance their forces and minimize their contact energy. The details for simulating these simultaneous deformations and relaxations can be found elsewhere [121, 128]. The low-frequency modulus G_0 is determined by the energy changes associated with these deformations.

3.3 Near-Equilibrium Radial Distribution Function

It is useful to understand the microstructure of soft particle dispersions before discussing the predictions of their elastic properties and the comparison of these with experimental data. The radial distribution function is a convenient form for characterizing the microstructure since one expects radial symmetry at near-equilibrium conditions. In Fig. 7, the radial distribution function computed for the undeformed system has been plotted against the scaled radial distance. But for the first peak (Fig. 7a), this radial structure is very similar to that observed for jammed hard-sphere packings [139]. For hard spheres, $g(r)$ shows a sharp rise from zero at the hard-sphere diameter, i.e., $r = 2R$. Here, since the soft-sphere packing is

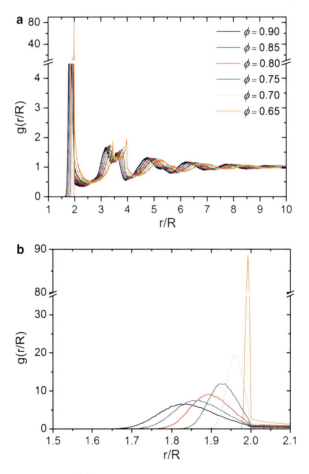

Fig. 7 Distribution function $g(r)$ for monodisperse packings versus the radial distance r scaled with particle radius R. The shift in $g(r)$ due to variation in the packing fraction is shown for (**a**) large and (**b**) small radial distances

compressed, the nearest-neighbor distance is much lower than the undeformed particle diameter. The first peak is thus located at radial distances $r < 2R$ and is also broader (Fig. 7b). This structure is similar to that reported by Chandler et al. for particles with soft repulsive potentials [140].

In Fig. 7b, we observe that each $g(r)$ curve has a first peak that resembles a Normal distribution. So, we fit each peak with a Gaussian curve of the form:

$$g(r) = a \exp\left(-\frac{(r-r_m)^2}{2s^2}\right), \tag{11}$$

where r_m is the mean radial distance, s is the spread, and a is the peak height. All three parameters have a systematic dependence on the volume fraction. As the volume fraction increases, the particles are packed closer together and r_m, which is the mean distance between neighbors, is lower. To quantify this dependence, consider one particle surrounded by N neighbors. The spherical volume of radius r_m surrounding the particle center contains approximately $(1+N/2)$ particles, assuming that this imaginary sphere cuts each neighboring particle into half (actually it is slightly less than half). The solid volume fraction is $\phi \sim (1+N/2)(1/r_m)^3$. At close-packing, where the particles just touch each other, r_m equals $2R$ and the fraction is $\phi_c \sim (1+N/2)/8$. Combining these two results, we obtain $r_m \sim (\phi_c/\phi)^{1/3}$. In the actual system, N is not constant but increases steadily from around 7 for close-packing up to 11 for high compressions. Thus $(1+N/2)$ is a weak function of ϕ. Upon fitting the observed r_m values we find that:

$$r_m(\phi) = 2.07 \left(\frac{\phi_c}{\phi}\right)^{\frac{1}{3}} \phi^{0.09}, \tag{12}$$

where the additional weak dependence on ϕ accounts for the slight increase in the number of contacts upon compression. The peak width s increases with the packing fraction. Since the right edge of the peaks reaches a minimum near $r = 2R$, s should be proportional to $(2-r_m)$. From the observed s values we obtain:

$$s \cong 0.478(2-r_m). \tag{13}$$

Finally, the peak height a also shows a distinct ϕ dependence. It is lower for greater compressions and appears singular at the hard-sphere contact. Upon fitting the data for a, we find that:

$$a(\phi) = 2.09(\phi - \phi_c)^{-0.87} \phi^{-0.56}, \tag{14}$$

where it indeed approaches a singularly high value as the packing fraction approaches ϕ_c.

These results point towards a universality in the particle radial distribution function from $r = 0$ to $2R$. This is shown in Fig. 8 where the first contact peaks, which are self-similar, are collapsed for all packing fractions onto the master Gaussian curve:

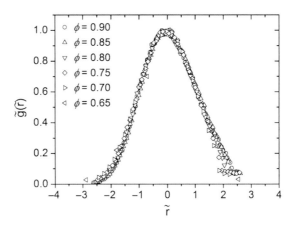

Fig. 8 Rescaled radial distribution function $\tilde{g}(\tilde{r})$ for monodisperse packings versus the modified radial distance $\tilde{r}$. Data for different volume fractions ϕ collapse on a single curve, which is well approximated by a standard Gaussian function

$$\tilde{g}(\tilde{r}) = \exp\left(-\frac{\tilde{r}^2}{2}\right), \tag{15}$$

where $\tilde{r}$ is the modified radial distance defined by:

$$\tilde{r} = \frac{1}{s(\phi)}\left(\frac{r}{2R} - r_m(\phi)\right), \tag{16}$$

and $\tilde{g}(\tilde{r})$ was calculated by scaling $g(r)$ with the individual peak heights as:

$$\tilde{g}(\tilde{r}) = \frac{g(r)}{a(\phi)}. \tag{17}$$

Thus, the parameters used for rescaling are solely functions of the packing fraction and its deviation from the close-packed value, using which we can predict the radial distribution function of any packing of soft particles a priori.

3.4 Shear Moduli and Osmotic Pressure

Figure 9 summarizes the simulation results for monodisperse packings. It shows a plot of G_0, G_∞ and π, all scaled with E^*, versus the packing fraction. These values of the shear moduli and the osmotic pressure were consistently reproduced over nine sample configurations for each fraction and show little spread, with error bars smaller than the symbols. Up to a packing fraction of around $\phi_c = 0.64$ (i.e., the close-packing density), there is no contact between particles, and both the moduli

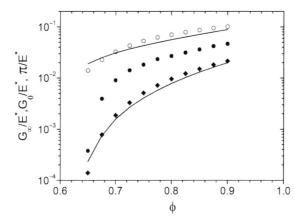

Fig. 9 Simulation results for monodisperse packings. The high-frequency shear modulus G_∞ (*open circles*), the low-frequency shear modulus G_0 (*closed circles*), and the osmotic pressure π (*diamonds*) are all scaled with the particle contact modulus E^* and plotted versus the packing fraction ϕ. The *lines* represent the predictions for π and G_∞ using (18) and (19)

and the osmotic pressure are zero. When compressed above this concentration, the spheres are deformed; the osmotic pressure grows rapidly at first and then continues to increase more slowly at higher volume fractions. The plot for G_∞ also shows a sharp rise at the critical packing density. As the packing fraction increases, the energy required to deform the packing also increases and hence G_∞ grows. G_0 shows the same trend, but is less stiff than G_∞. This is because the energy change accompanying deformation with intermediate relaxation is also lower.

Both the osmotic pressure and the high-frequency shear modulus can be calculated analytically from the radial distribution function derived above. The osmotic pressure π is related to the radial distribution function and the energy potential $u(r)$ as:

$$\pi = -\frac{4\pi n^2}{6} \int_0^{2R} r^3 \frac{du(r)}{dr} g(r) dr, \qquad (18)$$

where n is the number density of particles. The high-frequency shear modulus G_∞ is given by [141]:

$$G_\infty = \frac{2\pi}{15} n^2 \int_0^{2R} g(r) \frac{d}{dr}\left[r^4 \frac{du(r)}{dr}\right] dr. \qquad (19)$$

Figure 9, verifies the calculated osmotic pressure and high-frequency shear modulus against the values obtained from simulations. Both the calculated values and the simulation results agree well for the entire range of packing fractions. It is interesting to note that, unlike the high-frequency modulus, the low-frequency modulus cannot be obtained analytically.

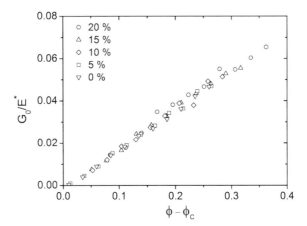

Fig. 10 Simulation results for the low-frequency shear modulus G_0 scaled with the contact modulus E^*, computed for 0, 5, 10, 15, and 20% polydisperse packings, as a function of $\phi - \phi_c$

We have found that small amounts of Gaussian polydispersity have little effect on the osmotic pressure and storage modulus. Surprisingly, even a 20% spread in the particle radius does not produce a significant effect on either G_0 or π. The low-frequency modulus G_0 is plotted versus the volume fraction ϕ in Fig. 10 for different values of the polydispersity. The results are very close and it is difficult to distinguish a definite trend due to polydispersity. Similar results are noted for the high-frequency modulus and osmotic pressure. The percolation threshold, which is the fraction at which the particles start touching each other and form force chains through the system volume, is not very different for each set and is fixed at around 0.64 [142–144] for this modest amount of polydispersity.

3.5 Comparison with Experimental Data

We have compared these theoretical predictions of the low-frequency modulus to experimental measurements on compressed emulsions and concentrated dispersions of microgels [121]. The emulsions were dispersions of silicone oil (viscosity: 0.5 Pa s) in water stabilized by the nonionic surfactant Triton X-100 [102, 121]. The excess surfactant was carefully eliminated by successive washing operations to avoid attractive depletion interactions. The size distribution of the droplets was moderately polydisperse with a mean droplet diameter of 2 µm. The interfacial energy Γ between oil and water was $4\,\text{mJ/m}^2$. The contact modulus for these emulsions was thus $E^* \approx 35\,\text{kPa}$. The volume fraction of the dispersed phase was easily obtained from weight measurements before and after water evaporation. Concentrated emulsions have a plateau modulus that extends to the lowest accessible frequencies, from which the low-frequency modulus G_0 was obtained. Figure 11 shows the variations of G_0/E^* with ϕ measured for the emulsions against the values calculated in the

Micromechanics of Soft Particle Glasses

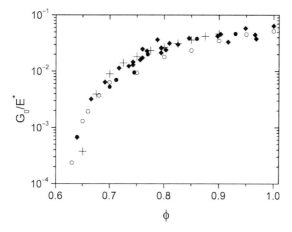

Fig. 11 Low-frequency shear modulus G_0 from simulations for elastic spheres with 15% polydispersity (*plus symbols*) is compared with experimental data for silicone oil/water/Triton X-100 emulsions (*closed circles*) and data from Mason et al. [124] for silicone oil/water/SDS emulsions (*open circles*). Data for microgel pastes in the presence of an excess of sodium chloride ($C_S = 0.1\,\text{mol/L}$) are also shown (*diamonds*)

simulations. On the same graph, we have also plotted data taken from Mason et al. [124] for silicone oil/water/SDS emulsions. All the sets of data are in good agreement, showing that the Hertzian potential of interaction, introduced in expressions (1) and (3), accounts reasonably well for the repulsive interactions between compressed droplets in concentrated emulsions.

The microgel pastes in Fig. 11 were made of alkali-swellable polyelectrolyte crosslinked microgels in water [47, 61]. At low pH, the particles are insoluble in water but at around pH 9, the carboxylic functions are ionized and they swell. The dispersions form pastes with appreciable shear moduli G_0 at large enough concentration. Unlike emulsion droplets, which do not change their size with concentration, microgels are polymeric micronetworks that change their sizes when the concentration is increased. This occurs by at least two mechanisms, which we refer to as steric deswelling and osmotic deswelling. Steric deswelling takes place at very high concentration, where the swelling of the microgels can be limited by the quantity of solvent available. Osmotic deswelling is due to the fact that the Donnan equilibrium, which controls the concentrations of counterions inside and outside the microgels, varies with the polymer concentration. In a previous study, we have shown that osmotic deswelling can be made negligible by the addition of salt. The volume fraction ϕ is then simply proportional to the polymer weight fraction C [61]. Another difficulty arises from the fact that the Young modulus and/or the contact modulus E^* cannot be easily estimated. We considered it as a free parameter that was adjusted to collapse the experimental data onto the numerical results at large volume fraction where G_0/E^* tends to a plateau. In practice, a bias in the choice of E^* simply lead to a vertical shift of the experimental data. In Fig. 11, we compare the variations of G_0/E^* versus ϕ for the microgel pastes to the predicted data. The agreement is reasonable given the indeterminacies in the determination of ϕ and E^*. In this particular example, the contact modulus is $E^* \approx 25$ kPa, from which we can estimate the shear modulus of individual microgels, obtaining $G_P \approx 10$ kPa ($\nu = 0.5$). This is comparable to the values measured for macroscopic polyelectrolyte gels [86, 145, 146].

In conclusion, the near-equilibrium properties of concentrated dispersions of soft particles dominated by elastic interactions, such as for emulsions and microgels, are well-described by assuming pairwise Hertzian contact. Furthermore, the radial distribution function for these materials appears to be self-similar at small distances, which is the relevant length scale for computing their elastic properties because the Hertzian repulsive force is short-ranged. This opens the possibility to predict the equilibrium properties of a great variety of soft glasses in terms of a limited amount of microscopic parameters such as the particle modulus and the volume fraction. The difficulty of measuring the properties of complex polymer–colloid particles urges further studies in that direction.

4 Wall Slip and Surface Rheology

4.1 Ubiquity of Wall Slip Phenomena in High-Solid Dispersions

The rheological behavior of high solid dispersions depends not only on bulk properties but also on the nature of the confining surfaces. Their interaction with solid surfaces – whether the material sticks or not – has a great importance. In real situations, the motion of weakly adhering suspensions is often dominated by wall slip [147]. Wall slip has been observed in a variety of solid particle dispersions such as concentrated particulate suspensions [148, 149], flocculated suspensions [150, 151], colloidal gels [152, 153], and suspensions of fillers in polymer matrices [154, 155]. Wall slip also occurs in concentrated dispersions of soft and deformable particles: concentrated emulsions [156–158], foams [159], microgel suspensions [102, 135, 160], synthetic gels [161], and many other complex formulations [162–164]. Although wall slip phenomena appear to be quite general, quantitative descriptions have remained scarce and fragmented until recently. It is now commonly agreed that slip is due to the presence of a thin layer of solvent near the wall, with the particles either not interacting with the wall or weakly interacting. The velocity and the local shear rate next to the wall are thus much greater than in the bulk. In extreme situations, the deformation of the material is entirely localized in the solvent layer near the wall and the bulk material does not deform at all.

Although this picture is remarkably generic, the mechanisms responsible for the formation of a particle-lean layer adjacent to the wall depend on the properties of the material under consideration. For the case of solid particle dispersions, wall depletion, particle migration, and solid–liquid separation are the most frequent sources of solvent layer lubrication. Wall depletion occurs whenever dispersions are brought into contact with smooth and solid surfaces because the suspended particles cannot penetrate rigid boundaries [147]. Particle migration is due to various forces arising from fluid inertia, fluid elasticity, and shear-induced diffusivity effects [165]. Solid–liquid separation, which frequently occurs in flocculated suspensions like slurries or cement pastes, is associated with the migration and exudation of the liquid phase though the solid network [166]. In concentrated soft particle dispersions, noncontact elastohydrodynamic interactions associated with particle deformability can be at the origin of low-viscosity layers that act as lubricants.

Until recently, the existence of slip has been inferred from indirect rheological observations or macroscopic visualizations. A very popular technique consists of injecting a line of a dye at the surface or in the bulk of the material [148, 152, 154, 158]. Notwithstanding the poor spatial resolution, the results are often affected by edge instabilities, which are ubiquitous in this type of material [102]. Recent important progress in the detection and characterization of wall slip has been made possible by the development of techniques that allow measuring velocity profiles at an excellent spatial resolution, even close to the particle scale. Magnetic resonance imaging techniques (MRI) have been applied to the investigation of surface phenomena in multiarm star polymer solutions [167] and in other pasty materials [163]. Heterodyne dynamic light scattering has been used to measure the local velocity profiles of slipping index-matched emulsions [168]. Recently, a high-frequency ultrasonic velocimetry technique, based on the ultrafast analysis of the ultrasonic speckle signal backscattered by particles following the flow, has been successfully applied to image the flow of complex fluids [169]. Last, but not least, a variety of techniques based on direct visualization have appeared recently. They consist in imaging the motion of individual particles in the bulk of a sheared suspension using simple videomicroscopy associated with particle image velocimetry [102, 170, 171] and confocal microscopy [172–174]. Isa et al. reviews recent advances in imaging the flow of concentrated suspensions to obtain time-resolved information on the particle scale level [175].

Wall slip has crucial implications with respect to the rheological characterization, transport, storage, and processing of concentrated suspensions. It is also the most serious difficulty experienced when testing solid dispersions. A direct manifestation of wall slip is that apparent motion can be detected below the bulk yield stress. This greatly affects the apparent yield stress values, the shape of the flow curves, and the linear storage and loss moduli. Over the years, different methods have been developed to reduce or suppress wall slip. This is generally achieved by modifying the surfaces of the tools by roughening or sticking a rough coating such as solvent-proof sandpaper [102, 176], or by using specific serrated tools [177]. The vane geometry, which is similar to the Couette geometry with the inner cylinder replaced by a vane and the outer cup profiled or roughened, is another useful and simple means of measuring the rheology of concentrated suspensions and other solid-like materials without the artifacts associated with wall slip or elastic instabilities [178, 179].

4.2 Generic Features of Wall Slip

4.2.1 Slip Regimes

Soft particle dispersions exhibit generic slip behavior when sheared near smooth surfaces. First, the magnitude of slip crucially depends on the applied velocity

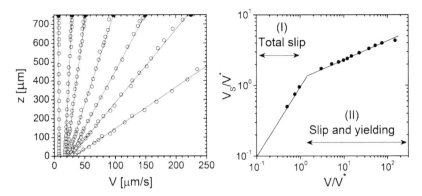

Fig. 12 *Left*: Steady-state velocity profiles measured for a microgel paste sheared between a rough moving surface ($z = 750$) and a smooth motionless surface ($z = 0$). Slip occurs at the smooth surface. The slip velocity is defined by the intercept of the velocity profile with $z = 0$. *Right*: Variations of the slip velocity (V_S) with the velocity applied at the moving surface (V); $V^*(10 \mu m/s)$ is the value of the slip velocity at the yield point. Different slip regimes occur at applied velocities below (regime *I*) and above (regime *II*) the yield point. Total slip leading to plug flow is observed below the yield point

and/or stress. Different regimes of slip can be distinguished, depending on whether the dispersion yields or not. This is illustrated in Fig. 12 for the case of concentrated emulsions, but essentially the same behavior has been found in microgel suspensions [160, 180], pasty materials [162], and hard-sphere suspensions [181].

Regime I is observed at low applied velocities, corresponding to stresses at and below the yield stress ($\sigma < \sigma_y$). Bulk flow is negligible and the motion is entirely due to slippage on the smooth surface, resulting in plug flow with zero-gradient velocity in the bulk (total slip). The plug velocity is equal to the applied velocity up to distances from the wall comparable to the particle size, indicating that the first layer of particles slips. The slip velocity is simply the plug velocity, i.e., the velocity of the mobile surface ($V_S = V$). This provides a simple way to determine the slip velocity from rheology.

Regime II occurs at applied velocities corresponding to stresses above the yield point ($\sigma > \sigma_y$). The apparent motion then involves a combination of slip and bulk deformation. In this regime, the increase of the slip velocity with the applied velocity is relatively slow ($V_S \propto V^{0.28}$), indicating that at large stresses, the effect of slip becomes negligible compared to the displacement associated with bulk flow. Some authors mention the possibility of a third regime when the applied velocity or stress is further increased [160, 180], and this point deserves more attention.

4.2.2 Surface Rheology

Each regime of slip is characterized by a specific stress/slip velocity relationship. Although this problem has been known for a long time, there exist relatively few

Micromechanics of Soft Particle Glasses

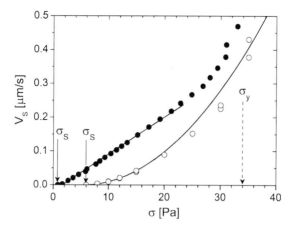

Fig. 13 Stress–slip velocity variations for a microgel paste sheared along hydrophilic (*closed circles*) and hydrophobic surfaces (*open circles*). The *dashed arrow* denotes the yield stress, $\sigma_y = 34$ Pa. The *solid arrows* denote the stress below which the paste adheres to the surface: $\sigma_S = 0.7$ Pa and 5.8 Pa for the hydrophilic and hydrophobic surface, respectively. The *solid lines* represent the quadratic ($V_S \propto \sigma^2$) and linear ($V_S \propto \sigma$) fits to the data for the hydrophobic and hydrophilic surfaces respectively

systematic studies and recent achievements have contributed to the understanding of the underlying phenomena. In Regime I, the surface rheology depends not only on the stress or applied velocity but also on the softness/deformability of the particles and on surface forces. For rigid particle suspensions, the stress/slip velocity relationship is generally linear [149, 181]. For soft particle suspensions, $V_S(\sigma)$ may be linear [160] or quadratic [102, 135, 168, 182]. For foams, the situation appears to be still more complex [183]. In Regime II, the stress/slip velocity relationship is linear and independent of the chemical nature of the shearing surface.

In practice, the slip properties of many materials can switch from one behavior to another when the surface chemistry is changed. This is illustrated in Fig. 13 where we present the stress/slip velocity relationships measured for a microgel paste sheared along two types of surfaces. For hydrophobic surfaces that are not wetted by water, the dependence of V_S on the applied stress σ is nearly quadratic. Slip stops at a finite value σ_S of the stress, which is termed the "sticking yield stress", which must not confused with the true yield stress σ_y. σ_S is much lower for hydrophilic wetting surfaces than for nonwetting surface. This indicates that short range forces and possible adhesion of dispersions onto the shearing surfaces have a great influence on the apparent flow properties. At small stresses, the slip velocity essentially increases linearly with stress except in the vicinity of the bulk yield stress where a quadratic dependence is recovered. We have proposed that these distinct stress/slip velocity relationships might simply reflect different slip mechanisms, which are ultimately determined by the attractive or repulsive nature of particle–wall interactions [160]. Recently, this approach, which was originally worked out for macroscopic gels [161], has been successfully applied to model the slip properties of microgel pastes and concentrated emulsions.

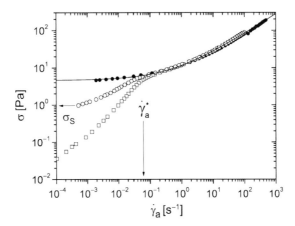

Fig. 14 Typical flow curves for a concentrated emulsion sheared between two rough surfaces (*closed circles*; the *line* is a fit to the Herschel–Bulkley equation), and between rough and smooth surfaces (*open circles* hydrophilic glass surface; *open squares* hydrophobic polymer surface). $\dot{\gamma}_a$ denotes the apparent shear rate, and $\dot{\gamma}_a^*$ is the value of the apparent shear rate at the yield point ($\sigma = \sigma_y$). σ_S is the sticking yield stress below which the emulsion adheres to the surface

4.2.3 Wall Slip and Macroscopic Rheology

The presence of wall slip dramatically impacts the apparent flow properties of solid dispersions. This is shown in Fig. 14, which represents the apparent flow curves measured using different shearing surfaces for a concentrated emulsion. In the absence of slip, the flow curve is well-represented by the Herschel–Bulkley equation, which characterizes the bulk flow properties of this class of materials (discussed in Sect. 5). A completely different behavior is observed when one of the shearing surfaces is smooth. The flow curve coincides with the bulk flow curve at high shear rates, but apparent motion continues to be detected well below the yield stress σ_y. Slip stops at the sticking yield stress σ_S. The resulting shape of the flow curve, with a sharp kink at a stress close to the yield stress, constitutes the unambiguous signature of wall slip, which can be observed in most concentrated dispersions [147]. An obvious consequence is that the yield stress can be underestimated when wall slip is present. Figure 14 shows that the apparent flow curve below the yield stress also depends on the nature of the shearing surface.

4.3 Theory of Soft Lubrication

4.3.1 General Formalism and Slip Equations

Here, we model concentrated dispersions of soft particles as elastic spheres of radius R closely packed into a disordered jammed configuration, as shown in Fig. 15. The volume fraction is higher than at close-packing so that the spheres are compressed

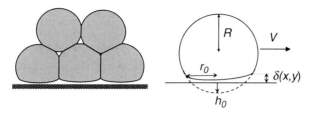

Fig. 15 Schematic of a soft concentrated dispersion near a solid surface (*left*) and detailed view of a particle squeezed against the surface (*right*)

and deform against their neighbors. Below the yield point, the stress is not sufficient to create large-scale rearrangements, and at a first approximation the particles are locked in their position. The first layer of particles adjacent to the wall experiences slip due to the formation of a lubricating film of solvent between particles and wall. To model slip, we focus on the behavior of a single particle of Young modulus E pressed against the wall moving at velocity V_S. At rest, the particle is compressed by a distance h_0 and develops a flat facet of radius r_0. Due to the proximity of the particle and the wall, the particles are sensitive to various short-range forces that eventually make them stick to the wall. Depending on the nature of the particles, there are many possible sources of interactions such as dispersive forces, steric hindrance, electrostatic contributions, and hydrophobic–hydrophilic forces [184]. During slip, the particle is dragged along the wall at a velocity V_S. The presence of neighbors is assumed to inhibit rotation so that the particle can only slide at the wall. The particle and the wall are separated by a lubricating film of solvent of thickness $\delta(x,y)$. The pressure field inside the film, $p(x,y)$ can further deform the elastic particle, which in turn maintains slip.

The equations governing the flow inside the lubricating film and the shape of the particle are:

$$\nabla \cdot [(\delta^3(x,y)\nabla p(x,y)] = -6\eta_S V \frac{\partial \delta(x,y)}{\partial x}, \tag{20}$$

$$\delta(x,y) = -h_0 + \frac{r^2}{2R} + w(x,y), \tag{21}$$

$$w(x,y) = \frac{1}{E^*} \int_{-\infty}^{+\infty}\int_{-\infty}^{+\infty} \frac{p(\xi,\eta) + \pi_d(\xi,\eta)}{\sqrt{(x-\xi)^2 + (y-\eta)^2}} d\xi d\eta, \tag{22}$$

where η_S is the solvent viscosity, $E^* = \pi E/(1-v^2)$ is the contact modulus for a particle against a rigid wall (v is the Poisson modulus of the particle). (x,y) and (ξ,η) are Cartesian coordinates in the plane of motion of the particle, $r = (x^2+y^2)^{1/2}$ is the radial coordinate parallel to the wall, $w(x,y)$ is the elastic deformation due to the hydrodynamic pressure, $p(x,y)$ and the disjoining pressure due to the short-range surface forces, $\pi_d(x,y)$. The first equation (20) is the Stokes flow describing the flow of solvent through the film (V is the velocity in the film). The third equation (22) expresses the elastic deformation of the particle due to the hydrodynamic pressure and the net disjoining pressure associated with the different surface forces at work [160]. These two equations are coupled through the second equation (21) that gives the geometrical shape of the particle. Once solved numerically, these equations nicely account for most of the experimental results observed so far. They provide a quantitative understanding of the importance of the surface forces, which definitely decide which slip mechanism is at work in real situations.

4.3.2 Case I: Elastohydrodynamic Slip

Case I is observed when the particle–wall interactions are attractive. At rest, the particles stick to the wall. If these contacts were to persist during motion, no-slip behavior would be expected. However, a lubricating slip layer can form and be maintained according to a specific mechanism, which we refer to as "elastohydrodynamic slip". The origin of elastohydrodynamic slip lies in the soft and deformable character of the particles. Indeed any elastic body that deforms asymmetrically as it is dragged along a solid wall is subjected to a lift force that pushes it away from the wall, generates a high pressure field underneath, and ultimately promotes slip. These forces are well-known to careless drivers who experience hydroplaning on a wet road, and the beauty of nature is that they also determine the behavior of mesoscopic dispersions slipping on rigid surfaces. This appears clearly in Fig. 16a, b where we present solutions of the lubrication equation presented earlier for a compressed elastic particle slipping under the influence of van der Waals attractive forces. The height contour lines show that the particle facet against the wall is asymmetrically deformed, with a crescent shape appearing at the lagging half of the particle. There is a large positive pressure zone along the leading edge, which dominates over the sharp negative pressure peak situated at the rear of the particle. This results in a net lift force that pushed the particle away from the wall, thus sustaining lubrication flow and wall slip.

Equations (20)–(22) have been analyzed using simple scaling arguments, allowing for analytical predictions [102, 160]. The strength of this approach is that it is general because it does not postulate a priori any particular shape for the particle. We reproduce here briefly the key steps of the derivation; the interested reader will find details in previous publications. At rest, the contact between the deformed particle and the rigid surface is treated as a simple Hertzian contact, providing simple expressions for the facet radius and the pressure underneath, respectively, $r_0 = R\xi_0^{1/2}$ and $p_0 = E^*\xi_0^{1/2}$ where $\xi_0 = h_0/R$ [102]. The pressure represents the contact stress that balances the osmotic pressure of the dispersion pushing the particle against the wall. Since the latter is proportional to the shear modulus of the dispersion, G_0, we express ξ_0 in terms of physical parameters, $\xi_0 = (G_0/E^*)^{2/3}$. Under flow, we assume that the particle compression is not much different to that at rest, indicating that $w \cong h_0$ and r_0 is not changed. The Stokes equation above yields directly an expression for the film thickness, $\delta \propto (\eta_S V_S R/E^*)^{1/2}$. We then predict the viscous drag on the contacting facet and the stress/slip velocity relationship:

Micromechanics of Soft Particle Glasses

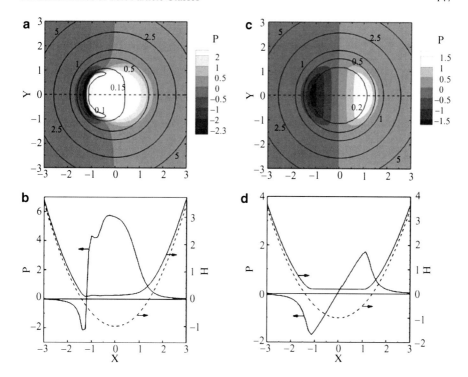

Fig. 16 Typical nondimensional height H and pressure P profiles for a translating, elastohydrodynamically lubricated sphere squeezed against a wall, in the presence of short-ranged attractive (*left panel*) and repulsive (*right panel*) forces. The *upper figures* (**a**) and (**c**) are contour plots of the pressure. The *lower graphs* (**b**) and (**d**) are plots of the pressure and height along the *center line* of the particles ($Y = 0$); the *dotted line* shows the undeformed particle shape (reprinted from [160])

$$f_d \propto \left(\eta_s V_s R^3 E^*\right)^{1/2} \left(\frac{h_{ij}}{R}\right), \tag{23}$$

$$\sigma \propto \left(\frac{\eta_s V_s G_0}{R}\right)^{1/2} \left(\frac{G_0}{E^*}\right)^{1/6}. \tag{24}$$

The second term on the right-hand side of (24) has a very weak dependence on the volume fraction so that the stress/slip velocity relationship has essentially a square root variation. This dependence reflects the fact that the lubrication film has itself a nontrivial dependence on the velocity. The value slip velocity V^* at the yield stress is:

$$\left(\frac{V}{V^*}\right) \propto \left(\frac{\sigma}{\sigma_y}\right)^2 \quad \text{with} \quad V^* \propto \gamma_y^2 \left(\frac{G_0 R}{\eta_s}\right)\left(\frac{E^*}{G_0}\right)^{1/3}. \tag{25}$$

It is interesting to note that the attractive forces do not play any significant role in this analysis, which assumes fully developed slip. As the slip velocity decreases, the lift force is not sufficient to prevent contact and the attractive forces cause the particles to adhere at the wall. The short-range forces can be included in the scaling analysis, thereby providing useful scaling predictions for the sticking yield stress σ_S at which slip stops. For van der Waals attractive forces, the latter is found to depend essentially on the shear modulus (which reflects the particle softness) and on the particle size, with a weak dependence on the Hamaker constant [160].

These predictions have been tested accurately in the case of concentrated emulsions and microgel pastes slipping along various substrates. The stress/slip velocity relationship and the onset of total slip V^* closely follow the predictions. The sticking yield stress is more difficult to reproduce quantitatively but the experimental results follow the predicted trends.

4.3.3 Case II: Hydrodynamic Slip

Case II refers to situations where the particle–wall interactions are purely repulsive. The particles are separated from the wall by a thin layer of solvent, even in the absence of any motion. Slip is thus possible for very slow flows, indicating that the sticking yield stress is vanishingly small. The residual film thickness for weak flows corresponds to a balance between the osmotic forces and the short-range repulsive forces, independently of any elastohydrodynamic contribution. This is clearly reflected in Fig. 16c, d, where we observe that the particle facet is nearly flat and symmetric. Since the pressure in the leading and rear regions of the facet are equal and opposite, the lift force is very small. The film thickness, which is set by the balance of the short-range forces, is constant so that the stress/velocity relationship is linear.

In conclusion, it is possible to rationalize wall slip in soft glasses as a consequence of the interplay of osmotic pressure and the various specific particle–wall interactions across the film of solvent that lubricates the contact between the particles and the wall. These results open up pathways for manipulating the flow of soft concentrated suspensions by making slight changes to the surface chemistry.

5 Shear Rheology of Soft Glasses

5.1 Generic Properties of the Nonlinear Rheology of Soft Glasses

Soft glasses are known to exhibit remarkable nonlinear shear rheology. They are yield-stress fluids that respond either like an elastic solid when the applied stress is zero or below the yield stress, or a like a viscoelastic fluid when a stress greater than the yield value of the material is applied [185]. Above their yield stresses, soft glasses are shear thinning fluids and very often the shear stress increases with the shear rate raised to the one-half power. This is well documented for the case of concentrated emulsions [102, 182, 186], microgel suspensions [31], and multilamellar vesicles [132]. Still more surprising, systematic rheological studies of concentrated microgel suspensions and compressed emulsions [13, 187] have shown that the flow properties of these materials are described by a universal flow curve. Recent investigations of other materials, namely diblock copolymer micellar solutions [70] and star polymers [188], have globally confirmed this result.

The generic flow properties of soft particle glasses are exemplified in Fig. 17, which shows the variations of the shear stress versus the shear rate measured at steady state for microgel pastes and compressed emulsions [187]. The flow curves in Fig. 17a obtained for microgel pastes with varying particle concentration, crosslink density, salt concentration, and solvent viscosity show the same qualitative behavior: a minimum shear stress, the yield stress of the material, below which the

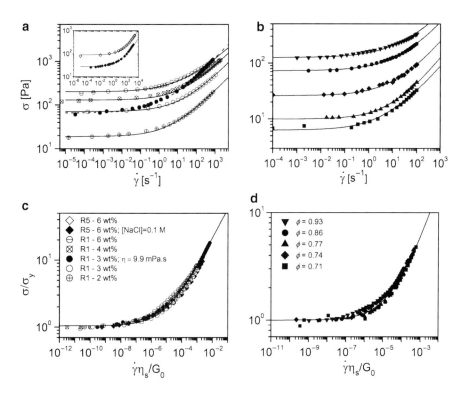

Fig. 17 Flow curves measured at steady state for microgel pastes (**a**) and concentrated emulsions of silicone oil in water (**b**). In (**a**) the data for microgel pastes are presented for varying particle concentration (wt%), crosslink density, salt concentration, and solvent viscosity. *Symbols* are the same as used in (**c**). R1 and R5 refer to two different crosslink densities, $N_x = 128$ and $N_x = 28$, where N_x is the average number of monomers between two crosslinks. In (**b**), data for emulsions are presented for varying packing fractions. *Symbols* are the same as used in (**d**). The *solid lines* in (**a**) and (**b**) are the best fits to the Herschel–Bulkley equation. Plots (**c**) and (**d**) show collapse of the different data sets when the shear stress is scaled by σ_y and the shear rate by η_S/G_0. The equations of the *solid lines* in (**c**) and (**d**) are of the form (26), where $m = 0.47$ and $K = 280$ for microgel pastes (**c**) and $m = 0.50$ and $K = 160$ for emulsions (**d**)

paste is solid-like and does not flow, and a power-law dependence between the shear stress and strain rate at higher stresses. Each curve can be described by a Herschel–Bulkley curve of the form $\sigma = \sigma_y + k\dot\gamma^m$. The yield stress σ_y is proportional to the low-frequency paste modulus G_0 through the yield strain γ_y. The latter is of the order of a few percent. k is a prefactor, which depends on the material properties (η_S, G_0). The exponent m is approximately equal to 0.5 for all samples. It is clear from Fig. 17a that the flow curves shift toward higher shear rates and yield stresses for samples with higher concentration of particles or those consisting of stiffer particles, due to either higher cross-link density or lower salt concentration. Similarly, the flow curves shift towards smaller shear rates for more viscous solvents. A similar trend is observed in Fig. 17b for concentrated emulsions consisting of silicone oil droplets dispersed in water using the nonionic surfactant Triton X-100. Again, the flow curves are described by the Herschel–Bulkley equation with the exponent m close to 0.5. Emulsions with larger packing fractions ϕ exhibit higher stresses and lower shear rates. As in microgel pastes, a higher continuous phase viscosity shifts the flow curve lower on the shear rate axis [187].

The flow curves of microgel pastes and emulsions can be rationalized in terms of the limited number of experimental parameters. When nondimensionalized using the yield stress σ_y and the characteristic time η_S/G_0, the flow curves presented in Fig. 17a, b collapse together on a single flow curve, as shown in Fig. 17c, d, respectively. The master curve is of the form:

$$\frac{\sigma}{\sigma_y} = 1 + K\left(\frac{\eta_S \dot\gamma}{G_0}\right)^m, \qquad (26)$$

where the coefficient K is only slightly dependent on the material and m is equal to 0.5 within the experimental accuracy. This universality of the shear rheology for these compositionally very different materials and its dependence on only the paste shear modulus G_0 and the solvent viscosity η_S is quite remarkable and is suggestive of an underlying common flow mechanism. Fundamentally, the common aspects of the structure of soft glasses (i.e., soft particles packed into in a dense and disordered structure) must be at the root of this behavior. In the literature, cage-effect dynamics in hard-sphere suspensions have been explained using phenomenological models [7, 189]. Several simulation studies via molecular dynamics [190, 191] and extensions of mode coupling theory [9, 12] for sheared glasses have also been used to understand the dynamics of molecular and colloidal glasses. These models and simulations have successfully captured some of the characteristic features observed in experiments. They predict a Herschel–Bulkley-like flow curve for the shear rheology of these materials. However, exact comparison with experimental data usually involves adjustable parameters that are not directly related to the properties of the materials.

A distinctive feature of soft particle glasses is the presence of specific interactions mediated by the suspending liquid. The packing is dense, and particles are compressed against their neighbors to form flat facets at contact. The solvent is present in thin films between the facets or in the interstitial volume. The relative motion between particles during shear creates a hydrodynamic pressure field in the films. As two particles are dragged past one another, a flow of solvent develops inside the liquid films separating the facets. This generates a high-pressure field and causes an additional elastic deformation of the particles, which self-consistently maintains the lubricating films and ultimately makes particle motion possible. This scenario is reminiscent of the elastohydrodynamic lubrication mechanism presented in Sect. 4 to explain the slip behavior of compressed particles at rigid boundaries. In this section, a model and simulation are described for the dynamics of dense suspensions of soft particles under shear that incorporates elastohydrodynamic lubrication interactions between particles of the amorphous suspension.

5.2 Model Description

To study the flow properties of soft glasses, we start from the model presented in Sect. 3.2. We consider a packing of N elastic spheres packed at a volume fraction ϕ, which is greater than random close-packing, $\phi_c = 0.64$. As shown in Fig. 6, the particles are compressed against one another because of the net compressive pressure, i.e., the osmotic pressure, acting on the dispersion. The dispersion is now subject to a dimensionless bulk shear flow in the plane (x,y), $\mathbf{u}^\infty = \mathbf{x} \cdot \mathbf{n}_x \mathbf{n}_y$ where $\mathbf{n}_x$ and $\mathbf{n}_y$ are the flow and gradient directions, respectively. Figure 18 shows a detail of two compressed particles i and j with radii R_i and R_j, centered at positions $\mathbf{x}_i$ and $\mathbf{x}_j$ and translating with velocities $\mathbf{u}_i$ and $\mathbf{u}_j$ respectively. For simplicity, we show the interaction between i and j only, but each of these particles has facets and interacts with several other neighbors. The total interaction between i and j involves a central repulsive force associated with the elastic contact at the facet between i and j, which is coupled to a frictional drag force due elastohydrodynamic lubrication inside the film of solvent separating the two particles.

Here, we model the elastic interactions using the Hertzian potentials presented in Sect. 2.4, which have been shown to describe the interactions between microgel

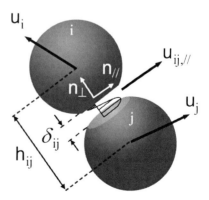

Fig. 18 Pairwise elastohydrodynamic lubrication interactions between two neighboring particles i and j

particles and emulsions droplets. Unlike in the static case, the deformations under shear can be much larger than the 10% limit above which the pure Hertz theory is no longer valid. Using the Hertzian potential, (1) leads to a significant underprediction of the shear stress and normal stress differences and large deformations beyond the accuracy of the model. Hence, we used the generalized potential given by (2). The elastic contact force acting on i is thus:

$$\mathbf{f}_{ij}^{el} = nCE^* \left(\frac{h_{ij}}{R_c}\right)^{n-1} R_c^2 \mathbf{n}_\perp, \qquad (27)$$

where R_c is the relative curvature and E^* the contact modulus. The values of n and C, which depend on the compression ratio h_{ij}/R_c, are specified in the definition of the generalized potential given by (2). The particles sliding over one another experience an elastohydrodynamic drag force coupled to the elastic deformation. To express this force, we generalized (23), standing for a single particle dragged along a flat surface, using the general form of the pair potential (2):

$$\mathbf{f}_{ij}^{EHD} = \left(nC\eta_S E^* u_{ij,//} R_c^3\right)^{1/2} \left(\frac{h_{ij}}{R_c}\right)^{(2n-1)/4} \mathbf{n}_{//}, \qquad (28)$$

where $u_{ij,//}$ is the relative velocity associated with sliding motion parallel to the contacting facet. These expressions characterize the pairwise interaction between two particles. Each particle has several randomly oriented facets and interacts with many others particles. We compute the net elastic, $\mathbf{f}_i^{el}$, and elastohydrodynamic, $\mathbf{f}_i^{EHD}$, forces on a particle by summing the pairwise forces due to all neighbours. Since inertia is negligible, the motion of particle i obeys to the equation:

$$\mathbf{u}_i = \mathbf{u}_i^\infty + \frac{f(\phi)}{6\pi\eta_S R_i}\left[\mathbf{f}_i^{el} + \mathbf{f}_i^{EHD}\right]. \qquad (29)$$

The coefficient $1/6\pi\eta_S R_i$ in the second term of the right-hand side of (29) represents the mobility coefficient for an isolated particle; $f(\phi)$ is the correction on the drag that accounts for the reduction of particle mobility at high concentration of particles. This equation can be made nondimensional by scaling lengths, time, and velocity by the average particle radius $R_0, \dot\gamma^{-1}$ and $\dot\gamma R_0$, respectively. The resulting dimensionless equation is then found to depend on the quantity $\lambda = \eta_S \dot\gamma/E^*$, which represents the ratio of viscous to elastic forces [187]. The dynamics and subsequent rheology of soft particle dispersions are thus characterized by λ and the degree of compression of particles that depends on ϕ. The equations describing the displacement of the particles form a set of N coupled equations, which can be solved numerically to give access to the spatial position and velocity of each particle, and to the different components of the stress tensor.

This model is implemented using a molecular-like simulation on random packings of N elastic spheres confined in a cubic box that is periodically replicated. $N = 10^3$ in the simulations reported here but the results with a much larger number of spheres ($N = 10^4$) were not significantly different. The radii of the spheres have a 10% polydispersity, similar to that in the experimental systems. We first created random packings of compressed force-free particles at prescribed volume fraction ϕ, as described in Sect. 3.2. Constant shear rate simulations were then performed using the open source LAMMPS code [192] assuming Lees–Edwards boundary conditions. The position and the velocity of each particle were obtained by solving the N equations of motion above using the Verlet integration algorithm [193]. The stress tensor σ was computed from the Kirkwood expression.

5.3 Simulation Results

Simulations were performed at nondimensional shear rates $\lambda = \dot{\gamma}\eta_S/E^*$ between 10^{-9} and 10^{-4} for five packing fractions between $\phi = 0.7$ and 0.9. For every (λ, ϕ) combination, the shear stress was calculated at regular time intervals until steady state was reached. Figure 19a shows the variations of the shear stress for the five volume fractions investigated. The flow curves exhibit a Herschel–Bulkley behavior, with the stress increasing with increasing volume fraction. At low shear rates, there is a yield stress and, at high shear rates, the shear stress increases as $\dot{\gamma}^{0.5}$ approximately. The simulations thus reproduce the experimental results presented in Fig. 17a. The inset shows the contributions to the total stress for $\phi = 0.9$ derived from the elastohydrodynamic and elastic forces acting on the particles. Clearly, the shear stress is primarily due to the elastic interactions, even at the highest shear rates.

In Fig. 19b, the predicted shear stresses at different volume fractions are collapsed onto a single master curve by normalizing the shear stress with the yield stress σ_y and rescaling the shear rate with time constant η_S/G_0, using the low-frequency modulus of the dispersion G_0 that was previously computed for these dispersions in Sect. 3. The simulations predict that the yield stress is proportional to G_0 through the yield strain, which is of the order of a few percent. These results agree well with the experimental data for concentrated microgels and compressed emulsions shown in Fig. 17c, d. The master curves are well described by (26), where the exponent m is very close to $1/2$ and the coefficient K agrees reasonably well with the experimental values. Thus, the micromechanical model and simulation presented here predict flow curves for concentrated dispersions of soft particles, which agree quantitatively with experimental measurements on concentrated suspensions of microgels and emulsions. It is crucial, however, that the non-Hertzian elastic contact model is used because an elastic Hertzian contact model would under-predict the stresses compared to those observed experimentally [187].

Although elasticity is found to make the dominant contribution to the stress for these soft particle glasses, viscous interactions also play a role, as evidenced by the importance of the solvent viscosity η_S in the rescaling used to collapse the experimental and simulation data in Figs. 17 and 19. It was originally speculated that the elastohydrodynamic interactions between compressed particles sliding by

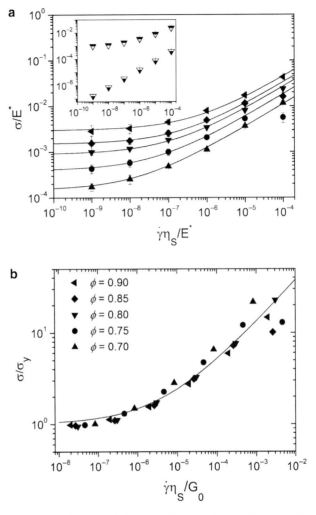

Fig. 19 (a) Shear stress σ from simulations: the *lines* are the best fits to the Herschel–Bulkley equation. The *inset* shows the elastic (*inverted triangles with open lower half*) and hydrodynamic component (*inverted triangles with open upper half*) of the stress for $\phi = 0.85$. (b) Collapse of flow curves; the *line* is of the form (26) with $m = 0.52$ and $K = 320$

one another leads directly to $\sigma \propto (\dot{\gamma}\eta_S/G_0)^{1/2}$ at high shear rates [102, 135], and similar predictions have been made for rows of sliding droplets in foams [194]. Indeed, the viscous component of the stress in the inset of Fig. 19a precisely follows this behavior, but its magnitude is much smaller than the component of stress from elasticity. The effect of the viscous interactions on the stresses at high shear rates is more subtle, and it occurs through the alteration of the pair distribution function of the dispersion during flow. As the particles are sheared by one another, the pair distribution function is deformed from that for the quiescent state, which was

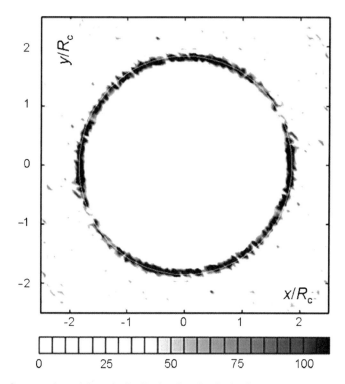

Fig. 20 Contour plots of the pair distribution function in the flow-gradient plane for $\dot{\gamma}\eta_S/E^* = 10^{-6}$ ($\phi = 0.85$). The *white circle* is a sphere of radius equal to the mean overlap between a pair of particles. *White* and *black* regions indicate lower and higher probabilities of finding other particles near a particle located at the center

analyzed in Sect. 3. In Fig. 20, we present the contour plots of the pair distribution function along the flow-gradient plane for $\dot{\gamma}\eta_S/E^* = 10^{-6}$. First, we observe that upon shearing particles accumulate along the compression axis and deplete along the extension axis. Second, the first-neighbor peak is distorted so that, on average, particles are more deformed along the compression axis than along the extension axis. It is important to note that these alterations of the pair correlation function induce preferential orientations of the facets and the interparticle solvent films. These effects together must lead to the observed increase in stress with the shear rate.

6 Outlook and Open Questions

We have shown that soft glasses encompass a broad range of materials made from soft and deformable particles dispersed in solvent at large volume fractions well above close-packing. These materials have several features in common with hard-sphere glasses, such as caged dynamics and nonergodicity. However, because the particles interact through potentials that are much weaker than in hard spheres, they develop repulsive elastic forces at contact when the volume fraction exceeds close-packing. The compressed particles then develop flat facets and adopt a generic amorphous structure. The solvent lubricates the contacts between the particles and plays an important role in transmitting the elastic interactions through the glass. In this chapter, we have shown that the static and dynamic properties of soft glasses result from an interplay between the disordered glass-like structure and solvent-mediated elastic interactions. We have incorporated these two basic ingredients into a micromechanical description that predicts several important rheological properties of soft glasses with purely repulsive interactions.

The near-equilibrium properties of soft glasses are essentially determined by the microstructure and the weak potentials existing between the particles. For undeformed glasses near equilibrium, the radial pair correlation function, which characterizes the statistical properties of the microstructure, is self-similar at small distances. This opens the possibility to predict the equilibrium properties of a great variety of soft glasses in terms of a limited number of microscopic parameters such as the particle modulus and the volume fraction. The predictions of the high-frequency shear modulus and of the osmotic pressure are in good agreement with the results of the simulations. The difficulty in predicting the low-frequency shear modulus from first principles urges for further work in that direction.

The existence of solvent-mediated elastohydrodynamic interactions between particles is central to the dynamic behavior of soft glasses. These interactions appear whenever a particle is dragged along a wall (surface flow) or a neighbor (bulk flow), as a result of the flow of solvent inside the liquid films between the particles. This generates a high-pressure field and causes an additional elastic deformation of the particles, which self-consistently maintains the lubricating films and ultimately sustains particle motion. We have described a micromechanical model, incorporating elastohydrodynamic interactions and the contribution of surface forces, which predicts the different slip regimes observed when soft glasses are sheared along a smooth wall. Similar ideas have been successfully extended to explain the remarkable universality of the bulk flow properties of soft glasses. Our approach predicts that the stress/shear rate relationship results from a balance between viscous dissipation and elastic interactions through the disordered network of compressed particles. Most of the results described here are for soft glasses interacting through Hertzian-like potentials but the method is general and can be applied to other materials. Star glasses, which interact through an ultrasoft potential and which, according to recent experiments, exhibit new and unexpected flow properties at low shear rates [188], deserve further investigation in this context.

Several important aspects of the rheology of soft glasses remain poorly understood. Important questions that merit attention concern the slow time evolution or aging [195], and other nonlinear phenomena associated with spatial heterogeneities such as shear banding [196]. The aging and slow dynamics of soft glasses is an inherent property of glasses that essentially affects the behavior of the materials at rest or under a stress smaller than the yield stress [197]. The most comprehensive studies of aging on soft colloids were carried out with laponite suspensions [198, 199], colloidal gels [200, 201], microgels [202, 203], sterically stabilized colloidal suspensions [204], and multilamellar vesicles [103, 205]. The waiting time (i.e., the rest time between the end of preparation and the beginning of a measurement) is the appropriate parameter for scaling the aging properties. While the soft glassy model has proven remarkably successful in reproducing several phenomenological aspects of aging [206], the underlying microscopic mechanisms remain poorly characterized. Aging seems to be related to the slow relaxation of internal stresses stored inside the dense disordered structure of the glasses [197, 202, 205]. The micromechanical model presented in this chapter will provide an ideal tool for characterizing in great detail the relaxation of internal stresses and the subsequent evolution of the microstructure.

Acknowledgments We would like to thank Dr. Jyoti Seth and Clementine Locatelli for their contribution to the work presented in this chapter. We are indebted to Lavanya Mohan for a critical reading of the chapter and for preparing Figs. 6 and 20. We are also grateful to Erik Zumalt for his help in preparing the graphics. RTB gratefully acknowledges financial support from National Science Foundation grant CBET 0854420.

References

1. Weeks E (1997) In: Maruyama S, Tokuyama M (eds) Statistical physics of complex fluids. Tohoku University Press, Sendai, Japan
2. Vlassopoulos D, Fytas G (2010) Adv Polym Sci doi 10.1007/12_2009_31
3. Bursac P, Lenormand G, Fabry B, Oliver M, Weitz DA, Viasnoff V, Butler JP, Fredberg JJ (2005) Nat Mater 4:557
4. Zhou EH, Trepat X, Park CY, Lenormand G, Oliver MN, Mijailovich SM, Hardin C, Weitz DA, Butler JP, Fredberg JJ (2009) Proc Natl Acad Sci 106:10632
5. Ancey C (2007) J Non-Newton Fluid Mech 142:4
6. Pusey PN (2008) J Phys Condens Matter 20:494202
7. Sollich P, Lequeux F, Hebraud P, Cates ME (1997) Phys Rev Lett 78:2020
8. Sollich P (1998) Phys Rev E 58:738
9. Fuchs M, Cates ME (2002) Phys Rev Lett 89:248304
10. Fuchs M, Cates ME (2009) J Rheol 53:957
11. Kobelev V, Schweizer KS (2005) J Chem Phys 123:164903
12. Fuchs M (2010) Adv Polym Sci doi 10.1007/12_2009_30
13. Cloitre M, Borrega R, Monti F, Leibler L (2003) Phys Rev Lett 90:068303
14. Russel WB, Saville DA, Schowalter WR (1989) Colloidal dispersions. Cambridge University Press, Cambridge
15. Lindsay HM, Chaikin P (1982) J Chem Phys 776:3774
16. Alexander S, Chaikin PM, Grant P, Morales GJ, Pincus P (1983) J Chem Phys 80:5776
17. Okubo T (1995) J Chem Phys 102:7721
18. Bonn D, Tanaka H, Wegdam G, Kellay H, Meunier J (1998) Europhys Lett 45:52
19. Bonn D, Kellay H, Tanaka H, Wegdam G, Meunier J (1999) Langmuir 15:7534
20. Levitz P, Ecolier E, Mourchid A, Delville A, Lyonnard S (2000) Europhys Lett 49:672
21. Bonn D, Coussot P, Huynh HT, Bertrand F, Debrégeas G (2002) Europhys Lett 59:786
22. Bonn D, Tanase S, Abou B, Tanaka H, Meunier J (2002) Phys Rev Lett 89:015701
23. Napper DH (1983) Polymeric stabilisation of colloidal dispersions. Academic, London
24. Castaing JC, Allain C, Auroy P, Auvray L, Pouchelon A (1996) Europhys Lett 36:153
25. Van Helden AK, Vrij A (1980) J Colloid Interface Sci 76:418

26. Mewis J, Frith WJ, Strivens TA, Russel WB (1989) AICHE J 35:415
27. Neel O, Ducouret G, Lafuma FJ (2000) J Colloid Interface Sci 229:244
28. Nicholson JW (2006) The chemistry of polymers. RSC, London
29. Bucknall CB, Paul DR (2000) Polymer blends: Formulation and Performance. Wiley, New York
30. Bagheri R, Marouf BT, Pearson RA (2009) Polymer Rev 49:201
31. Wyss H, Fernandez de Las Nieves A, Mattson J, Weitz DA (eds) (2010) Microgel based-materials. Wiley-VCH, New York
32. Das M, Zhang H, Kumacheva E (2006) Annu Rev Mater Res 36:117
33. Morris GE, Vincent B, Snowden MJ (1997) J Colloid Interface Sci 190:198
34. Xu S, Zhang J, Paquet C, Lin Y, Kumacheva E (2003) Adv Funct Mater 13:471
35. Saunders BR, Vincent B (1999) Adv Colloid Interface Sci 80:1
36. Cloitre M (2005) Macromol Symp 229:99
37. Pelton R (2000) Adv Colloid Interface Sci 278:972
38. Wolfe MS, Scopazzi C (1989) J Colloid Interface Sci 122:265
39. Raquois C, Tassin JF, Rezaiguia S, Gindre AV (1995) Prog Org Coat 26:239
40. Paulin SE, Ackerson BJ, Wolfe MS (1996) J Colloid Interface Sci 178:251
41. Meyer S, Richtering W (2005) Macromolecules 38:1517
42. Pich A, Tesssier A, Boyko V, Lu Y, Adler H-JP (2006) Macromolecules 39:7701
43. Ketz RJ, Prud'homme RK, Graessley WW (1998) Rheol Acta 27:531
44. Bromberg L, Temchenko M, Hatton TA (2002) Langmuir 18:4944
45. Eichenbaum GM, Kisetr PF, Shah D, Meuer WP, Needham D, Simon DA (2000) Macromolecules 33:4087
46. Saunders BR, Crowther HM, Vincent B (1997) Macromolecules 30:482
47. Cloitre M, Borrega R, Monti F, Leibler L (2003) C R Physique 4:221
48. Fernandez-Nieves A, Fernandez-Barbero A, Vincent B, de las Nieves FJ (2003) J Chem Phys 119:10383
49. Frith WJ, Stokes JR (2004) J Rheol 48:1195
50. Caggioni M, Spicer PT, Blair DL, Lindberg SE, Weitz DA (2007) J Rheol 51:851
51. Dingenouts N, Norhausen Ch, Ballauff M (1998) Macromolecules 31:8912
52. Stieger M, Pedersen JS, Lindner P, Richtering W (2004) Langmuir 20:7283
53. Rodriguez BE, Wolfe MS, Fryd M (1994) Macromolecules 27:6642
54. Berndt I, Richtering W (2003) Macromolecules 36:8780
55. Meng Z, Cho J-K, Breedveld V, Lyon A (2009) J Phys Chem B 113:4590
56. Fernandez-Nieves A, Fernandez-Barbero A, de las Nieves FJ (2001) J Chem Phys 115:7644
57. Routh AF, Ferandez-Nieves A, Bradley M, Vincent BJ (2006) Phys Chem B 110:12721
58. Kiefer J, Naser M, Kamel A, Carnali (1993) J Colloid Polym Sci 271:253
59. Saunders BR, Vincent B (1997) Prog Colloid Polym Sci 105:11
60. Fernandez-Nieves A, Fernandez-Barbero A, Vincent B, de las Nieves FJ (2003) J Chem Phys 119:10383
61. Borrega R, Cloitre M, Betremieux I, Ernst B, Leibler L (1999) Europhys Lett 47:729
62. Roovers J, Zhou L-L, Toporowski PM, van der Zwan M, Iatrou H, Hadjichristidis N (1993) Macromolecules 26:4324
63. Likos CN (2001) Phys Rep 348:267
64. Likos CN (2006) Soft Matter 2:478
65. Alexandridis P, Lindman B (2000) Amphiphilic block copolymers: self-assembly and applications. Elsevier, Amsterdam
66. Hamley IW (2005) Block copolymers in solution: fundamentals and applications. Wiley, Chichester
67. Hadjichristidis N, Pispas S, Floudas G (1998) Block copolymers: synthetic strategies, physical properties, and applications. Wiley Interscience, New York
68. Stellbrink J, Rother G, Laurati M, Lund R, Willner L, Richter D (2004) J Phys Condens Matter 16:S3821
69. Laurati M, Stellbrink J, Rother G, Lund R, Willner L, Richter D (2005) Phys Rev Lett 94:195504

70. Merlet-Lacroix N (2008) Auto-assemblage de copolymères à blocs acryliques en solvent apolaire: un exemple de verre colloïdal attractif. Ph.D. thesis, University Paris VI
71. Lacroix-Merlet N, Di Cola E, Cloitre M (2010) Soft Matter 6:984
72. Becher P (1983) Encyclopedia of emulsion technology. Marcel Dekker, New York
73. Mabille C, Schmitt V, Gorria Ph, Leal Calderon F, Faye V, Deminière B, Bibette J (2000) Langmuir 16:422
74. Rhutesh K, Shah, RK, Cheung Shum H, Rowat AC, Lee D, Agresti JJ, Utada AS, Chu L-Y, Kim J-W, Fernandez-Nieves A, Martinez CJ, Weitz DA (2008) Mater Today 11:18
75. Ekwall P (1975) In: Brown GM (ed) Advances in liquid crystals. Academic, New York, p 1
76. Porte G (1992) J Phys Condens Matter 4:8649
77. Ramos L, Molino F (2000) Europhys Lett. 51:320
78. Diat O, Roux D, Nallet F (1983) J Phys II France 3:1427
79. Princen HM (1986) J Colloid Interface Sci 112:427
80. De Gennes PG, Prost J (1994) Physics of liquid crystals. Oxford University Press, Oxford
81. Van der Linden E, Dröge JHM (1993) Physica A 193:439
82. Panizza P, Roux D, Vuillaume V, Lu C-YD, Cates ME (1996) Langmuir 12:248
83. Rubinstein M, Colby R (2003) Polymer physics. Oxford University Press, Oxford
84. Obukhov SP, Rubinstein M, Colby RH (1994) Macromolecules 27:3191
85. Rubinstein M, Colby R, Dobrynin AV, Joanny J-F (1996) Macromolecules 29:398
86. Skouri R, Schosseler F, Munch JP, Candau SJ (1995) Macromolecules 28:197
87. Senff H, Richtering W (1999) J Chem Phys 111:1705–1711
88. Kim JU, Matsen MW (2008) Macromolecules 41:4435
89. Rodriguez BE, Wolfe MS, Fryd M (1994) Macromolecules 27:6642
90. Eichenbaum GM, Kiser PF, Dobrynin AV, Simon SA, Needham D (1999) Macromolecules 32:4867
91. Pusey PN, van Megen W (1986) Nature 320:340
92. Pusey PN (1991) In: Hansen JP, Levesque D, Zinn-Justin J (eds) Liquids, freezing and glass transition. North Holland, Amsterdam
93. Sciortino F, Tartaglia P (2005) Adv Phys 54:471
94. Frenkel J (1946) Kinetic theory of liquids. Clarendon, Oxford
95. Cates ME (2003) Ann Henri Poincare 4:S647
96. Gang H, Krall AH, Cummins Z, Weitz DA (1999) Phys Rev E 59:715
97. Bartsch E, Frenz, V, Baschnagel J, Schartl W, Sillescu H (1997) J Chem Phys 106:3743
98. Crassous JJ, Régissier R, Ballauff M, Willenbacher N (2005) J Rheol 49:851
99. Crassous JJ, Siebenbürger M, Ballauff M, Drechsler M, Henrich O, Fuchs M (2006) J Chem Phys 125:204906
100. Purnomo EH, van den Ende D, Mellema J, Mugele F (2006) Europhys Lett 76:74
101. Crassous JJ, Siebenbürger M, Ballauff M, Drechsler M, Hajnal D, Henrich O, Fuchs M (2008) J Chem Phys 128:204902
102. Meeker SP, Bonnecaze RT, Cloitre M (2004) J Rheol 48:1295
103. Ramos L, Cipelletti L (2001) Phys Rev Lett 87:245503
104. Sciortino F (2002) Nat Mater 1:146
105. Asakura A, Oosawa F (1958) J Polym Sci 33:183
106. Poon WCK (2002) J Phys Condens Matter 14:R859
107. Eckert T, Barsh E (2002) Phys Rev Lett 89:125701
108. Eckert T, Barsh E (2004) J Phys Condens Matter 16:2004
109. Pham KN, Puertas AM, Bergenholtz J, Egelhaff SU, Moussaïd A, Pusey PN, Schofield AB, Cates ME, Fuchs M, Poon WCK (2002) Science 296:104
110. Pham KN, Egelhaff SU, Pusey PN, Poon WCK (2004) Phys Rev E 69:2004
111. Stradner A, Sedgwick H, Cardinaux F, Poon WCK, Egelhaaf U, Schurtenberger P (2004) Nature 432:492
112. Chen S-H, Chen W-R, Mallamace F (2003) Science 300:619
113. Bhatia S, Mourchid A (2002) Langmuir 18:2002
114. Grandjean J, Mourchid A (2004) Europhys Lett 65:712
115. Grandjean J, Mourchid A (2005) Phys Rev E 72:041503

116. Narayanan T, Sztucki M, Belina G, Pignon F (2006) Phys Rev Lett 96:258301
117. Mayer C, Zaccarelli E, Stiakakis E, Likos CN, Sciortino F, Munam A,Gauthier M, Hadjichristidis N, Iatrou H, Tartaglia P, Löwen H, Vlassopoulos D (2008) Nat Mater 7:780
118. Johnson KL (1985) Contact mechanics. Cambridge University Press, Cambridge
119. Liu KK, Williams DR, Briscoe BJ (1998) J Phys D 31:294
120. Lacasse M-D, Grest GS, Levine D (1996) Phys Rev E 54:5436
121. Seth R, Cloitre M, Bonnecaze RT (2006) J Rheol 50:353
122. Dagastine DR, Manica R, Carnie SL, Chan DYC, Stevens GW, Grieser F (2006) Science 313:210
123. Likos CN, Löwen H, Watzlawek M, Abbas B, Jucknischke O, Allgaier J, Richter D (1998) Phys Rev Lett 80:4450
124. Mason TG, Bibette J, Weitz DA (1995) Phys Rev Lett 75:2051
125. Mason TG, Bibette J, Weitz DA (1996) J Colloid Interface Sci 179:439
126. Mason TG, Lacasse M-D, Grest GS, Levine D, Bibette J, Weitz DA (1997) Phys Rev E 56:3150
127. Lacasse M-D, Grest GS, Levine D, Mason TG, Weitz DA (1996) Phys Rev Lett 76:3448
128. Lacasse M-D, Grest GS, Levine D (1996) Phys Rev E 54:5436
129. Buzza DMA, Cates ME (1994) Langmuir 10:4503
130. Adams S, Frith WJ, Stokes JR (2004) J Rheol 48:1195
131. Leng J, Nallet F, Roux D (2001) Eur Phys J E 4:337
132. Fujii S, Richtering W (2006) Eur Phys J E 19:139
133. Mason TG, Bibette J, Weitz DA (1995) Phys Rev Lett 75:2051
134. Mason TG, Bibette J, Weitz DA (1996) J Colloid Interface Sci 179:439
135. Meeker SP, Bonnecaze RT, Cloitre M (2004) Phys Rev Lett 92:198302
136. Larson RG (1999) The structure and rheology of complex fluids. Oxford University Press, New York
137. Landau LD, Lifshitz EM (1986) Theory of elasticity. Pergamon, New York
138. Lubachevsky BD, Stillinger FH (1990) J Stat Phys 60:561
139. Donev A, Torquato S, Stillinger, FH (2005) Phys Rev E 71:011105
140. Chandler D, Weeks JD, Andersen HC (1983) Science 220:787
141. Zwanzig R, Mountain RD (1965) J Chem Phys 43:4464
142. Lorenz B, Orgzall I, Heuer H-OJ (1993) Phys A 26:4711
143. Mecke KR, Seyfried A (2002) Europhys Lett 58:28
144. Phan SE, Russel WB, Zhu J, Chaikin PM (1998) J Chem Phys 108:9789
145. Nisato G, Skouri R, Schosseler F, Munch J-P, Candau SJ (1995) Faraday Discuss 101:133
146. Nisato G, Schosseler F, Candau SJ (1996) Polym Gels Netw 4:481
147. Barnes HA (1995) J Nonnewton Fluid Mech 56:221
148. Aral BK, Kalyon DM (1994) J Rheol 38:957
149. Walls HJ, Caines SB, Sanchez AM, Khan SA (2003) J Rheol 47:847
150. Buscall R, McGowan JI, Mortonjones AJ (1993) J Rheol 37:621
151. Russell WB, Grant MC (2000) Colloids Surf A 161:271
152. Persello J, Magnin A, Chang J, Piau J-M, Cabane B (1994) J Rheol 38:1845
153. Pignon F, Magnin A, Piau J-M (1996) J Rheol 40:587
154. Kalyon DM, Yaras P, Aral B, Yilmazer U (1993) J Rheol 37:35–53
155. Kalyon DM (2005) J Rheol 49:621
156. Plucinski J, Gupta RK Chakrabarti S (1998) Rheol Acta 37:256–269
157. Pal R (2000) Colloids Surf A 162:55
158. Princen HM (1985) J Colloid Interface Sci 105:150
159. Denkov ND, Subramanian V, Gurovich D, Lips A (2005) Colloids Surf A 263:129–145
160. Seth J, Cloitre M, Bonnecaze RT (2008) J Rheol 52:1241
161. Gong JP, Osada Y (2010) Adv Polym Sci doi 10.1007/12_2010_91
162. Franco JM, Gallegos C, Barnes HA (1998) J Food Eng 36:89
163. Bertola V, Bertrand F, Tabuteau H, Bonn D, Coussot P (2003) J Rheol 47:1211
164. Meeten GH (2004) Rheol Acta 43:6–16
165. Leighton D, Acrivos A (1987) J Fluid Mech 181:415

166. Louge A (1996) C R Acad Sci Paris IIb:785
167. Holmes WM, Callaghan PT, Vlassopoulos D, Roovers J (2004) J Rheol 48:1085
168. Salmon J-B, Bécu L, Manneville S, Colin A (2003) Eur Phys J E 10:209
169. Manneville S, Bécu L, Colin A (2004) Eur Phys J 28:361
170. Tapadia P, Wang S-Q (2006) Phys Rev Lett 96:016001
171. Degré G, Joseph P, Tabeling P, Lerouge S, Cloitre M, Ajdari A (2006) Appl Phys Lett 89:024104
172. Derks D, Wisman H, van Blaaderen, A, Imhof A (2004) J Phys Condens Matter 16:S3917
173. Cohen I, Davidovitch, B, Schofield AB, Brenner MP, Weitz DA (2006) Phys Rev Lett 97:215502
174. Isa L, Besseling, R, Poon WCK (2007) Phys Rev Lett 98:198305
175. Isa L, Besseling R, Schofield AB, Poon WCK (2010) Adv Polym Sci doi 10. 1007/12_2009_38
176. Khan SA, Schnepper CA, Armstrong RC (1988) J Rheol 1:69
177. Nickerson CS, Kornfield JA (2005) J Rheol 49:865
178. Stokes JR, Telford JH (2004) J Non-Newton Fluid Mech 124:137
179. de Vicente J, Stokes JR, Spikes HA (2006) Food Hydrocolloids 20:483
180. Davies GA, Stokes JR (2008) J Non-Newton Fluid Mech 148:2008
181. Ballesta P, Besseling R, Isa L, Petekidis G, Poon WCK (2008) Phys Rev Lett 101:258301
182. Goyon J, Colin A, Ovarlez G, Ajdari A, Bocquet L (2008) Nature 454:84
183. Marze S, Langevin D, Saint-James A (2008) J Rheol 52:1081
184. Israelachvili J (1991) Intermolecular and surfaces forces. Academic, San Diego
185. Barnes J (1999) J Non-Newton Fluid Mech 81:133
186. Bécu L, Manneville S, Colin A (2006) Phys Rev Lett 96:138302
187. Seth JR (2008) On the rheology of dense pastes of soft particles. Ph.D. thesis, The University of Texas
188. Erwin BM, Cloitre M, Gauthier M, Vlassopoulos D (2010) Soft Matter doi:10.1039/b926526k
189. Picard G, Ajdari A, Lequeux F, Bocquet L (2005) Phys Rev E 71:010501
190. Yamamoto R, Onuki A (1997) Europhys Lett 40:61
191. Varnik F, Bouquet L, Barrat L, Berthier L (2003) Phys Rev Lett 90:095702
192. Plimpton SJ (1995) J Comp Phys 117:1–19
193. Rapaport DC (2004) The art of molecular dynamics simulation. Cambridge University Press, Cambridge
194. Denkov ND, Tcholakova S, Golemanov K, Ananthapadmanabhan KP, Lips A (2008) Phys Rev Lett 100:138301
195. Barrat J-L, Dalibard J, Feigelman M, Kurchan J (eds) (2003) Slow relaxations and nonequilibrium dynamics in condensed matter. Springer, Berlin
196. Manneville S (2008) Rheol Acta 47:301
197. Cipelletti L, Ramos L (2005) J Phys Condens Matter 17:R253
198. Abou B, Bonn D, Meunier J (2001) Phys Rev E 64:021510
199. Bellour M, Knaebel, A, Harden JL, Lequeux F, Munch JP (2003) Phys Rev E 97:031405
200. Cipelletti L, Manley S, Ball RC, Weitz DA (2000) Phys Rev Lett 84:2275
201. Chung B, Ramakrishnan S, Bandyopadhyay R, Liang D, Zukoski CF, Harden JL, Leheny RL (2006) Phys Rev Lett 96:228301
202. Cloitre M, Borrega R, Leibler L (2000) Phys Rev Lett 85:4819
203. Purnomo EH, van den Ende D, Mellema J, Mugele F (2008) Phys Rev Lett 101:238301
204. Derec C, Ducouret G, Ajdari A, Lequeux F (2003) Phys Rev E 67:061403
205. Ramos L, Cipelletti L (2005) Phys Rev Lett. 94:158301
206. Fielding S, Sollich P, Cates ME (2000) J Rheol 44:323

Quantitative Imaging of Concentrated Suspensions Under Flow

Lucio Isa, Rut Besseling, Andrew B. Schofield, and Wilson C.K. Poon

Abstract We review recent advances in imaging the flow of concentrated suspensions, focussing on the use of confocal microscopy to obtain time-resolved information on the single-particle level in these systems. After motivating the need for quantitative (confocal) imaging in suspension rheology, we briefly describe the particles, sample environments, microscopy tools and analysis algorithms needed to perform this kind of experiment. The second part of the review focusses on microscopic aspects of the flow of concentrated 'model' hard-sphere-like suspensions, and the relation to non-linear rheological phenomena such as yielding, shear localization, wall slip and shear-induced ordering. We describe both Brownian and non-Brownian systems and show how quantitative imaging can improve our understanding of the connection between microscopic dynamics and bulk flow.

Keywords Colloidal dispersion · Confocal imaging · Glass transition · Nonlinear rheology

Contents

1 Introduction 164
2 Experimental Methods 168
   2.1 The Colloidal Particles 168
   2.2 Imaging 170
   2.3 Flow Geometries 173
3 Image Analysis 178
   3.1 General Methods 178
   3.2 Locating Particles 179
   3.3 Tracking Algorithms 181

L. Isa, R. Besseling, A.B Schofield, and W.C.K. Poon (✉)
Scottish Universities Physics Alliance (SUPA) and The School of Physics and Astronomy,
The University of Edinburgh, Kings Buildings, Mayfield Road, Edinburgh EH9 3JZ,
United Kingdom
e-mail:wckp@ph.ed.ac.uk

4	Imaging of Systems Under Deformation and Flow	183
	4.1 Disordered Systems	184
	4.2 Ordered Systems and Ordering Under Flow	192
5	Conclusion and Outlook	197
References		198

1 Introduction

Understanding the deformation and flow, or rheology, of complex fluids in terms of their constituents (colloids, polymers or surfactants) poses deep fundamental challenges, and has wide applications [1]. Compared to the level of understanding now available for the rheology of polymer melts, the rheology of concentrated suspensions lags considerably behind. In the case of polymer melts, it is now possible to predict with some confidence flow fields in complex geometries starting from a knowledge of molecular properties [2]. No corresponding fundamental understanding is yet available for concentrated colloids. The reasons for this are as follows [3]; see the illustration in Fig. 1.

In a polymer melt, each chain moves under the topological constraints imposed by many other chains. The number of these constraints, typically of order 10^3,

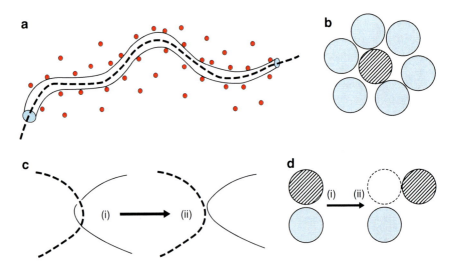

Fig. 1 A schematic comparison between polymer melt rheology and colloid rheology. (**a**) In a polymer melt, a typical chain (*dashed curve*) is constrained by many (in reality, $\sim 10^3$) other chains, here represented by *small circles*. This gives rise to the fruitful mean-field concept of a 'tube' in which the chain has to move. (**b**) In a concentrated colloidal suspension, a typical particle (*hatched*) is surrounded (in 3D) by ~ 10 neighbours. This number is too small for mean-field averaging to be meaningful. (**c**) Large deformations in polymer melts, such as the process (i)→(ii), involves breaking covalent bonds, and so do not ordinarily occur. (**d**) There are no covalent constraints on order unity deformations, such as (i)→(ii), in a colloidal suspension

is sufficiently large that a mean field picture, the so-called 'tube model', can be successfully applied [1]. Moreover, the topological entanglement between chains means that the breaking of covalent bonds is needed to impose large deformations, so that strains often remain small ($\ll 1$). In contrast, the maximum number of neighbours in a (monodisperse) suspension of spheres is about 10, so that 'mean-field' averaging of nearest-neighbour 'cages' will not work, and local processes showing large spatio-temporal heterogeneities are expected to be important. Moreover, no topological constraints prevent the occurrence of strains of order unity or higher, so that very large deformations are routinely encountered. These two characteristics alone render suspension rheology much more difficult. Added on top of these difficulties is the fact that concentrated suspensions are in general 'non-ergodic', i.e. they are 'stuck' in some solid-like, non-equilibrium amorphous state. On the other hand, in a monodisperse system or in a mixture with carefully chosen size ratios, highly-ordered (crystalline) states can occur. In either case, any flow necessarily entails non-linearity (yielding, etc.). Moreover, a suspension is a multiphase system, so that the relative flow of particles and solvent can, and often does, become important. This complication does not arise in polymer melts. Finally, compared to the considerable effort devoted to the synthesis and experimental study of very well characterized model materials in polymer melts, corresponding work in colloid rheology remains relatively rare. It is therefore not surprising that the rheology of concentrated suspensions is not nearly as well understood as that of polymer melts.

Advances in this field will pay rich dividends – the successful processing and application of concentrated colloids more often than not depends on understanding, tuning and exploiting their unique flow properties [4]. But to build and validate predictive theories of bulk rheological properties, we need microstructural information. A 'case study' of this claim comes from recent work on the qualitatively distinct rheologies of so-called 'repulsive' and 'attractive' colloidal glasses.

Colloidal glasses are concentrated suspensions in which long-range diffusion effectively vanishes [5]. These dynamically-arrested amorphous states have finite shear moduli in the low-frequency limit. In slightly polydisperse hard-sphere like suspensions, the transition from an 'ergodic fluid' (where long-range diffusion is possible) to a glass occurs at a particle volume fraction of $\phi \approx 0.58$ [6, 7]. The cause of dynamical arrest is crowding. As ϕ increases, each particle spends longer and longer being 'caged' by its nearest neighbours, until at $\phi \approx 0.58$, the lifetime of these cages becomes longer than any reasonable experimental time window. Each particle can still undergo Brownian motion within its nearest-neighbour cage, but its root mean-square displacement saturates at just over $0.1a$ (where a is the particle radius) [7], this quantity being a measure of the time-averaged fluctuating 'cage size'. When oscillatory shear strain is applied to a hard-sphere colloidal glass, yielding occurs in a single-step process at a strain amplitude of just over 10% [8]. This has been interpreted in terms of 'cage breaking': the glass yields when nearest-neighbour cages are strained beyond their 'natural' (thermally induced) deformation.

When a strong enough short-range attraction is present (where 'short' means a few per cent of a), a second type of glassy state occurs, the 'attractive glass', in which arrest is due to particles being trapped by nearest neighbour 'bonds' [9].

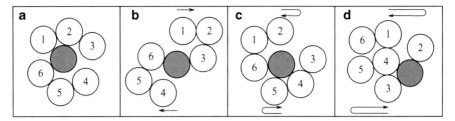

Fig. 2 Cartoon of two different yielding mechanisms in an attractive glass under oscillatory shear (motion indicated by *arrows*). (**a**) Un-sheared configuration of particles at time zero. The shaded particle's bonding neighbours are particles 1, 2 and 6 and topological neighbours are particles 1–6. At small shear amplitude γ_0 the particles retain the same topology during shear. (**b**) Above the lower yield strain, bonds break. (**c**) After one period of oscillatory shear, the shaded particle retains the same topological neighbours but different bonding neighbours, cf. (**a**). (**d**) If the strain amplitude exceeds the higher yield strain, the identity of the topological neighbours around the shaded particle is changed after one period of shear, cf. (**a**), the glass is melted and becomes liquid-like

It turns out that, under oscillatory strain, an attractive glass yields in a two-stepped process [10]. The first step occurs at a strain amplitude of a few per cent, matching the range of the interparticle attraction, and then at a second step when the strain amplitude reaches a few tens of per cent. The first step has been interpreted as the breaking of interparticle 'bonds'; see Fig. 2. Once these are broken many times, however, each particle 'realises' that it is still in a topological cage formed by its neighbours. A second, 'cage breaking', step is still necessary for complete yielding to occur. This 'two step' signature is discernible in other rheological tests applied to attractive colloidal glasses [11].

For the purposes of this review, the important point is that the explanation proposed for the observed qualitative difference between repulsive and attractive glass rheology has been couched in microstructural terms; see Fig. 2. Thus, conclusive validation of these explanations depends on studying yielding at the single-particle level. Such investigation can, in principle, be performed using computer simulations. But simulating particles under flow with realistic system sizes and including hydrodynamic interactions is a major challenge, although progress is being made (see the useful introduction in [12] and references therein). Experimentally, the situation is also difficult. Until comparatively recently, the method of choice for obtaining microstructural information in colloids is scattering (X ray, neutron or light), whether in quiescent or flowing systems. But scattering can provide information only on the average structure; obtaining local information is difficult, if not impossible. The ideal method for gaining local information is direct imaging.

The appropriate imaging method of imaging colloids depends, of course, on a number of factors, principally perhaps the size and concentration of the particles concerned. The size imposes a basic constraint, so that, for example, visible light can only be used to image particles (at least in the 'far field') with diameter $\gtrsim 300$ nm. The concentration is also an important factor, because any multiple scattering by concentrated systems will degrade image quality. In this review, we focus on the use of *confocal* optical microscopy to study colloidal systems under flow. Confocal imaging rejects out-of-focus information using suitably placed pinholes, thus permitting the study of concentrated systems. While the use of confocal microscopy to study quiescent concentrated colloids is, by now, well established [13–16], the extension to flow is relatively recent. The quantitative use of this methodology depends on developments in both hardware and software. First, to minimise image distortion, *fast* confocal scanning is necessary; we review recent developments in this area in Sect. 2.2. Moreover, sample geometries are required for simultaneous flow/rheometry and imaging, which we review in Sect. 2.3. In terms of software, algorithms must be developed to track particles from image sequences distorted by flow; such algorithms are explained in Sect. 3. After discussing these methodological issues, we turn to review a number of applications (Sect. 4).

To date, the use of quantitative confocal microscopy to image concentrated colloids under flow has been applied mostly to the study of very well characterised, model systems, particularly systems of particles that interact as more or less perfect hard spheres (HS). The focus on model systems is partly dictated by the desire to obtain as high quality images as possible, which requires the matching of the refractive indices of the particles and the suspending medium (see Sect. 2.1). The degree of control necessary, both in the particle synthesis and in solvent choice, is only really achievable in model systems. The reason for choosing to study HS is that these represent the simplest possible (classical) strongly-interacting particles. The properties of a quiescent collection of HS are well understood in statistical mechanical terms, much of this understanding having been obtained in the last few decades through the study of model hard-sphere colloids. It is therefore natural to extend such studies into the area of *driven* systems.

We first briefly summarize the quiescent behaviour of ideal HS. As first shown via numerical simulations [17, 18], below a volume fraction of $\phi_F = 0.494$, the thermodynamically stable phase is a fluid of colloidal particles. Long-ranged order is absent and the particles are able to explore all available space (the system is ergodic). For $0.494 < \phi < 0.545$, the equilibrium state is a coexistence of a fluid phase and a crystalline phase. At $\phi_C = 0.545$ the entire system must be crystalline to minimise the free energy, and this remains the case up to the (crystalline) closed packing fraction of $\phi_{CP} = 0.74$. This behaviour was largely confirmed by the experiments of Pusey and van Megen [6] using sterically-stabilised poly-methyl-methacrylate (PMMA) particles (for which see the next section). But, they also found that their system failed to crystallise for volume fractions $\phi \gtrsim \phi_G = 0.58$, remaining 'stuck' in a non-ergodic (or glassy) state up to the maximum possible concentration for amorphous packing (which, for monodisperse HS, is $\phi_{RCP} \simeq 0.64$). We have already introduced such colloidal glasses. Our review of microscopic phenomena will cover both the flow of amorphous states (Sect. 4.1) and of ordered (or ordering) states (Sect. 4.2) of concentrated colloids. While our focus will be on hard-sphere colloids, we will also mention various imaging studies on the the flow of non-Brownian suspensions.

2 Experimental Methods

2.1 The Colloidal Particles

One of the key ingredients for high-quality confocal imaging of concentrated suspensions is a colloidal system that allows identification of the individual spheres, does not suffer from rapid photobleaching and gives good depth penetration into the bulk of the dispersion. Fluorescently-labelled PMMA particles in hydrocarbon solvents show these characteristics.

The synthesis of hard-sphere PMMA particles was first described by Barrett [19] in 1975 and subsequently by Antl et al. [20] in 1986. It is a two-stepped dispersion polymerization reaction, yielding particles made of PMMA cores kept stable by a thin ($\simeq 10$ nm) outer layer of poly-12-hydroxystearic acid chains, which act as a steric barrier to aggregation. In the first step of the polymerization the spheres are made by growing PMMA chains in solution, which become insoluble when they reach a certain size. At this point they come out of solution and clump together to form particles that are kept stable by physisorbed poly-12-hydroxystearic acid chains. The second step of the preparation involves chemically linking the stabiliser chains to the spheres. These chains are tightly packed on the particle surface [21] and stretch out in good solvents such as various hydrocarbons, which causes the particles to interact as nearly-perfect hard-spheres [22]. The resulting particles can have polydispersities as low as a few percent, and can be made with radii ranging from $\simeq 50$ nm up to and above 1 μm.

Fluorescent labelling of the particles for confocal microscopy may be achieved in three ways. The first involves the use of polymerisable dyes. These dyes have been chemically modified to include a reactive group that can be chemically attached to the particle as they are produced. The advantage of this procedure is that the dye will not leave the particle once it is incorporated. For sterically-stabilised PMMA particles this involves adding a methacrylate group to the dye, and several such procedures [23–27] have been described in the literature. The most commonly used dye is 7-nitrobenzo-2-oxa-1,3-diazole-methyl methacrylate (NBD-MMA) [26, 27], which is excited at 488 nm and emits at 525 nm, while the red end of the spectrum is well served by (rhodamine isothiocyanate)-aminostyrene (RAS) [27].

The second way to dye the PMMA spheres is to add an unreactive species during particle formation. Here the dye has no polymerisable group and is just dissolved in one of the reaction reagents with the hope that it will become incorporated into the growing particle. The advantage of this method is that no chemistry, which may alter the dye's physical properties, is required prior to use; it also allows for a wider range of dyes. The disadvantage is that the dye may leak out when, for example, solvency conditions are changed, or migrate within the particle. This technique was employed by Campbell and Bartlett [28], who examined how four different red dyes affected particle formation. They found optimum properties when using 1,1-dioctadecyl-3,3,3,3-tetramethylindocarbocyanine (DiIC18), which had a significantly slower photobleaching rate than other dyes tested, did not affect particle preparation and did not interfere with the hard-sphere behaviour. However, this dye degrades at temperatures around 100°C and therefore the reaction that chemically links the stabiliser to the spheres cannot be performed [29].

The third method to stain the PMMA is to add the fluorescent dye after particle synthesis. This is achieved by finding a solvent that will dissolve the dye and also be taken up by the particles. Thus, an acetone/cyclohexanone mixture can be used to deliver rhodamine perchlorate dye to preformed PMMA spheres [13]. The advantage of this method is that once a suitable delivery system is found, many possible dyes, and even multiple dyes, may be added to the spheres. The disadvantage is that the solvent mixture may attack the spheres, swell them or alter their physical properties.

An advantage of PMMA spheres is that the polymer may be cross-linked [30]. This is achieved via a molecule with two polymerisable groups which is used to bind chemically all the individual PMMA polymer chains in a particle together into a network. This can stabilise the particles in solvents which would normally dissolve them such as aromatics. An additional benefit is that this method allows the fluorescent dye to be kept within one particular area of the particle, usually its core. The preparation of such core-shell particles [27, 31–33] involves modifying the usual reaction by adding the cross-linking agent and dye at the start of the procedure and creating particles as usual. However, instead of going on to link chemically the stabiliser to the particles, they are cleaned to remove excess dye and then more methyl methacrylate monomer and cross-linking agent are reacted with them to produce an undyed shell. The stabiliser is then chemically attached. Such core-shell particles allow for more accurate detection of the particle centres from microscopy in concentrated systems, as the fluorescent cores are well separated from each other (e.g. [34]).

To achieve good imaging conditions in the bulk of the sample and reduce scattering of the laser and excited light, the refractive index (RI) of the particles and the solvent should be closely matched. For PMMA spheres the RI is around 1.5 and can be matched using a mixture of *cis*-decahydronaphthalene (RI = 1.48) and tetrahydronaphthalene (RI = 1.54) in a ratio of approximately 2:1. In earlier work [6] a decalin-carbon disulphide mixture was used (ratio 2.66:1) but this is problematic due to the volatility and toxicity of the carbon disulphide. The particle RI may also be modified by a few per cent by adding different monomers during preparation [35].

Another concern when studying colloidal dynamics is sedimentation. To counter this phenomenon, solvent mixtures have been sought which not only index-match the colloids but also allow density matching. Adding carbon tetrachloride [36] to the *cis*-decahydronaphthalene/tetrahydronaphthalene system was tried but had only limited success as it enhanced photobleaching and imparted a charge to the particles. Another mixture used consists of *cis*-decalin and cycloheptylbromide (CHB) [36, 37] (or cyclohexylbromide [38]). This also imparts a charge to the colloids, likely due to photo-induced cleavage of the Br–C bond and subsequent solvent acidification, but this may be screened by the addition of salt [39, 40], (partly) restoring the nearly-hard-sphere nature of the system.

For both RI and density matching, it is important to note that the extra solvent components can swell the PMMA particles, and the solvent uptake may take up to weeks to saturate. Time-dependent monitoring is required to ensure that the particle size has stabilised. Otherwise, significant errors in volume fraction estimation may result.

Besides PMMA particles, various other systems have been considered in the literature for use in confocal microscopy. One of the best known is silica spheres. They are prepared by the hydrolysis and condensation of alkyl silicates in ethanol using ammonia as catalyst as first described by Stöber et al. [41]. They can be made fluorescent by adding dyes that have been reacted with silane coupling agents which make them affix to the silica [42–45]. The advantage of this system is that various dyes can be used and added at any time during the synthesis [42]. The spheres are stabilized in organic solvents via a dense layer of organophilic material grafted onto their surface [46–49], and the RI can be matched with the solvent (RI = 1.45 for Stöber silica spheres). However, the main problem with silica spheres is that they are quite dense (reported values range from 1.51 [49] to 2.2 g cm^{-3} [50]) so that sedimentation problems may be severe with all but the smallest spheres.

Water-based fluorescent particles can also be used, but their RI differs considerably from that of water, limiting observation in the bulk. The RI of water can be modified by adding salts [51], but the concentrations required are exceptionally high, making most particles unstable against van der Waals attraction.

A final interesting development is that of quantum-dot-loaded particles. Quantum dots [52] are themselves very small semiconducting colloids (1–10 nm) which fluoresce due to quantum confinement. They can be trapped within a polymer or silica [53] colloid. Since quantum dots photobleach much less than organic dyes, they can be used for experiments involving long-term observation.

2.2 Imaging

Direct imaging dates back to the work of Robert Brown [54], who used it to discover and study Brownian motion, which is *the* defining characteristic of colloids. Subsequently, Perrin [55] used direct imaging to great effect in his Nobel-Prize-winning work on sedimentation equilibrium and diffusion in *dilute* suspensions. However, direct imaging has flourished as a major research tool in colloidal research only in the last few decades, mainly due to the increase in imaging and computing power.

Our main emphasis in this chapter is on high-resolution imaging and reconstruction of the position of all particles in some volume of a concentrated colloidal suspension under flow in real-time. Before reviewing this subject in detail, we point out the use of conventional, low-resolution imaging to track the position of a number of *tracer* particles. Perhaps most important for our purposes here, the well-established technique of particle imaging velocimetry (PIV) [56] can be used to shed light on complicating factors in conventional rheological measurements such as wall slip [57, 58] and flow non-uniformities such as shear banding [59]. Thus this technique has recently been used in a rheometer to elucidate the physics of wall slip in concentrated emulsions under shear [60, 61]. The correlation techniques used in PIV to measure tracer velocity can also be used to track the shear-induced diffusion of tracers in concentrated suspensions of non-Brownian spheres [62, 63]. PIV and related techniques are, of course, limited to transparent samples.

For completeness, we mention other methods for velocimetry that have no requirement for transparency, such as heterodyne light-scattering [64] and ultrasonic velocimetry [65]. The latter has been applied to characterise slip and flow non-linearities in micelles and emulsions [66, 67]. We also mention Nuclear Magnetic Resonance Imaging (NMRI) [68–71], another velocimetry method independent of transparency, which can also provide information on local density. The technique has spatial resolution down to $\sim 20\,\mu$m and has been combined with rheometric set ups to relate velocity profiles to macroscopic rheology [72, 73], to give insight on the occurrence of shear bands [74] and shear thickening [75] in concentrated suspensions.

While all of these techniques give additional insight unavailable from bulk rheology alone, building up a complete picture of colloidal flow requires dynamic information on the single-particle level. We now turn to microscopic methods that give precisely such information. We focus on single-particle imaging in 3D, but mention that the imaging of a single layer of colloids has been used to great effect to study fundamental processes in 2D (e.g. [76–78]). While perhaps somewhat less complex than 3D imaging, 2D imaging nonetheless presents some challenges, e.g. when the imaged objects come into very close proximity [79].

The use of conventional (non-confocal) optical microscopy to study concentrated colloidal suspensions in 3D has been reviewed before [80]. In nearly index-matched suspensions, contrast is generated using either phase contrast or differential interference contrast (DIC) techniques. One advantage of conventional microscopy is speed: image frames can easily be acquired at video rate. Conversely, it has poor 'optical sectioning' due to the presence of significant out-of-focus information, so that particle coordinates in concentrated systems cannot be reconstructed in general, although structural information is still obtainable under special circumstances [81].

Compared to conventional microscopy, confocal microscopy delivers superior 'optical sectioning' by using a pinhole in a plane conjugate with the focal (xy) plane. It allows a crisp 3D image to be built from a stack of 2D images even for somewhat turbid samples, but each 2D image is acquired by scanning, which imposes limits on acquisition speed. The technique has been described in detail before [82].

The use of confocal microscopy to study concentrated colloidal suspensions was pioneered by van Blaaderen and Wiltzius [83], who showed that the structure of a random-close-packed sediment could be reconstructed at the single-particle level. Confocal microscopy of colloidal suspensions in the absence of flow has been recently reviewed [13–16]. We refer the reader to these reviews for details and references. Here, we simply note that this methodology gives direct access to *local* processes, such as crystal nucleation [84] and dynamic heterogeneities in hard-sphere suspensions near the glass transition [34, 37].

Our main interest is the use of microscopy to study the flow of concentrated suspensions at single-particle resolution in 3D. It is possible to use conventional (non-confocal) video microscopy for this purpose [85–88], but the poor optical sectioning hinders complete, quantitative image analysis. Conversely, crisp confocal images in principle permit the extraction of particle coordinates, but due to slow scanning and acquisition rates, early observations in real time (i.e. during shear) produced blurred images that again limited the potential for quantitative analysis [89]. A common solution was to apply shear, and then image immediately after the cessation of shear, both in 2D [90, 91] and in 3D [89, 92–94]. In the next subsection we review developments in confocal microscopy that permit faster acquisition and hence time-resolved 3D imaging of particulate systems under flow.

Confocal Microscopy

Confocal images are built by scanning a laser beam across the field of view and collecting the emitted fluorescent light through a pinhole Laser Scanning Confocal Microscopy (LSCM). Traditionally, the laser is scanned across the specimen by two galvo-mirrors which gives maximum acquistion rates of the order of 1 Hz, depending on image size. Technical advances such as the use of resonant galvo-mirrors, spinning (Nipkow) discs (possibly extended with an array of micro-lenses) and acousto-optic deflectors (AODs) have significantly improved the acquisition rates.

AODs are crystals which act as diffraction gratings. By sending a standing sound wave at radio-frequency across the crystal the local index of refraction is changed, creating a grating which deflects laser beams passing through the crystal. By changing the frequency of the sound wave, the diffraction angle is changed and therefore the field of view can be scanned extremely rapidly. The main problem associated with AODs is that the grating deflects light of different wavelengths by different angles and therefore obstructs the fluorescent light travelling back along the optical path. This is partly resolved by combining an AOD with a galvo-mirror and a slit instead of a pinhole. The galvo-mirror positions the beam in one direction, the AOD scans in the orthogonal direction and the fluorescent signal is detected through the slit. The slit slightly reduces the rejection of out-of-focus light and causes a slight anisotropy in the in-plane point-spread function, but the method offers frame rates $\gtrsim 100$ Hz for 512×512 pixels images.

Another technique that improves the scanning speed is the use of a spinning (Nipkow) disc [82]. This solution operates by illuminating the sample through an array of pinholes printed onto a spinning disc, thus achieving the scanning of the sample during the disc's revolution. Spinning discs can achieve full scanning of the field of view by fractions of a revolution and therefore yield frame rates of hundreds of Hz. Disadvantages of this method are the fixed size of the pinholes, limiting the use of different objectives for optimum operation, and also the strong loss of intensity occurring at the disk. Recently, the last disadvantage has mostly been overcome by using laser illumination in combination with an additional array of micro-lenses, strongly increasing the efficiency and considerably reducing photobleaching [95].

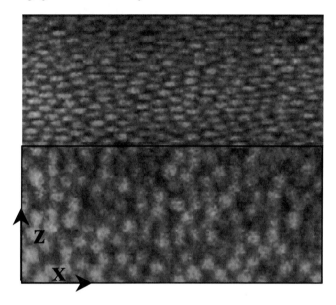

Fig. 3 Reprinted from [96]. Projection of a raw 3D image stack of 1.7-μm PMMA particles, acquired during shear at a local flow rate of $\dot{\gamma} = 3.2 \times 10^{-3}$ s^{-1}, starting 15 μm above the coverslide. The image volume, $x \times y \times z \sim 29 \times 29 \times 15 \mu m^3$ (256 × 256 × 76 voxels), was scanned in ~1 s

Regardless of the scanning principle, the confocal microscope can be operated either in a 2D mode, i.e. capturing time sequences of images at a fixed focal depth z, or in a 3D mode, i.e. capturing image stacks obtained by scanning the sample volume; see Fig. 3. For rapid 3D acquisition, the best method of scanning along z is to use a piezo-element for focus control of the objective. The two operation modes impose different limits on the acquisition rates for successful quantitative imaging (see Sect. 3.2 and [96]).

2.3 Flow Geometries

Microscope studies of flowing particulate suspensions, and soft matter systems in general, require well-defined flow geometries to facilitate data interpretation and allow, in principle, for a mapping of the observed dynamics to the rheological properties of the system. To date, many studies have employed flow cells which impose the deformation (rate) but do not have the ability to measure directly the (shear) stress. Recently this limitation has been overcome by combining confocal imaging with a rheometer or by inferring (indirectly) stress from pressure drops in channel flows.

In all geometries, wall properties play a crucial role in the application of shear. The most direct manifestation of wall effects is that of (apparent) slip in many complex fluids, in particular colloid gels or pastes driven along smooth surfaces

[60, 61, 97]. While the physics of slip in suspensions and pastes is an interesting topic in its own right (see Sect. 4.1), in most flow geometries the practical goal is to minimise slip and transfer shear to the suspension. The remedy for wall slip differs between various systems; for moderate to very dense particle suspensions, a coating consisting of a sintered monolayer of similar size particles generally provides stick boundary conditions.

Parallel Plate Shear Cells

Planar shear, or planar Couette flow, is simply implemented by placing the material between two parallel plates much larger than their separation z_{gap}, and translating them relative to each other. This can be achieved either by fixing one of the plates and moving the other one, or by moving them in opposite directions so that, in the laboratory frame, the zero velocity plane is situated somewhere between the two plates. For a Newtonian fluid without slip and sufficiently far from the edges of the plates (a few z_{gap}), the shear rate in the fluid is constant, $\dot{\gamma} = (v_T - v_B)/z_{gap}$, with v_T (v_B) the top (bottom) plate velocity. In the following, we denote the shear velocity direction by x, the vorticity direction by y and the velocity gradient direction by z.

By construction, parallel plate shear cells can only achieve finite strains after which the direction of motion must be reversed; they are thus particularly suited for oscillatory strain studies. However, in some designs, extremely large strains are possible (~20–50) so that, with constant drive velocity, steady shear is effectively achieved in each half-cycle. Using a microscope glass (cover) slide as the bottom plate allows imaging of the suspension during shear. An important requirement is that the plates should be strictly parallel; this condition is particularly stringent for small plate separations, e.g. [98]. Non-parallel plates induce non-uniform shear as well as drifts in the sample due to capillarity effects. Other considerations relate to component weight, mechanical resonances, the driving motor and minimising sample evaporation. Depending on the desired operation range, compromises may be required to optimise either long-time stability or high-frequency behaviour.

Planar shear cells were initially coupled to conventional microscopes [85, 86] or scattering set ups [86, 99–101]. With the advance of confocal imaging they are now preferentially used in combination with an inverted confocal microscope.

Cohen et al. [102, 103] used a simple design with a movable microscope cover slide as lower plate and fixed top plate. The maximum plate separation was 100 μm and they were parallel to within 1μm. The lower plate was driven by a piezoelectric actuator with displacements up to 90 μm at frequencies ≤100 Hz. The sample between the plates was in contact with a reservoir of un-sheared bulk suspension.

A similar design but allowing for larger strains and higher shear rates is described by Solomon et al. [94]. Two tilt goniometers allow one to tune the parallelism of the plates; a gap of 150 μm is set by a linear micrometer. Oscillatory shear was produced by applying a sinusoidal displacement with a linear stepper motor. The shear rates were in the range 0.01–100 s^{-1} and strain amplitudes in the range 0.05–23.

A cell optimised for slow shear and large amplitudes was developed by the Edinburgh group [96, 104]. The gap size ranged from $\sim$200 to 1000 µm with plates parallel to $\pm$5 µm over the shear region. The top plate was driven at 0.05–10 µm s^{-1} by a mechanical actuator with magnetic encoder. The maximum translation was $L_s \sim$1 cm, allowing steady shear up to a total strain $\Delta\gamma = L_s/Z_{\text{gap}} \gtrsim 1000\%$. The cell could be operated either with the bottom plate fixed or with the plates counter-propagating via an adjustable lever system to tune the zero velocity plane.

A recent, 'state of the art', parallel plate cell is the design by Wu et al. [98], which combines high mechanical stability and a modular construction. The plate separation ranges between 20 and 200 µm, and a special pivot system was designed to align the plates parallel to the highest degree. A piezo-stepper motor provided plate velocities ranging from $\sim$2 µms^{-1} to 10 mm s^{-1}. As in the Edinburgh design, the relative plate motion can reach up to $\sim$1 cm, allowing for large accumulated strain.

Rotational Geometries

Application of continuous shear is achieved in rotational geometries such as cylindrical Couette, plate–plate or cone–plate devices, the standard geometries used in traditional rheology [105]. Of these, cone–plate and plate–plate geometries are most suitable for microscopic observation. Various ways of conventional (microscopic) imaging in such geometries have been used, including an early direct observation of crystallisation in a plate–plate rheometer [106] and a microscopic study of a confined, charged suspension under continuous shear in a rotational plate–plate set-up [87]. Another technique is to image tracer particles in the system from the side of a cone–plate or plate–plate geometry to obtain the deformation or velocity profile, either qualitatively [107] or quantitatively [61].

For completeness we also mention geometries used to study non-Brownian suspensions, where the typical dimensions are much larger (particle radius $a \gtrsim 50$ µm). One is an annular channel formed by two concentric glass cylinders where a ring-shaped top plate drives the suspension [108], the other consists of a special truncated counter-rotating cone–plate cell [63]. Both set ups allow for imaging in multiple directions using different camera positions. As an exponent of imaging with multiple cameras, we mention the recent work in [109], where two orthogonally positioned cameras were used simultaneously to image tracers in an immersed granular packing sheared in a cylindrical Couette cell. Analysis of the data from the two cameras provided the full 3D trajectory of the tracers.

The first combination of a rotational shear cell with confocal microscopy was described in [110, 111]. It consists of a cone–plate, similar to Fig. 4b, where the cone and the plate (a large glass cover slide) could be rotated independently. In contrast to Fig. 4b, the objective was located at a fixed radial position where the gap height was 1.7 mm. This exceeds the working distance of high numerical aperture objectives so that only the lower part of the gap could be imaged at single-particle resolution, a general limitation of any large-gap geometry. The shear rates obtained were in the range of 10^{-2} to 10^2 s^{-1}. The height of the zero velocity plane could be adjusted

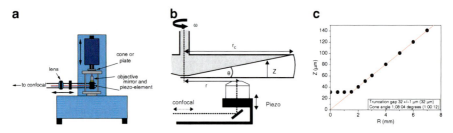

Fig. 4 Reprinted from [96]. (**a**) Schematics of the confocal rheoscope of the Edinburgh group [96]. The *top arrow* marks translation of the rheometer head to adjust the geometry gap, the *horizontal arrow* indicates translation of the arm supporting the objective to image at different radial positions r. (**b**) Close up of the central part of the rheoscope, similar to the cone–plate imaging system of Derks [111] except that in the latter the lower plate can also be rotated, while in the former the microscope objective radial position r can be varied. (**c**) Gap profile of a 1° cone–plate geometry, measured in the confocal rheoscope with fluorescent particles coated on both surfaces

by tuning the ratio of the cone and plate rotational speed, while keeping the shear rate constant. Due to the weight of the cone and plate, oscillatory shear experiments were limited to low frequencies. The main limitation of this set up is the acquisition rate of the confocal scanner: particle tracking could be achieved only near the zero velocity plane and in systems where the out-of-plane motion was slower than the acquisition rate, such as s colloidal crystal.

A more recent development of confocal imaging in a 'rheometric' geometry, which also permits simultaneous measurement of the rheological properties, is the set up of Besseling et al. [96]; see Fig. 4. Here a rheometer was equipped with a custom-built, open base construction, providing space for imaging optics. An adjustable plate with an imaging slit at the top of the base is covered with a circular cover slide forming the bottom surface of the geometry. This allows for imaging at various radial distances r. The stress-controlled rheometer (AR2000, TA instr.) enables all classical rheological tests either in plate–plate or cone–plate geometry. In practice a cone with an angle of 1° was used; see Fig. 4b,c. The imaging optics under the plate (piezo-mounted objective to scan along z, mirror and lenses) are coupled to a confocal scanner (VTEye, Visitech) to give full 3D imaging capabilities as in a standard microscope mount. A major advantage is the acquisition rate (200 fps) of the confocal scanner, which allows 3D imaging of single particle dynamics up to rates of $\sim 0.1\,\mathrm{s}^{-1}$ and measurements of in the xy plane up to $100\,\mathrm{s}^{-1}$. The surface of the cone and the bottom glass plate can be treated (e.g. coated with a sintered layer of colloids) for the study of slip-related problems.

Capillaries and Micro-Channels

Micro-channels and capillaries are flow geometries that occur in many practical applications. They are also interesting from a fundamental perspective, offering insight on issues such as flow instabilities [112], confinement [113] and particle migration effects [114–117]. When the pressure drop over the channel is measured, it is possible, in principle, to relate microscopic observations to bulk rheological properties (see e.g. [118]), but this is non-trivial as it requires steady, uniform flow along the channel and absence of entrance or confinement effects [119].

From the 1970s, a large number of imaging studies of non-Brownian suspensions flowing in mm- to cm-sized channels have been performed via Laser Doppler Velocimetry [120–123] and NMRI [124, 125], but little information has been obtained at the single-particle level. Optical microscopy experiments on channel flows of colloids have only recently started to appear, often in relation to microfluidics applications [126]. To avoid image distortions, channels with square or rectangular cross sections are preferred to cylindrical capillaries.

Haw [112] studied the jamming of concentrated hard-sphere colloids at the entrance of mm-sized, cylindrical capillaries using conventional microscopy. The channel shape and the imaging method did not allow him to obtain detailed microscopic information.

Confocal studies in the group of Weeks [116, 117] used rectangular capillaries (50 μm × 500 μm) coupled to a syringe as a reservoir to drive suspensions of intermediate volume fractions. Due to the high flow rates involved in their experiments, particle tracking was impossible. Instead the flows were studied using image correlation techniques and intensity measurements. A very similar geometry, but with square channels was used in [127] to study flow of attractive gels of silica particles.

Isa and co-workers [96, 128, 129] used a slightly different geometry to study the flow of very dense colloidal pastes. Figure 5 shows a sketch of the setup: square or rectangular, micron-sized glass capillaries are connected to a glass reservoir at one end and flow is driven by suction at the other end of the channel. The inner walls of the capillaries are either untreated and smooth, or coated with colloids.

The recent development of micro-fabrication techniques (soft lithography) has opened up the possibility of studying flow in microfluidics geometries. Degrè et al. [118] have performed PIV studies of polymer solutions seeded with tracers in micro-fabricated geometries made by adhering a moulded block of polydimethylsiloxane (PDMS) onto a glass cover slide, obtaining channels with a thickness of tens of microns. PDMS has many practical advantages, but swells in most organic

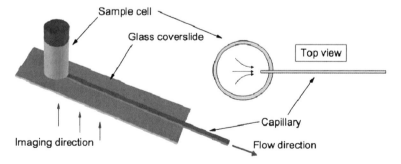

Fig. 5 A possible sample cell for capillary flow. The capillary is not drawn to scale. The construction is placed on the microscope stage and imaged from below

solvents [130], making it incompatible with many colloidal model systems. It also has a relatively low elastic modulus causing deformation under high pressures. A solution to these problems is to substitute PDMS with a photo-curable monomer, as described in [131]. It is clear that soft lithography offers great flexibility in designing channels of almost any geometry, paving the way for high-resolution imaging studies of colloidal flows in a variety of complex micro-environments. Such studies will be relevant for microfluidic applications, as well as for modelling flows in porous and other complex materials.

3 Image Analysis

To extract the maximum amount of quantitative information from (confocal) images, detailed image analysis is necessary. The central components of such analysis are particle location and tracking.

3.1 General Methods

The basic method to obtain coarse-grained information on the flow without details on the dynamics at the particle level is Particle Image Velocimetry (PIV) or related correlation techniques [56]. Using a sequence of images of tracer particles in the suspension or confocal images of the full microstructure during flow, a map of the advected motion between consecutive 2D images, $i-1$ and i, or between parts of these images ('tiles'), is obtained as the shift, $(\Delta X, \Delta Y)$, which maximizes the correlation between these images or regions. By repeating the procedure over a sequence of frames, one obtains the displacement as function of time, $(\Delta X(t_i), \Delta Y(t_i))$. In general, PIV yields a discrete vector field, $\Delta \mathbf{R}(\mathbf{r}_{pqr})$, with displacement $\Delta \mathbf{R}$ in each element $\mathbf{r}_{pqr}$ of a 3D image ($\mathbf{r}_{pq}$ in 2D). In many practical cases however, the flow field has a simpler structure; see the examples in Fig. 6.

For 2D images with uniform motion in the xy plane, the procedure is applied to the entire image, Fig. 6a, giving $\Delta \mathbf{R}(\mathbf{r}, t_i) = \Delta X(\mathbf{r}, t_i) = \Delta X(t_i)$. For advective motion which depends on the position y transverse to the flow, Fig. 6b, the procedure is performed on image strips, yielding $\Delta X(y_q)$ discretised at the strip centres y_q. For 3D images of simple shear, Fig. 6c, the motion is a function of z only. The image stacks are then decomposed in xy slices at different z and the procedure is applied to the individual 2D slices. In channel flow, Fig. 6d, the stacks are first decomposed in xy slices and then each slice is decomposed in y-bins for which the motion is analysed.

These methods have been used to measure the velocity profiles in various dense suspensions. Meeker et al. [60] imaged tracers embedded in pastes from the edge of a cone–plate geometry. A number of studies have been reported in which micro-PIV

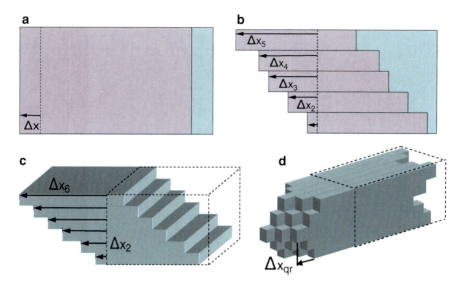

Fig. 6 Reprinted from [96]. Advection profiles in various geometries. (**a**) A uniform 2D shift ΔX across the entire field of view maximises the correlation. (**b**) 2D case where the advected motion is a function of y; the image is then decomposed in bins centred at y_q, each of which is shifted by $\Delta X(y_q)$ to obtain maximum correlation. (**c**) 3D image where the motion is a function of z only. The stack is decomposed in slices centred at z_r, each of which is shifted by $\Delta X(z_r)$ to maximize the correlation. (**d**) 3D case with y and z dependent motion. Decomposition into y and z bins yields the advection profile $\Delta X(y_q, z_r)$

is applied to the flow in channels [118, 132] or to obtain displacement profiles in parallel plate geometries [102].

A related application of the correlation method has been reported by Derks and co-workers [111]. In their study of dense colloids flowing in a cone–plate geometry, a single confocal scan in the velocity-gradient plane ($y - z$ plane in their notation, see Fig. 7) was performed. This was then analysed by shifting and correlating *lines* (along y) between consecutive z-values in the image, rather than performing PIV between consecutive frames. Via this procedure the distortion of the particle images could be quantified, yielding parabolic displacement profiles, Fig. 7, corresponding to linear velocity profiles from which the shear rate was extracted.

3.2 Locating Particles

To go further in the analysis, one first needs to determine the location of the particle centres from the images. The standard method was developed over a decade ago by Crocker and Grier (CG) [133], and has since been used in numerous studies on colloidal dynamics.

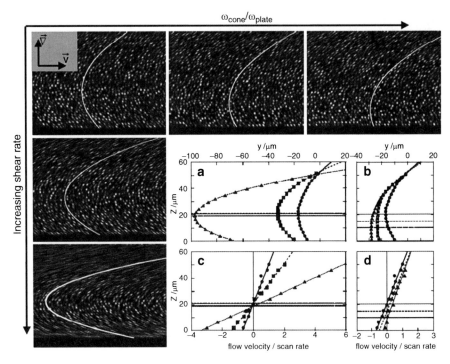

Fig. 7 Reprinted with permission from [111], copyright (2004), Institute of Physics Publishing. Confocal images (yz, 75 μm×56 μm, 512×512 pixels^2) of a colloidal fluid at various shear conditions taken in a counter-rotating cone–plate shear cell [111]; see Sect. 2.3. The applied shear rates are 1.67, 3.36 and 8.39 s^{-1} (*top to bottom*); the ratios of the applied cone to plate rotation speeds are 84, 129 and 175 (*left to right*). Graphs (**a**) and (**b**) show displacement profiles, $y(z)$, measured from these images via cross-correlation of scanned lines. The appropriate profile is overlaid on each image (*white curves*). The velocity profiles (dy/dz) calculated from these displacement profiles are shown in the graphs (**c**) and (**d**). The particle diameter is 1.50 μm

While some of the concepts in their method may also apply to identifying objects with varying shapes [134], the method is primarily aimed at identifying circular (2D) or spherical (3D) objects that appear bright on a dark background. Since experimental images have unavoidable pixel noise and often also undesired intensity modulations, the images are first treated with a spatial bandpass filter, which eliminates long-wavelength contrast gradients and pixel-to-pixel noise. Next, the coordinates of the centres of the features are obtained by locating the local intensity maxima in the filtered images. These coordinates are then refined to a higher accuracy by applying a *centroiding* algorithm, which locates the brightness-weighted centre of mass (centroid) of the particles; with this refinement the coordinates of the particle centre can be obtained with a typical resolution of $1/n$ of the pixel size [133] where n is the particle diameter in pixels. However, the method has some limitations in concentrated systems, as individual particle images may start to overlap. Unless one uses the core-shell particles mentioned in Sect. 2.1, an alternative method to the coordinate refinement may be required. A very useful technique for doing this, based on optimizing the overlap between the measured intensity profile of each particle and the so-called 'sphere spread function', has been described recently by Jenkins and Egelhaaf [135].

In addition to the intrinsic sub-pixel accuracy from the above refinement methods, additional errors arise from particle motion during image acquisition. These errors may be considerable when images are obtained by scanning, i.e. when pixels are not acquired instantaneously. For an acquisition time $1/f_{scan}$ for a 2D image with n lines and particle radius (in pixels) of $\tilde{a}$, the time to image a particle is

$$t_{im}^{2D} = 2\tilde{a}/(nf_{scan}). \quad (1)$$

For a 3D image (a z-stack of 2D slices), the voxel size in the z direction may differ from that in the x and y direction. When the particle radius in z-pixels is given by $\tilde{a}_z$, the acquisition time for a 3D particle image is

$$t_{im}^{3D} = 2\tilde{a}_z/f_{scan}. \quad (2)$$

Typical parameters ($f_{scan} = 90$ Hz in a fast confocal, $n = 256$, $\tilde{a} = \tilde{a}_z = 5$) yield $t_{im}^{2D} \simeq 0.4$ ms and $t_{im}^{3D} \simeq 0.1$ s.

The additional errors resulting from this finite acquisition time are easily estimated. We do so here for HS. The short time diffusion leads to an error $\delta_{HS} = \sqrt{D_s(\phi)t_{im}/2}$. Here $D_s(\phi)$ is the volume fraction (ϕ) dependent short time diffusion constant for HS: $D_s(\phi) = (k_BT)/(6\pi\eta a)H(\phi)$ with $H(\phi) < 1$ a hydrodynamic correction [7, 136–139]. Note that this is an upper bound applicable to HS; for colloids with softer interactions, $\delta < \delta_{HS}$. With a solvent viscosity $\eta \simeq 3 \times 10^{-3}$ Pa s, and frame rates as above, typical values are $\delta_{2D} \simeq 2$ nm and $\delta_{3D} \simeq 35$ nm for a colloid with radius $a = 1\,\mu$m. For 3D imaging, this error exceeds the intrinsic sub-pixel accuracy. Further errors due to flow-induced distortion of the particle image are estimated by comparing the imaging time t_{im} with the time t_f required for the flow to displace the particle over its own diameter. For a flow velocity $\tilde{V}$ (in pixels per second), $t_f = 2\tilde{a}/\tilde{V}$. We consider the particle significantly distorted if $t_{im}/t_f \geq 0.1$. Hence, using (1, 2), the maximum velocities are

$$\tilde{V}_{2D}^{max} = 0.1nf_{scan}, \qquad \tilde{V}_{3D}^{max} = 0.1f\tilde{a}/\tilde{a}_z. \quad (3)$$

Using the above parameters, a typical maximum velocity in 2D is $V^{max} \simeq 500\,\mu$ms^{-1}, while for 3D images with $\tilde{a}/\tilde{a}_z = 1$, we obtain $V^{max} \sim 2\,\mu$ms^{-1}. In both cases, further improvement could be achieved by removing the distortion prior to locating the particles by correlating scanned images or lines at different z [111].

3.3 Tracking Algorithms

Merging particle coordinates from subsequent frames into single-particle trajectories is the optimal route to analyse colloidal dynamics. However, in some cases this is difficult [140], but quantitative information may still be obtained via alternative methods. For example, Breedveld et al. [62, 140, 141] measured the distribution of all possible displacement vectors in images of tracer particles in dense non-Brownian suspensions. From these distributions, the contribution due to cross correlations, corresponding to vectors connecting the position of particle i in one frame to that of particle j in the next frame, could be subtracted. From the resulting autocorrelation part of the distribution, the full shear-induced self-diffusion tensor could be obtained. While this method also has potential for analysis of coordinate ensembles obtained from (confocal) microscopy during flow, we will focus on a complete analysis of the particle dynamics via tracking.

Conventional Tracking

The most widely used algorithm to track particles from ensembles of coordinates in consecutive frames is that of Crocker and Grier [133]. It is based on the dynamics of dilute non-interacting colloids. Given the position of a particle in a frame and all possible new positions in the next frame, within a 'tracking range' R_T of the old position, the algorithm chooses the identification with the minimum mean squared frame to frame displacement (MSFD). The algorithm has been used to analyse particle dynamics in a wide variety of 2D and 3D images of quiescent systems; see, e.g. [15, 16]. Its main limitation is that, when the particle motion between frames is excessive, misidentifications can occur. Such motion can be due to diffusion, flow-advection or both.

In quiescent systems, the ability to track particles between frames is limited by experimental constraints such as the acquisition rate of the (confocal) microscope, image dimensionality, particle size and concentration, and solvent viscosity. After locating the particles, the relevant quantity which relates to the tracking performance of the CG algorithm is the root mean squared frame-to-frame displacement relative to the mean interparticle distance ℓ. If particles move on average a substantial fraction of ℓ, then the algorithm starts to misidentify them. This has recently been quantified [96] by testing the CG algorithm on computer-generated hard-sphere or hard-disk ensembles at different concentrations. These tests showed that the algorithm can handle larger mean squared displacements (MSDs) for more concentrated systems, since the higher concentration prevents particles from coming into close proximity of each other between frames.

The tests were also performed on computer-generated data in which additional uniform or non-uniform motion was added, to study how far the CG algorithm could be pushed beyond its original design parameters. For uniform motion, CG tracking was as successful as in the quiescent case for small drifts but failed for drifts of the order of half the particle-particle separation. For non-uniform (linear shear) flows with small strains between frames the identification worked correctly, but large non-uniform displacements caused major tracking errors.

Correlated Image Tracking

Some important limitations to particle tracking in 2D or 3D images with large drift or non-uniform motion (see also [142]) can be overcome by the method of correlated image tracking (CIT), described in detail in [96]. The main extension compared to conventional (CG) tracking is that, prior to tracking, the time- and position-dependent advective motion is obtained from a PIV-type correlation analysis as in Sect. 3.1. This advected motion is then subtracted from the raw particle coordinates, shifting them to a 'locally co-moving' ('CM') reference frame where the particles can be tracked with the CG algorithm.

Because particle coordinates are distributed continuously, the advection profile $\Delta \mathbf{R}(x_p, y_q, z_r, t_i)$ from the PIV analysis is first interpolated to obtain a *continuous* profile $\Delta \mathbf{R}(x, y, z, t_i)$. Using the latter, the transformation of the position $\mathbf{r}_k(t_i) = [x_k(t_i), y_k(t_i), z_k(t_i)]$ of particle k in the laboratory frame to its position $\bar{\mathbf{r}}_k(t_i)$ in the CM reference frame is

$$\bar{\mathbf{r}}_k(t_i) = \mathbf{r}_k(t_i) - \sum_{j=1}^{i} \Delta \mathbf{R}(\mathbf{r_k(t_i)}, t_j), \qquad (4)$$

with $\Delta \mathbf{R}(\mathbf{r}_k(t_i), t_j)$ the *past* motion between frame j and $j-1$, at the *current* particle location $\mathbf{r}_k(t_i)$. In the CM frame, the average particle motion is essentially zero over the entire image. Therefore the classic CG algorithm tracks particles successfully in this reference frame. The limitations on the CG tracking performance are essentially the same as discussed in Sect. 3.3 for quiescent systems. Finally, the resulting trajectories can be restored in the laboratory frame by inverting 4.

The main limitation to CIT originates from the failure of the image correlation procedure when the *relative* particle motion between frames is excessive, rather than from failure due to large absolute shifts between images [96]. This apart, CIT has the same limitations of CG tracking with significant failure for mean squared frame-to-frame displacements in the CM reference of $\sim(0.3\ell)^2$ in 2D. Two direct applications of CIT will be reviewed in Sect. 4.1.

4 Imaging of Systems Under Deformation and Flow

This section reviews the application of (confocal) imaging to study the flow response of concentrated suspensions, with emphasis on quantitative analysis as described in the previous sections. We separately discuss disordered systems and systems in which order is present either before or as a result of flow. In each case, the focus is on Brownian systems, but we briefly review non-Brownian systems at the end of each subsection.

4.1 Disordered Systems

Slow Glassy Flows

A significant motivation for recent developments in imaging concentrated colloidal suspensions under flow is to investigate the behaviour of samples that are so dense that structural rearrangements are arrested in the quiescent (unsheared) state, i.e. colloidal glasses [5]. Elucidating the mechanisms for the deformation and flow of colloidal glasses is currently one of the 'grand challenges' in soft matter science. Much insight to this problem can be expected to come from the detailed study of model systems. The simplest model systems are concentrated hard sphere suspensions, which are glassy at volume fractions $\phi \gtrsim 0.58$. (This claim has recently been disputed [143]; but however this controversy eventually resolves, it remains true that, for almost all practical rheological purposes, amorphous colloidal hard-sphere suspensions behave as soft solids for $\phi \gtrsim 0.58$.)

The first time resolved studies of the flow of HS colloidal glasses at single particle level have been described in [104] and [96, 97]. Both the 3D particle dynamics and global flow were investigated in steady shear under various boundary conditions using both a planar shear cell and 'confocal rheoscope' [97, 104] (geometries with uniform stress in the flow gradient direction z). Standard fluorescent PMMA-PHS particles ($a = 850$ nm, $\phi = 0.62$) in a charge-screened, CHB-decalin mixture were used for observations at the particle scale; for measurements on a global scale, fluorescent PMMA tracers in a host suspension of non-fluorescent, smaller PMMA colloids were used.

We first focus on the phenomenology observed for flow between plates coated with colloid, where the micron scale roughness provides stick boundary conditions. At high concentrations $\phi \gtrsim 0.62$, the HS glasses shear non-uniformly, exhibiting rate-dependent shear localization [97, 104, 144] which is most pronounced close to the yield stress: at the lowest *applied* rates $\dot{\gamma}_a$, the local shear rate $\dot{\gamma}$ near (one of) the walls considerably exceeds $\dot{\gamma}_a$ and vanishes smoothly into the bulk where the system moves as a solid [144]. This continuous decrease of the flow rate differs from observations in other yield stress fluids, e.g. [145, 146], possibly due to thixotropic effects in those systems. The characteristic length, along z, characterizing the decrease of $\dot{\gamma}$, starts from ~25 particle diameters and increases with the applied rate or global stress, i.e. the wall 'fluidization' propagates into the sample and eventually leads to fully linear flow profile at the largest rate. Recent theories and simulations [147, 148] have suggested spatial correlations as well as boundary effects in the 'fluidity' of soft glassy materials near yielding, in line with these experimental observations, as well as with observations in confined emulsions under flow [119].

The microscopic dynamics inside a shear band was then studied by focussing on a $30 \times 30 \times 15\,\mu m^3$ volume (~3,000 particles) the bottom layer of which was ~10 particles away from the wall. A series of undistorted 3D snapshots of the entire microstructure was acquired, as shown in Fig. 3, from which 3D particle positions and dynamics were obtained as described in Sects. 3.2 and 3.3. Particle tracking using the new CIT method allowed one to study the flow for *local* shear rates up

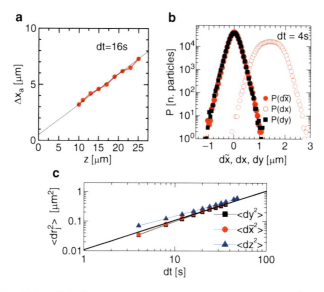

Fig. 8 Reprinted from [96]. 3D analysis of a sheared glass at $\dot{\gamma} = 0.019\,\mathrm{s}^{-1}$. (**a**) (Redbullet) accumulated displacement $\Delta X_a(z, \Delta t)$ (see Fig. 6) from image correlation over $\Delta t = 16\,\mathrm{s}$ (4 frames). *Line*: linear fit giving an accumulated strain $d\Delta X_a/dz = 0.28$. (**b**) Distribution of frame-to-frame displacements $P(\Delta x)$ and $P(\Delta y)$ from CIT with coordinates restored in the laboratory frame. Also shown is $P(\Delta \tilde{x})$ of the non-affine x-displacements, using Equation 5 and $\dot{\gamma} = 0.019\,\mathrm{s}^{-1}$. (**c**) The (non-affine) MSD in the three directions. *Line*: $\langle \Delta y^2(\Delta t) \rangle = 2Dt$ with $D = 5.4 \times 10^{-3}\,\mu\mathrm{m}^2\mathrm{s}^{-1}$

to $\dot{\gamma} \simeq 0.05\,\mathrm{s}^{-1}$. The results are shown in Fig. 8, where (a) presents the advected motion $\Delta X(z, \Delta t)$ (from image correlation) for $\Delta t = 16\,\mathrm{s}$ and a shear rate $0.02\,\mathrm{s}^{-1}$ [96]. *Local* velocity profiles such as these are linear on this scale and extrapolate to zero within the resolution, showing the absence of slip in this case. Figure 8b shows the distributions $P(\Delta x)$ and $P(\Delta y)$, which are frame-to-frame particle displacements obtained from CIT. The laboratory frame data for the displacements in the velocity direction (x) show a large contribution from the advected motion. To focus on the more interesting non-affine part of the displacements, $\Delta \tilde{x}$, the z-dependent advected motion was subtracted via

$$\Delta \tilde{x}(t) = x(t) - x(0) - \dot{\gamma} \int_0^t z(t')dt', \qquad (5)$$

with $\dot{\gamma}$ measured from the data. The results for $P(\Delta \tilde{x})$, Fig. 8b, showed that non-affine motion for x and y were very similar. Analysis of the full 3D dynamics further revealed that it is indeed nearly isotropic in all directions, as shown by the long time behaviour of the MSDs for x, y and z in Fig. 8c. The linearity of these curves at long times also indicates that, after a sufficient number of shear induced cage-breaking events, particle dynamics in the system becomes diffusive, with long-time diffusion coefficients varying by <20% in the three directions.

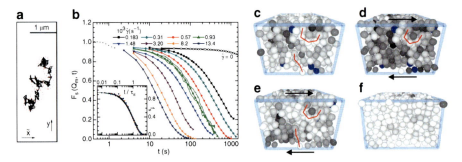

Fig. 9 Partly reprinted from [104]. (**a**) Trajectory of a single particle over 800 s in the velocity-vorticity plane in the co-moving frame as defined in Equation 5 for a sheared glass at a local shear rate $\dot{\gamma} = 9.3 \cdot 10^{-4}$ s^{-1}. (**b**) Selected incoherent scattering functions $F_s(Q_m, t)$, with $\dot{\gamma}$ increasing from *right to left*. The *dashed line* schematises initial relaxation. *Inset*: data collapse using $f_s(Q_m, t/\tau_\alpha)$, with the line showing that $f_s \propto \exp(-t/\tau_\alpha)$. (**c–f**) Snapshots of the structure for $\dot{\gamma} = 0.0015$ s^{-1}, (**c**) at $t_0 = 80$ s after start of imaging. The grey-scale measures the change in local environment of each particle ($\propto C_6^i(t_0, dt)$) over the past $dt = 40$ s. (**d**) at $t_0 = 120$ s and $dt = 80$ s. (**e**) at $t_0 = 120$ s and $dt = 40$ s. *Red lines* show local deformations, *arrows* mark the shear direction. (**f**) A quiescent glass, with $dt = 200$ s

These experiments give information on individual plastic cage-breaking events as well as rate-dependent structural relaxation. The first phenomenon is illustrated by the non-affine motion of a single particle in the $\tilde{x} - y$ plane; see Fig. 9a. Intermittent jumps, reflecting 'plastic' breaking of the particle cages, are observed between periods of 'rattling' in which the cage is deformed. An average accumulated strain of ~10% is found between these events, in reasonable agreement with the yield strain obtained from bulk rheology or light scattering [8, 100]. The changes in the particles' local environment during shear were also studied via the change in the bond order parameters $Q_{6,m}^i$ for each particle i [149] over a time dt, measured by the quantity $C_6^i(t, dt)$, where $C_6 = 0$ reflects no change. Figure 9c–e show snapshots of the sheared microstructure at two times, where the grey-scale is proportional to $C_6^i(t, dt)$. In Fig. 9c, the changes over 6% accumulated strain are shown. Clusters of strong rearrangements, termed 'Shear Transformation Zones' (STZs) in earlier theoretical studies [150], can be observed, where the local strain appears much higher than elsewhere. Over the following 6% accumulated strain, Fig. 9e, STZs appear in different locations while the earlier ones remain essentially locked.

The average structural relaxation was examined via the incoherent scattering function, $F_s(Q, t) = \langle \cos(Q[y_i(t_0 + t) - y_i(t_0)]) \rangle_{i, t_0}$ (using $Q = Q_m \simeq 3.8 a^{-1}$). As shown in Fig. 9b, the long-time dynamics is essentially frozen at rest ($\dot{\gamma} = 0$). At short times and small shear rates, F_s exhibits a plateau corresponding to caging at small accumulated strain. At longer times, F_s decays to zero due to repeated cage-breaking events in line with the diffusive dynamics in Fig. 9c. This decay accelerates strongly on increasing rate. When scaling time by the characteristic (α) relaxation time τ_α, F_s collapsed onto a single exponential curve (see the inset), which directly confirmed the theoretically predicted time-shear superposition principle [151].

The rate dependence of the inverse relaxation time and diffusivity D exhibited a non-trivial scaling: $D \sim 1/\tau_\alpha \sim \dot{\gamma}^{0.8}$. Such 'power-law fluid' scaling contrasts the yield-stress behaviour observed in the global rheology (see [104]) or predictions from Mode Coupling Theory [152]. (For further discussion of these data in the context of Mode Coupling Theory, see the article by Fuchs in this issue.) However, these observations are consistent with a stochastic non-linear Langevin equation treatment [153, 154]. In this approach to hard-sphere dynamics, particles become more deeply trapped in effective free-energy (or, equivalently, entropic) wells as the concentration increases beyond a certain critical value; the effect of applied stress is to lower the entropic barrier that particles have to surmount in order to escape [155]; the predicted shear-induced relaxation time indeed shows power-law scaling with shear rates in an intermediate regime with exponents close to 0.8 [156].

Subsequently, Schall et al. studied the behaviour of dense colloidal packings under very small shear deformations [157]. The packing was formed by sedimentation of silica-spheres ($a = 0.75\,\mu$m) in a lower density solvent, so that the volume fraction varied as a function of height. The shear geometry consisted of a movable, coated bottom cover slide, while a metal grid positioned on top confined the packing within a layer of 42 µm ($\sim$30 particle diameters). The analysis focussed on the microscopic strain variations resulting from either thermal fluctuations or applied strain. From the thermal distribution of strain energies, an estimate for the shear modulus was obtained, but volume fraction gradients complicate the analysis. Large accumulated local strains were observed at long times in the quiescent sample, indicating aging. The behaviour under application of shear is shown in Fig. 10. In this work, the total accumulated strain was $\sim$3% on average, Fig. 10a, well below the typical bulk yield strain in colloidal hard sphere glasses [8, 100]. Thus, it is probable that the data represent creep rather than yielding and flow.

Figure 10b shows a spatial map of the shear strain ε_{yz} in a slice normal to the shear gradient. The authors identified the localized regions with high strain with STZs, the first published images of STZs in colloidal glasses. Interestingly, the highlighted

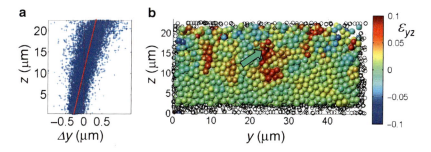

Fig. 10 Reprinted from [157], with permission from AAAS. (**a**) Shear-induced displacements of particles in a colloidal packing between $z = 0$ (the cover slide) and $z = 23\,\mu$m after 50 min of shear (3% average accumulated strain). (**b**) Strain distribution and STZs in a 7 µm thick section in the shear-displacement gradient plane. Particle colour indicates the value of the local shear strain ϵ_{yz} (see colour scale), accumulated over 50 min of shear. The *arrow* indicates a shear transformation zone

region revealed a fourfold symmetric strain field just after its (earlier) formation, consistent with theoretical predictions for a single plastic event. Schall et al. further identified the STZ-cores and, using somewhat uncontrolled assumptions, calculated the STZ formation energy, activation volume and the activation energy. From this it was concluded that STZ formation was mainly thermally activated, although some strain assistance was involved.

Data such as those shown in Figs. 10b and 9c–e show that direct imaging is a very powerful method for studying inhomogeneous local responses to applied stress. In particular, the discovery of STZs in model colloids means that the physics of these systems may have relevance beyond soft matter physics, e.g. in the study of large-scale deformation of metallic glasses [158]. An important open question in the flow of these dense disordered systems is if and how the STZs are related to the previously mentioned (larger) length scale for correlations in fluidity [148], associated with shear localization. This issue deserves strong emphasis in future research.

Particulate suspensions and other concentrated soft materials (e.g. emulsions) often exhibit wall slip during flow along smooth boundaries. This is important in practical applications, and strongly relevant when interpreting bulk flow measurements. Slip in particulate suspensions has been seen mostly for high Péclet numbers (non-Brownian suspensions), both in solid- [159, 160] and liquid-like [161] systems. In a recent study, Ballesta et al. addressed the effect of Brownian motion and the glass transition on slip in hard-sphere colloids using confocal imaging and simultaneous rheology [97] (Sect. 2.3). The main results are shown in Fig. 11. The rheology for rough walls in Fig. 11a shows the traditional change from a shear thinning fluid ($\phi = 52\%$) to a yielding solid ($\phi = 59\%$) associated with a glass

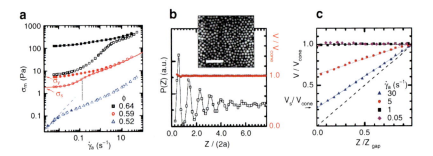

Fig. 11 Reprinted from [97]. (**a**) Measured stress σ_m vs applied rate $\dot{\gamma}_a$ for particles with $a = 138$ nm, at different volume fractions for coated (*black squares, red bullets, blue triangles*) and un-coated geometries (*open squares, red open circles, blue open triangles*). *Dashed line*: linearly viscous behaviour. The *full line* is a flow curve predicted using the model in [97]. *Dotted line*: applied rate at which yielding starts at the cone-edge. (**b**) Density profile $P(z)$ (*open squares*), from 3D-imaging of the $a = 652$ nm system at $\phi = 0.63$, during slip at $r = 2.5$ mm and $\dot{\gamma}_a = 0.01$ s^{-1}. (*Red bullets*) Corresponding velocity profile from particle tracking, showing full plug flow. *Inset*: Slice of one of the 3D-images, showing the first colloid layer. Scale bar: 10 μm. (**c**) Velocity profiles for the $\phi = 0.59$ data in (**a**), in units of the cone velocity $v_{\text{cone}} = \dot{\gamma}_a \theta r$, as function of the normalized hight $z/z_{\text{gap}} = z/\theta r$, for various applied rates $\dot{\gamma}_a$ at $r = 2.5$ mm. The *arrow* marks the slip velocity for $\dot{\gamma}_a = 30$ s^{-1}; the *dashed line* is the behaviour without slip $v = \dot{\gamma}_a z$ as observed for $\phi < \phi_G$

transition at $\phi \sim 58\%$ [8, 162]. For a smooth wall and index-matching solvent (which prevents van der Waals attraction and sticking of particles to the glass) the results for a dense liquid remained unchanged, indicating no slip. This is confirmed by imaging (Fig. 11c, dashed line) and also agrees with the measurements for dense colloidal liquids by Derks et al. [111]. However, for the colloidal glass, a slip branch developed in the flow curve at small rates. This branch was described by a Bingham relation between the slip stress σ_{slip} and slip velocity v_s, $\sigma_{\text{slip}} = \sigma_s + \beta v_s$, associated with solid-body motion of the suspension along the glass plate. This 'solid' structure during slip was shown to extend down to the first particle layer, Fig. 11b and the inset, giving microscopic insight to the physical origin of the parameters σ_s and β in the above Bingham-form. Furthermore, the transition from pure slip at low rates to yielding at high rates was probed, as illustrated in Fig. 11c. It was shown that this transition depended on the local gap, i.e. the radial position r, illustrating the possibility of non-uniform (r-dependent) stress in a cone–plate geometry.

Interestingly, this slip behaviour of hard-sphere glasses is different in nature from that found earlier by Meeker et al. in jammed systems of emulsion droplets [60]. There, a non-linear elasto-hydrodynamic lubrication model, appropriate for deformable particles, could quantitatively account for their observations. It therefore appears that, while slip is ubiquitous for yield stress fluids flowing along smooth walls, the mechanism for its occurrence can be highly system dependent.

Fast Channel Flow of Fluids and Pastes

A different regime of flow in hard-sphere suspensions, either in dense fluids or glasses, occurs when the shear rate becomes comparable to the rate for thermal relaxation of the particles within their cages, which is of the order of the inverse Brownian time. In these situations jamming or shear thickening may start to occur, while in geometries with non-uniform stress, pronounced shear-induced migration may take place. Here we describe recent studies of imaging in this regime, focussing on the specific geometry of capillaries and micro-channels.

The effects of shear-induced migration in intermediate volume fraction hard-sphere suspensions ($\phi < 0.35$) was studied by Frank et al. [116] and Semwogerere et al. [117]. In these experiments, confocal imaging was used to measure both velocity and concentration profiles, $\phi(y)$, of colloids across $50\,\mu\text{m} \times 500\,\mu\text{m}$ glass channels. Flow velocities up to $8\,\text{mm}\,\text{s}^{-1}$ were probed, corresponding to Péclet numbers up to ~ 5000. The gradient in shear rate $\dot{\gamma}(y)$ causes particles to migrate from the boundaries (with large $\dot{\gamma}$) to the centre of the capillary (small $\dot{\gamma}$), resulting in development of a steady state concentration profile beyond a certain 'entrance length'. Frank et al. found profiles (transverse to the flow) where the concentration in the centre increased considerably with the average volume fraction or on increasing flow rate (Péclet number, Pe). The data were interpreted via a model which included a shear rate and local volume fraction dependence of normal stresses in the sample. The rapid rise of these stresses with ϕ and Pe was responsible for the observed behaviour. Semwogerere et al. [117] studied the entrance length over which the flow became fully developed and found that it increases strongly with Pe, in marked contrast to non-Brownian flows for which it is flow-rate independent [163]. The entrance length also decreased with increasing volume fraction and the data could be successfully described within their model.

Few imaging studies have been carried out in higher concentration suspensions. Haw [112] observed intriguing behaviour in traditional microscopy studies of PMMA suspensions ($\phi \gtrsim 55\%$) sucked into mm-sized capillaries with a syringe. Strikingly, the volume fraction of the suspension in the capillary was as much as 5% lower than the initial concentration, depending on the particle size. Such unexpected "self-filtration" was attributed to jamming of the particles at the capillary entrance, leading to higher flow rate of the pure solvent under the applied pressure difference.

Motivated by these experiments, Isa et al. conducted further studies of flows in similar geometries at the single particle level using confocal microscopy [129]. The system consisted of a hard-sphere suspension (PMMA spheres, radius $1.3 \pm 0.1\,\mu\text{m}$) at nearly random close packing, 'a paste', in a 20-particle-wide square capillary. The motion of individual colloids was tracked via CIT and velocity profiles were measured in channels with both smooth and rough walls. Despite the colloidal nature of the suspension, significant similarities with granular flow [164, 165] were found.

The bulk flow curve of the system studied by Isa et al. fitted a Herschel–Bulkley (HB) form for a yield-stress fluid [8] at small to moderate flow rates. The velocity profile predicted from a HB constitutive relation consists of a central, unsheared "plug" and shear zones adjacent to the channel walls, which fits qualitatively with what was observed; see Fig. 12a. At the same time, however, the size of the central

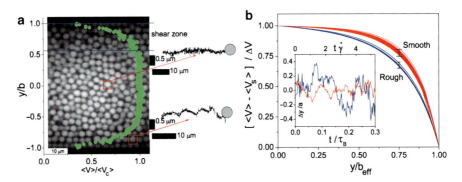

Fig. 12 (a) Velocity profile of a 63.5% volume fraction suspension of PMMA hard-spheres (radius $a = 1.3\,\mu\text{m}$) in a square, glass micro-channel (side $2b = 50\,\mu\text{m}$). The velocity $\langle V \rangle$ is normalized by the centreline velocity $\langle V_c \rangle$. The profile is overlaid onto a confocal image of the suspension in the channel. The velocity profile consists of a central, unsheared plug and of lateral shear zones, whose width is highlighted by the horizontal *blue lines*. The right side of the image shows two examples of particles trajectories in the plug and in the shear zone respectively. The *x* and *y* scales of the trajectories are different and the colloids are not drawn in scale. (b) Reprinted from [129]. Normalised velocity profiles at $z = 17\,\mu\text{m}$ vs y/b_{eff} ($b_{\text{eff}} = b - a$) for a wide range of flow rate. For clarity, fits to the data rather than the raw data are shown; the *error bars* show the spread in the measurements. *Inset*: normalized transverse fluctuations of individual particles in the shear zone for the two boundary conditions, vs normalized time t/τ_B or *local* accumulated strain $t \cdot \partial \langle V(y) \rangle / \partial y = t \cdot \dot{\gamma}(y)$, top axis

plug is expected to decrease with increasing flow rate, vanishing beyond some critical velocity above which the yield-stress fluid is shear-melted everywhere (see, e.g. [105, 166]). But Isa et al. found that the width of the central plug was independent of applied flow rate; see Fig. 12b. This behaviour is analogous to that observed in the pipe flow of dense dry grains [165, 167]. The data for the colloidal flow could be captured by a theory of stress fluctuations originally developed for the chute flow of dry granular media [167, 168]. Presumably, this model can be successfully applied to nearly-close-packed colloidal systems because at high flow rates, interparticle contacts rather than Brownian motion dominate. Direct imaging showed indeed that the trajectory of a particle in the shear zone was determined by 'collisions' with neighbouring particles; see Fig. 12. Moreover, the presence of rough boundaries enhanced such fluctuations, causing wider shear zones.

In more recent work on the same system [113], using smooth walls, it was shown that confinement can induce instabilities in the flow. Upon decreasing the width of the channels from ∼40 to ∼20 particle diameters, the flow developed oscillations above a threshold applied flow rate. Such oscillations consisted of cyclical jamming and un-jamming of the suspended particles and led to filtration effects similar to those reported by Haw [112]. Single-particle imaging was used to demonstrate the presence of a concentration profile across the channel, which was well correlated with the local velocity profile.

Non-Brownian Suspensions

Other recent imaging experiments address the flow of non-Brownian suspensions (i.e. Pe → ∞) at the single particle level. In a series of papers [62, 63, 140, 141], Breedveld et al. investigated the steady shear-induced self diffusion of non-Brownian spheres. They used PMMA particles (radius $a \sim 45\,\mu\text{m}$) in an RI and density-matching mixture (water, zinc chloride, and Triton X-100, viscosity ∼3.4 Pa s) and studied suspensions with volume fractions ranging from 20% to 50%. By studying the correlated motion of a small fraction of the particles which had been dyed (Sect. 3.3), the self and 'off-diagonal' MSDs could be extracted without the use of explicit particle tracking.

For these non-Brownian suspensions, the long-time shear-induced MSDs depend on accumulated strain $\dot{\gamma}\Delta t$ only. The 'long-time' diffusion for $\dot{\gamma}\Delta t > 1$ is due to the chaotic nature of (hydrodynamically mediated) multiple particle collisions. The associated self-diffusion constants D_{xx}, D_{yy}, D_{zz} (in the velocity, vorticity and gradient direction respectively) were found to be anisotropic with $D_{zz}/D_{yy} < 2$, while D_{xx} was almost an order of magnitude larger. The data were in qualitative agreement with hydrodynamic theories and simulations, although quantitative discrepancies remained. In contrast, for the *slowly* sheared glasses in Sect. 4.1 [104], which exhibited a non-trivial rate dependence, $D_{yy}/(a^2\dot{\gamma})$ scales as $\dot{\gamma}^{-0.2}$, with a value ranging from 0.7 at low rates to ∼0.3 at the largest rates, exceeding the result for non-Brownian suspensions. A last notable observation in the non-Brownian system in [140] was a regime of reduced diffusion for small accumulated strains

$\dot{\gamma}\Delta t \simeq \Delta\gamma_{\text{cage}} \ll 1$. While its nature remains unclear, it was shown that the typical strain $\Delta\gamma_{\text{cage}}$ roughly corresponded to the affine deformation δ/a at which particles come into direct contact (on average), which varies with volume fraction as $(\phi_{rcp}/\phi)^{1/3} - 1$, and vanishes as the average surface-surface separation between particles on approaching close packing.

In a very recent study by Wang et al. [109], the 3D motion and diffusion of heavy tracer particles in a granular packing ($\phi \sim 0.6$), immersed in a density and RI matching liquid in a cylindrical Couette-cell was studied. Among many other remarkable results, the data in [109] showed direct evidence that $D \propto \dot{\gamma}$ in non-Brownian systems. Interestingly, the value for shear diffusion in the vorticity direction, $D_{yy}/(\dot{\gamma}a^2) \simeq 0.1$, extracted from the data in [109], was close to the values found by Breedveld for slightly lower density.

Another study on non-Brownian suspensions [169] focussed on a transition from reversible to irreversible particle dynamics in oscillatory shear. While the creeping (Stokes) flow equations for simple shear flow are in principle reversible, particle roughness, three particle collisions or repulsive forces (ignoring Brownian motion) may render the dynamics irreversible. Pine et al. showed that there exists a volume-fraction-dependent critical strain amplitude beyond which the dynamics become diffusive. In a follow-up study [170], they demonstrated that the state below the critical strain is in fact an 'absorbing state', where the particles self-organize into a structure where collisions are avoided. They also demonstrated an increase in the dynamic viscosity of the suspension on crossing the critical strain amplitude, as was also observed previously by Breedveld et al. [140]. An interesting question is whether any relation exists between the absorbing states measured in these non-Brownian systems and the behaviour below the yield strain in colloidal glasses.

To date, flow studies of Brownian and non-Brownian systems have largely been conducted independently. It is clear that much can be gained by detailed comparison of such studies [171] and that single-particle imaging can provide crucial insights in this exercise [129].

4.2 Ordered Systems and Ordering Under Flow

HS colloids undergo an entropy driven fluid-crystal phase transition when the suspension's volume fraction reaches $\phi_f = 0.494$. This behaviour was first confirmed experimentally using quasi-monodisperse hard-sphere-like PMMA colloids [6]. Static light scattering [172] shows (and later, conventional microscopy [81] confirms) that the resulting colloidal crystals consist of hexagonally-packed layers more or less randomly stacked on top of each other; i.e. these crystals have a mixture of face-centred cubic (fcc) and hexagonal close packed (hcp) structures. Starting with a paper in 1998 [173], van Megen and co-workers have published a series of studies in the kinetics of crystallisation in PMMA suspensions using time-resolved dynamic and static light scattering, investigating in particular the effect of polydispersity [174] in some detail.

Gasser and co-workers [84] pioneered the use of confocal microscopy to study the kinetics of colloidal crystallisation. Using buoyancy-matched PMMA particles, they observed nucleation and growth of colloidal crystals with single-particle resolution. These authors identified the size of critical nuclei, 60–100 particles, in rough agreement with computer simulations [175], and measured nucleation rates and the average surface tension of the crystal-liquid interface. For completeness we mention that Palberg and coworkers have pioneered the use of non-confocal microscopy methods to image ordering in highly-deionized suspensions [87, 176], which crystallise at very low volume fractions (10^{-3} or lower).

The relative mechanical weakness of colloidal crystals compared to their atomic and molecular counterparts means that modest shear will have large, non-linear effects on crystallisation as well as crystal structure. In the case of hard-sphere-like PMMA colloids, these effects have been studied using light scattering [177] almost immediately after the discovery of crystallisation in this system [6]. Much more recently, confocal microscopy has been used to image shear effects in PMMA colloids. Below, we focus on reviewing such studies, but we note that imaging studies of crystallisation in other colloids have also been performed (see, e.g. [87]). At the end of this section we have also review briefly experimental work on ordering and ordered structures in non-Brownian suspensions.

Hard-Sphere Colloids

It is known from static light scattering [177], diffusing wave echo spectroscopy [86] and conventional optical microscopy [85] that the application of oscillatory shear to dense colloidal fluids or glasses can drive crystallisation. Rheological studies [178] show that shear-crystallised samples have lower elastic and viscous moduli than their glassy counterparts at the same volume fraction, in agreement with the enhanced effective 'cage' volume in the ordered case, leading to smaller entropic stiffness.

A detailed confocal imaging study of shear-induced crystallisation in hard-sphere colloids was performed recently by Solomon and Solomon [94]. They imaged crystals formed under an oscillatory shear field at a particle volume fraction of 52% in slightly-charged PMMA particles (diameter 1.15 µm, polydispersity 4%) in a planar shear cell (Sect. 2.3). Particles were identified using the conventional Crocker and Grier algorithm but, due to the low acquisition rate (0.6 frames/second), imaging was only possible after the cessation of shear. Consistent with previous work [85, 177], they observed ordering of the initially amorphous suspension into close-packed planes parallel to the shearing surface; see Fig. 13. Upon increasing the amplitude of the applied oscillatory strain, γ, the close-packed direction of these planes was observed to shift from an orientation parallel to the vorticity direction to one parallel to the flow direction, and the quality of the layer ordering decreased. In addition, they studied shear-induced stacking faults and reported their three dimensional structure. For large strain amplitudes ($\gamma = 300\%$), ordering in the flow-vorticity plane only persisted for 5–10 layers, while in the gradient direction the crystal consists of alternating sequences of fcc and hcp stacking.

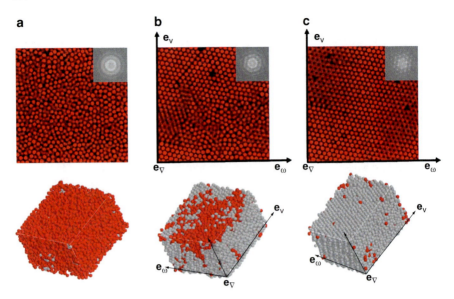

Fig. 13 Reprinted with permission from [94]. Copyright (2006), American Institute of Physics. Confocal images (*top row*, with *insets* showing the Fourier transform) and reconstructed 3D stacks (*bottom row*) showing the time dependent effect of oscillatory shear (frequency = 3 Hz, strain amplitude = 1) on a 48% volume fraction PMMA suspension. Time progresses from *left to right*: quiescent (24 h after preparation, **a**), after 2 min (**b**), and after 5 min (**c**) of shear. In (**b**) and (**c**), unit vectors (*e*) in the velocity (*v*), vorticity (*ω*) and shear gradient (∇) directions are shown. The light grey particles in the 3D renderings (*bottom row*) represent those with a high degree of local crystalline ordering

Once formed, whether it be under quiescent or sheared conditions, colloidal crystals show a complex response to applied stress. In particular, they exhibit banding under simple shear. We have already discussed flow localization in colloidal glasses in Sect. 4.1. The first evidence for the occurrence of this phenomenon in ordered structures was reported in a study of hard-sphere colloidal crystals under steady shear by Derks et al. [111]. They studied the velocity profiles and single-particle dynamics via correlation techniques and particle tracking in a cone–plate system (Sect. 2.3) around the zero-velocity plane. Since particles were locked in their lattice positions and their motion was limited, they could be tracked using the conventional Crocker-Grier algorithm. The close-packed direction in hexagonally-ordered planes was observed to be parallel to the velocity, in agreement with high strain amplitude oscillatory shear results [85, 94]. The study also showed direct evidence that colloids in adjacent layers move in a zig-zag fashion, Fig. 14a, as previously inferred from light scattering [179] and conventional imaging [85]. The velocity profile measured over the accessible z range were linear, but the measured shear rates were much larger than the applied rates, Fig. 14b, indicating considerable shear banding in the sample. In a more recent study [180], Derks et al. analysed in more detail the particle dynamics in the flow-vorticity plane of a sheared

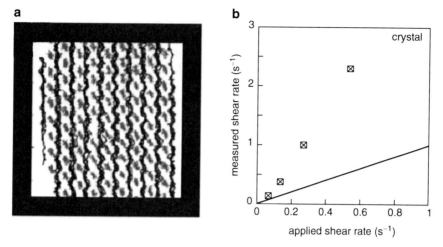

Fig. 14 Reprinted with permission from [111], copyright (2004), Institute of Physics Publishing. (**a**) Superposition of particle tracks in the zero-velocity plane plus one adjacent plane in a sheared colloidal crystal, showing zig-zag motion. The image size is 18.75 μm × 18.75 μm (256 × 256 pixels). The colloids have a diameter of 1.50 μm. (**b**) Measured shear rate vs applied shear rate for the colloidal crystal (points), the continuous line shows equality of these two rates

crystal. They evidenced increasing fluctuations in the crystal layers on increasing rate, eventually leading to shear melting when the short-time mean square displacements due to shear attained a 'Lindemann' value of ∼12% of the particle separation.

More direct evidence of shear banding in colloidal crystals under *oscillatory* shear was found by Cohen and co-workers [102]. In this confocal microscopy study, the oscillatory motion across the gap was measured at various applied amplitudes and frequencies. For low applied deformation, γ_{app}, the authors claim that the crystal is linearly strained at all frequencies, f. At higher γ_{app}, the crystal yields, and two different regimes are observed. At low frequencies, a high-shear band appears close to the upper (static) plate. As the applied strain and frequency increase, the band becomes larger until the whole gap is sheared at $f > 3$ Hz. Moreover, large slip is observed in all cases, with the measured strain always below the applied one. Cohen et al. proposed a model based on the presence of two coexisting, linearly-responding phases to explain their observations. This model is attractive for its simplicity, but questions remain, in particular concerning the lack of bulk strain for small-amplitude, low-frequency data and some significant deviations between the modelled and measured profiles.

Simple shear is, of course, a highly idealised deformation geometry. Motivated partly by the possibility of shedding light on the deformation, fatigue and fracture of metals in more 'real-life' situations, Schall and co-workers carried out a confocal imaging study of a more complex form of deformation: indentation in defect-free colloidal crystals grown by slowly sedimenting silica particles on templated surfaces [181]. Indentation (or nano-indentation in the case of metallic crystals) con-

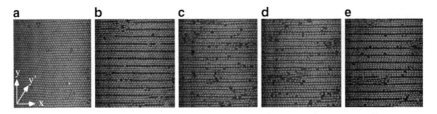

Fig. 15 Reprinted with permission from [103], copyright 2004 by the American Physical Society. Structure of a sheared suspension (diameter 1.42 μm) with strain amplitude $\gamma_0 = 0.38$, frequency $f = 30$ Hz and $\phi = 0.61$. The plate moves in the x direction. (**a**) Confocal micrograph of a sheared suspension forming hcp layers when the gap between the plates $D = 80$ μm. (**b**)–(**e**) Images of the suspension in the buckled state. The gap is set slightly below the height commensurate with confinement of four flat hcp layers. (**b**) An xy image slice of the suspension near the upper plate. (**c**)–(**e**) Slices that are, respectively, 1.3, 2.6, and 3.9 μm below the slice in (**b**)

sists of pressing a sharp tip (diamond for metals, and a needle for colloidal crystals) against a surface to probe its mechanical properties by relating the applied pressure and the measured penetration. It is also a method for the controlled introduction of defects into a crystal.

Schall et al. indented colloidal crystals using a needle with an almost hemispherical tip of diameter 40 μm, inducing a strain field in which the maximum shear strain lies well below the contact surface. The tip diameter, particle radius and crystal thickness in their experiments were chosen to be similar to parameters in typical metallic nano-indentation experiments. The authors discussed their observations using a model that addresses the role played by thermal fluctuations in the nucleation and growth of dislocations.

We have already discussed confinement effects in the channel flow of colloidal glasses. Such effects are also seen in hard-sphere colloidal crystals sheared between parallel plates. Cohen et al. [103] found that when the plate separation was smaller than 11 particle diameters, commensurability effects became dominant, with the emergence of new crystalline orderings. In particular, the colloids organise into "z-buckled" layers which show up in xy slices as one, two or three particle strips separated by fluid bands; see Fig. 15. By comparing osmotic pressure and viscous stresses in the different particle configurations, the cross-over from buckled to non-buckled states could be accurately predicted.

Non-Brownian Suspensions

Shear induced structuring has been predicted in the flow of non-Brownian suspensions (e.g. [182]), but to date there have been few imaging studies. In a series of experiments by Tsai and co-workers [108, 183, 184], laser sheet imaging was employed to study the dynamics of confined glass sphere packings (diameter 600 μm) in an index-matching (non-density matching) solvent in an annular shear cell with load applied from the top shearing surface. Using simultaneous measurements of the

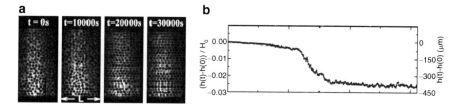

Fig. 16 Reprinted with permission from [108], copyright 2003 by the American Physical Society. (**a**) Time evolution of the structure of an index matched suspension of glass beads in [108], imaged in the flow-gradient plane after the start of shearing at $t = 0$. The observed layering is parallel to the flow direction. (**b**) Fractional volume compaction vs time, based on the change of height of the upper driving boundary. The actual change in microns is indicated at the right

volume (and torque in [183]), they discovered a transition from a disordered packing to an ordered packing under steady shear, Fig. 16a, accompanied by compaction steps, Fig. 16b, and reduction of the shear force. They observed strongly non-linear shear profiles with a rapid reduction of the rate away from the top driving plate, varying significantly with the order of the packing. Due to these non-linear profiles, the delay observed before the ordering transition was strongly dependent on the layer thickness for fixed boundary speed, since global ordering requires an accumulated strain $\Delta\gamma \sim 1$ through the entire sample.

5 Conclusion and Outlook

Recent advances in imaging and data analysis techniques have enabled time resolved imaging and tracking of individual particles in colloidal systems under flow. We have reviewed these techniques, as well as examples of their application, primarily on concentrated systems of hard-sphere colloids. Despite what has been achieved, we believe that this area is only in its infancy. Results already obtained raise many more questions than they answer. Many phenomena and systems remain little explored, and existing results invite further investigation using imaging techniques. For example, computer simulations of crystal nucleation [185] as well as proposed explanations of complex yielding behaviour in glassy states of systems of particles with short-range attraction [10, 11] can both be directly tested using such methods. In other cases, the use of imaging may shed light on well-known and long-studied phenomena that remain incompletely understood, e.g. shear thickening of concentrated colloidal suspensions [75]. The detailed comparative study of distinct but related systems using imaging techniques, e.g. Brownian and non-Brownian suspensions, or particulate suspensions and emulsions, should also be a fruitful area for exploration.

Further technical developments will enhance the power of the methodology. Faster scanning methods will obviously extend the upper limit of flow rates that can be studied. The development of algorithms for identifying [186] and tracking particles in polydisperse systems will significantly extend the range of systems that can be studied – there has been no quantitative study to date of polydispersity effects in suspensions under flow by imaging. Further use of confocal imaging combined with (simultaneous) bulk rheometry will give a wealth of information on the relation between micro-scale dynamics and bulk flow and provide a crucial test ground for various theories. The coupling of 3D imaging with microrheological techniques [187] utilising optical trapping ('laser tweezers' [188]) should also yield many new discoveries and insights; confocal microscopy in two dimensions has already been used to explore the structure of the 'wake' downstream from a probe particle being dragged through a concentrated suspension [189]. Studying flow in microfluidic geometries using time-resolved single-particle imaging is of direct relevance to many emergent lab-on-a-chip type applications. And the parallel use of optical microscopy and other imaging techniques (perhaps especially NMR) to study the same flow system should open up many possibilities.

Finally, the example applications reviewed here pertain mostly to well characterised, model particles. Confocal microscopy can fruitfully be used to image the (static) microstructure of many 'real-life' systems, such as foods [190]. To date, there have been few confocal microscopy and rheological studies of such systems [110]. Given the importance of rheological properties in the processing and utilisation of a wide range of industrial products (controlling the 'mouth feel' of food is but one example), we may expect increasing application of this methodology to applied colloidal systems.

Acknowledgements We would thank Eric Weeks for discussions. Rut Besseling, Andrew Schofield and Wilson Poon were funded respectively by EPSRC grants GR/S10377/01, EP/E030173/1 and EP/D067650. Lucio Isa acknowledges support from EU network MRTN-CT-2003-504712.

References

1. Larson RG (1999) The structure and rheology of complex fluids. Oxford University Press, Oxford
2. Bent J, Hutchings LR, Richards RW, Gough T, Spares R, Coates PD, Grillo I, Harlen OG, Read DJ, Graham RS, Likhtman AE, Groves DJ, Nicholson TM, McLeish TCB (2003) Science 301:1691
3. Cates ME (2003) Annales Henri Poincaré 4 (Suppl. 2), S647
4. Wilson DI, Rough SL (2006) Chem Engin Sci 61:4147
5. Sciortino F, Tartaglia P (2005) Adv Phys 54:471
6. Pusey PN, van Megen W (1986) Nature 320:340
7. van Megen W, Mortensen TC, Williams SR, Müller J (1998) Phys Rev E 58:6073
8. Petekidis G, Vlassopoulos D, Pusey PN (2004) J Phys Condens Mat 16:S3955
9. Pham KN, Puertas AM, Bergenholtz J, Egelhaaf SU, Moussaid A, Pusey PN, Schofield AB, Cates ME, Fuchs M, Poon WCK (2002) Science 296:104
10. Pham KN, Petekidis G, Vlassopoulos D, Egelhaaf SU, Pusey PN, Poon WCK (2006) Europhys Lett 75:624
11. Pham KN, Petekidis G, Vlassopoulos D, Egelhaaf SU, Poon WCK, Pusey PN (2008) J Rheol 52:649

12. Padding JT, Louis AA (2006) Phys Rev E 74:031402
13. Dinsmore AD, Weeks ER, Prasad V, Levitt AC, Weitz DA (2001) Appl Optics 40:4152
14. Habdas P, Weeks ER (2002) Curr Opin Colloid Interface Sci 7:196
15. Semwogerere D, Weeks ER (2005) In: Encyclopedia of biomaterials and biomedical enigineering. Taylor and Francis, New York
16. Prasad V, Semwogerere D, Weeks ER (2007) J Phys Condens Matter 19:113102
17. Wood WW, Jacobson JD (1957) J Chem Phys 27:1207
18. Hoover WG, Ree FH (1968) J Chem Phys 49:3609
19. Barrett KEJ (1975) Dispersion polymerization in organic media. Wiley, London
20. Antl L, Goodwin JW, Hill RD, Ottewill RH, Owens SM, Papworth S, Waters JA (1986) Colloid Surf 17:67
21. Cebula DJ, Goodwin JW, Ottewill RH, Jenkin G, Tabony J (1983) Colloid Polym Sci 261:555
22. Bryant G, Williams SR, Qian L, Snook IK, Perez E, Pincet F (2002) Phys Rev E 66:060502
23. Tse AS, Wu Z, Asher SA (1995) Macromol 28:6533
24. Liu R, Winnik MA, Stefano FD, Vanketessan J (2001) J Polym Sci A Polym Chem 39:1495
25. Tronc F, Li M, Winnik MA, Kaul BL, Graciet JC (2003) J Polym Sci A Polym Chem 41:766
26. Jardine RS, Bartlett P (2002) Colloids Surf A 211:127
27. Bosma G, Pathmamanoharan C, de Hoog EHA, Kegel WK, van Blaaderen A, Lekkerkerker HNW (2002) J Colloid Interface Sci 245:292
28. Campbell AI, Bartlett PJ (2002) Colloid Interface Sci 256:325
29. Hu H, Larson RG (2004) Langmuir 20:7436
30. Pathmamanoharan C, Groot K, Dhont JKG (1997) Colloid Polym Sci 275:897
31. Dullens RPA, Claesson EM, Derks D, van Blaaderen A, Kegel WK (2003) Langmuir 19:5963
32. Dullens RPA, Claesson EM, Kegel WK (2004) Langmuir 20:568
33. Dullens RPA (2006) Soft Matter 2:805
34. Kegel WK, van Blaaderen A (2000) Science 287:290
35. Underwood SM, van Megen W (1996) Colloid Polym Sci 274:1072
36. de Hoog E (2001) Ph.D. thesis, Interfaces and crystallization in colloid-polymer suspensions, University of Utrecht
37. Weeks ER, Crocker JC, Levitt AC, Schofield A, Weitz DA (2000) Science 287:627
38. Leunissen ME, Christova CG, Hynninen AP, Royall CP, Campbell AI, Dijkstra AIM, van Roij R, van Blaaderen A (2005) Nature 437:235
39. Yethiraj A, van Blaaderen A (2003) Nature 421:513
40. Sedgwick H, Egelhaaf SU, Poon WCK (2004) J Phys Condens Mat 16:S4913
41. Stöber W, Fink A, Bohn E (1968) J Colloid Interface Sci 26:62
42. van Blaaderen A, Vrij A (1992) Langmuir 8:2921
43. Nyffenegger R, Quellet C, Ricka J (1993) J Colloid Interface Sci 159:150
44. Verhaegh NAM, van Blaaderen A (1994) Langmuir 10:1427
45. Imhof A (1996) Ph.D. thesis, Dynamics of concentrated colloidal suspensions, University of Utrecht
46. van Helden AK, Jansen JW, Vrij A (1981) J Colloid Interface Sci 81:354
47. Pathmamanoharan C (1988) Colloid Surf 34:81
48. Pathmamanoharan C (1990) Colloid Surf 50:1
49. van Blaaderen A, Vrij A (1993) J Colloid Interface Sci 156:1
50. Vrij A, Jansen JW, Dhont JKG, Pathmamanoharan C, Kops-Werkhoven MM, Fijnaut HM (1983) Faraday Discuss 76:19
51. Hendriks F, Aviram A (1982) Rev Sci Inst 53:75
52. Murphy CJ (2002) Anal Chem 74:520A
53. Ma Q, Wang C, Su X (2008) J Nanosci Nanotech 8:1138
54. Brown R (1866) The miscellaneous botanical works of Robert Brown. R. Harwicke, London
55. Perrin JB (1913) Les atomes. Alcan, Paris
56. Raffel M, Willert C, Kompenhans J (1998) Particle image velocimetry, a practical guide. Springer, Berlin
57. Russel W, Grant M (2000) Colloid Surf A 161:271
58. Buscall R, McGowan JI, Morton-Jones AJ (1993) Colloids Surf A 37:621

59. Olmsted PD (2008) Rheol Acta 40:283
60. Meeker SP, Bonnecaze RT, Cloitre M (2004) Phys Rev Lett 92:198302
61. Meeker SP, Bonnecaze RT, Cloitre M (2004) J Rheol 48:1295
62. Breedveld V, van den Ende D, Tripathi A, Acrivos A (1998) J Fluid Mech 375:297
63. Breedveld V, van den Ende D, Bosscher M, Jongschaap RJJ, Mellema J (2001) Phys Rev E 63:021403
64. Salmon J-B, Manneville S, Colin A, Pouligny B (2003) Eur Phys J Appl Phys 22:143
65. Manneville S, Bécu L, Colin A (2004) Eur Phys J Appl Phys 28:361
66. Bécu L, Manneville S, Colin A (2006) Phys Rev Lett 96:138302
67. Bécu L, Anache D, Manneville S, Colin A (2007) Phys Rev E 76:011503
68. Fukushima E (1999) Ann Rev Fluid Mech 31:95
69. Callaghan PT (1999) Rep Prog Phys 62:599
70. Bonn D, Rodts S, Groenink M, Rafaï S, Shahidzadeh-Bonn N, Coussot P (2008) Ann Rev Fluid Mech 40:209
71. Gladden LF, Alexander P (1996) Meas Sci Technol 7:423
72. Ovarlez G, Bertrand F, Rodts S (2006) J Rheol 50:259
73. Raynaud JS, Moucheront P, Baudez JC, Bertrand F, Guilbaud JP, Coussot P (2002) J Rheol 46:709
74. Huang N, Ovarlez G, Bertrand F, Rodts S, Coussot P, Bonn D (2005) Phys Rev Lett 94:028301
75. Fall A, Huang N, Bertrand F, Ovarlez G, Bonn D (2008) Phys Rev Lett 1:018301
76. Zahn K, Maret G (2000) Phys Rev Lett 85:3656
77. Zahn K, Wille A, Maret G, Sengupta S, Nielaba P (2003) Phys Rev Lett 90:155506
78. Ebert F, Keim P, Maret G (2008) Eur Phys J E 26:161
79. Baumgartl J, Bechinger C (2005) Europhys Lett 71:487
80. Elliot MS, Poon WCK (2001) Adv Colloid Interface Sci 92:133
81. Elliot MS, Bristol BTF, Poon WCK (1997) Physica A 235:216
82. Wilson T (1990) Confocal microscopy. Academic, San Diego
83. van Blaaderen A, Wiltzius P (1995) Science, 270:1177
84. Gasser U, Weeks ER, Schofield AB, Pusey PN, Weitz DA (2001) Science 292:258
85. Haw MD, Poon WCK, Pusey PN (1998) Phys Rev E 57:6859
86. Haw MD, Poon WCK, Pusey PN, Hebraud P, Lequeux F (1998) Phys Rev E 58:4673
87. Biehl R, Palberg T (2004) Europhys Lett 66:291
88. Smith P, Petekidis G, Egelhaaf SU, Poon WCK (2007) Phys Rev E 76:041402
89. Tolpekin VA, Duits MHG, van den Ende D, Mellema J (2004) Langmuir 20:2614
90. Hoekstra H, Vermant J, Mewis J (2003) J Langmuir 21:9134
91. Stancik EJ, Gavranovic GT, Widenbrant MJO, Laschtisch AT, Vermant J, Fuller GG (2003) Faraday Discuss 123:145
92. Varadan P, Solomon MJ (2003) J Rheol 47:943
93. Cohen I, Mason TG, Weitz DA (2004) Phys Rev Lett 93:046001
94. Solomon T, Solomon MJ (2006) J Chem Phys 124:134905
95. Wang E, Babbey CM, Dunn KW (2005) J Microscopy 218:148
96. Besseling R, Isa L, Weeks ER, Poon WCK (2009) Adv Colloid Interface Sci 146:1
97. Ballesta P, Besseling R, Isa L, Petekidis G, Poon WCK (2008) Phys Rev Lett 101:258301
98. Wu YL, Brand JHJ, van Gemert JLA, Verkerk J, Wisman H, van Blaaderen A, Imhof A (2007) Rev Sci Instrum 78:103902
99. Hebraud P, Lequeux F, Munch JP, Pine DJ (1990) Phys Rev Lett 78:4657
100. Petekidis G, Vlassopoulos D, Pusey PN (2003) Faraday Discuss 123:287
101. Petekidis G, Moussaïd A, Pusey PN (2002) Phys Rev E 66:051402
102. Cohen I, Davidovitch B, Schofield AB, Brenner MP, Weitz DA (2006) Phys Rev Lett 97:215502
103. Cohen I, Mason TG, Weitz DA (2004) Phys Rev Lett 93:046001
104. Besseling R, Weeks ER, Schofield AB, Poon WCK (2007) Phys Rev Lett 99:028301
105. Steffe JF (1996), Rheological methods in food processing engineering. Freeman, East Lansing, MI, USA
106. Rodriguez B, Wolfe M, Kaler E (1993) Langmuir 9:12

107. Aral B, Kalyon D (1994) J Rheol 38:957
108. Tsai J-C, Voth GA, Gollub JP (2003) Phys Rev Lett 91:064301
109. Wang P, Song C, Briscoe C, Makse HA (2008) Phys Rev E 77:061309
110. Nicolas Y, Paques M, van den Ende D, Dhont JKG, van Polanen RC, Knaebel A, Steyer A, Munch JP, Blijdenstein TBJ, van Aken GA (2003) Food Hydrocolloids, 17:907
111. Derks D, Wisman H, van Blaaderen A, Imhof A (2004) J Phys Cond Mat 16:S3917
112. Haw MD (2004) Phys Rev Lett 92:185506
113. Isa L, Besseling R, Morozov AN, Poon WCK (2009) Phys Rev Lett 102:058302
114. Abbott JR et al. (1991) J Rheol 35:773
115. Leighton D, Acrivos A (1987) J Fluid Mech 181:415
116. Frank M, Anderson D, Weeks ER, Morris JF (2003) J Fluid Mech 493:363
117. Semwogerere D, Morris JF, Weeks ER (2007) J Fluid Mech 581:437
118. Degre G, Joseph P, Tabeling P, Lerouge S, Cloitre M, Ajdari A (2006) Appl Phys Lett 89:024104
119. Goyon ACJ, Ovarlez G, Ajdari A, Bocquet L (2008) Nature 454:84
120. Koh CJ, Hookham P, Leal LG (1994) J Fluid Mech 266:1
121. Averbakh A, Shauly A, Nir A, Semiat R (1997) Int J Multiphase Flow 23:409
122. Shauly A, Averbakh A, Nir A, Semiat R (1997) Int J Multiphase Flow 23:613
123. Lyon MK, Leal LG (1998) J Fluid Mech 363:25
124. Sinton SW, Chow AW (1991) J Rheol 35:735
125. Hampton RE, Mammoli AA, Graham AL, Tetlow N, Altobelli SA (1997) J Rheol 41:621
126. Psaltis D, Quake SR, Yang CH (2006) Nature 442:381
127. Conrad JC, Lewis JA (2008) Langmuir 24:7628
128. Isa L, Besseling R, Weeks ER, Poon WCK (2006) J Phys Conf Ser 40:124
129. Isa L, Besseling R, Poon WCK (2007) Phys Rev Lett 98:198305
130. Lee JN, Park C, Whitesides GM (2003) Anal Chem 75:6544
131. Bartolo D, Degre G, Nghe P, Studer V (2008) Lab Chip 8:274
132. Roberts MT, Morhaz A, Christensen KT, Lewis JA (2007) Langmuir 23:8726
133. Crocker JC, Grier DG (1996) J Colloid Interface Sci 179:298
134. Brangwynne CP, Koenderink GH, Barry E, Dogic Z, MacKintosh FC, Weitz DA (2007) Biophys J 93:346
135. Jenkins M, Egelhaaf SU (2008) Adv Colloid Interface Sci 136:65
136. Beenhakker C, Mazur P (1984) Physica A 126:249
137. Pusey PN, Tough RJA (1983) Farad Discuss 76:123
138. Tokuyama M, Oppenheim I (1994) Phys Rev E 50:R16
139. Brady J (1993) J Chem Phys 567:99
140. Breedveld V, van den Ende D, Jongschaap RJJ, Mellema J (2001) J Chem Phys 114:5923
141. Breedveld V, van den Ende D, Bosscher M, Jongschaap RJJ, Mellema J (2002) J Chem Phys 116:10529
142. Xu H, Reeves AP, Louge MY (2004) Rev Sci Instrum 75:811
143. Brambilla G, Masri DE, Pierno M, Berthier L, Cipelletti L, Petekidis G, Schofield AB (2009) Phys Rev Lett 102:085703
144. Besseling R, Ballesta P, Isa L, Petekidis G, Poon WCK, in preparation
145. Moller PCF, Rodts S, Michels MAJ, Bonn D (2008) Phys Rev E 77:041507
146. Coussot P, Raynaud J, Bertrand F, Moucheront P, Guilbaud J, Jarny HHS, Lesueur D (2002) Phys Rev Lett 88:218301
147. Picard G, Ajdari A, Lequeux F, Bocquet L (2005) Phys Rev E 71:010501
148. Bocquet L, Colin A, Ajdari A (2009) Phys Rev Lett 103:036001
149. ten Wolde PR, Ruiz-Montero MJ, Frenkel D (1996) J Chem Phys 104:9932
150. Falk ML, Langer JS (1998) Phys Rev E 57:7192
151. Berthier L, Barrat J-L, Kurchan J (2000) Phys Rev E 61:5464; Berthier L (2003) Journal of Physics: Condensed Matter, 15:S933
152. Fuchs M, Cates ME (2009) J Rheol 53:957
153. Schweizer KS, Saltzman EJ (2003) J Chem Phys 119:1181
154. Saltzman EJ, Schweizer KS (2003) J Chem Phys 119:1197

155. Kobelev V, Schweizer KS (2005) Phys Rev E 71:021401
156. Saltzman EJ, Yatsenko G, Schweizer KS (2008) J Phys Cond Mat 20:244129
157. Schall P, Weitz DA, Spaepen F (2007) Science 318:1895
158. Langer JS (2005) Scr Mater 54:375
159. Yilmazer U, Gogos CG, Kaylon DM (1989) Polymer Compos 10:242
160. Kalyon DM (2005) J Rheol 49:621
161. Jana SC, Kapoor B, Acrivos A (1995) J Rheol 39:1123
162. Fuchs M, Cates ME (2002) Phys Rev Lett 89:248304
163. Nott PR, Brady JF (1994) J Fluid Mech 275:157
164. Campbell CS (2006) Powder Tech 162:208
165. MiDi GDR (2004) Eur Phys J E 14:341
166. Huilgol RR, You Z (2005) J NonNewton Fluid Mech 128:126
167. Pouliquen O, Gutfraind R (1996) Phys Rev E 53:552
168. Gutfraind R, Pouliquen O (1996) Mech Mater 24:273
169. Pine DJ, Gollub JP, Brady JF, Leshansky AM (2005) Nature 438:997
170. Corte L, Chaikin PM, Gollub JP, Pine DJ (2008) Nat Phys 4:420
171. Coussot P, Ancey C (1999) Phys Rev E 59:4445
172. Pusey PN, van Megen W, Bartlett P, Ackerson BJ, Rarity JG, Underwood SM (1989) Phys Rev Lett 63:2753
173. Henderson SI, van Megen W (1998) Phys Rev Lett 80:877
174. Schope HJ, Bryant G, van Megen W (2006) Phys Rev E 74:060401
175. Auer S, Frenkel D (2001) Nature 409:1020
176. Biehl R, Palberg T (2004) Rev Sci Instrum 75:906
177. Ackerson BJ, Pusey PN (1988) Phys Rev Lett 61:1033
178. Koumakis N, Schofield AB, Petekidis G (2008) Soft Matter 4:2008
179. Ackerson BJ (1990) J Rheol 34:553
180. Derks D, Wu YL, van Blaaderen A, Imhof A (2009) Soft Matter 5:1060
181. Schall P, Cohen I, Weitz DA, Spaepen F (2006) Nature 440:319
182. Sierou A, Brady JF (2002) J Rheol 46:1031
183. Tsai J-C, Gollub JP (2004) Phys Rev E 70:031303
184. Tsai J-C, Gollub JP (2005) Phys Rev E 72:051304
185. Blaak R, Auer S, Frenkel D, Lowen H (2004) Phys Rev Lett 93:068303
186. Penfold R, Watson AD, Mackie AR, Hibberd DJ (2006) Langmuir 22:2005
187. Wilson LG, Harrison AW, Schofield AB, Poon WCK (2009) J Phys Chem B 113:3806
188. Waigh TA (2004) Rep Prog Phys 68:685
189. Meyer A, Marshall A, Bush BG, Furst EM (2006) J Rheol 50:77
190. Lorén N, Langton M, Hermansson AM (2007) In: Understanding and controlling the microstructure of complex foods. CRC, Cambridge, p 232

Soft and Wet Materials: From Hydrogels to Biotissues

Jian Ping Gong and Yoshihito Osada

Abstract This chapter describes recent progress in the study and development of hydrogels with tough mechanical strength, low frictional coefficient, and wear-resisting properties. Furthermore, examples of application of gels as cell scaffolds and substitutes of biological tissues, such as artificial articular cartilage, will be introduced.

Keywords Artificial cartilage · Cell scaffold · Double network · Hydrogel · Low friction · Strength · Toughness

Contents

1	General Introduction	204
2	Tough Double Network Gels	206
	2.1 Optimized Structure for Tough DN Gels	206
	2.2 Necking and Hysteresis of Tough DN Gels	209
	2.3 Proposed DN Gel Toughening Mechanism: Local Damage Zone Models	212
	2.4 Structure of DN Gels Characterized by Small-Angle Neutral Scattering	213
	2.5 Tough DN Gels from Neutral Polymers	214

J.P. Gong (✉)
Faculty of Advanced Life Science, Hokkaido University, Sapporo 060–0810, Japan
e-mail: gong@sci.hokudai.ac.jp

Y. Osada
Faculty of Advanced Life Science, Hokkaido University, Sapporo 060–0810, Japan
and
Riken, Saitama, Japan
e-mail: osadayoshi@riken.jp

3	Friction and Lubrication of Gels: An Approach to Low Friction Materials	216
	3.1 Rich and Complex Friction Behavior of Hydrogels	216
	3.2 Proposed Gel Friction Mechanism: A Repulsion–Adsorption Model	219
	3.3 Comparison of the Model with Observation	223
4	Gels with Low Friction	225
	4.1 Friction Reduction by Template Effect	225
	4.2 Friction Reduction by Dangling Chains	226
	4.3 Friction Reduction by Dilute Polymer Solution	228
	4.4 Friction Reduction by Substrate Roughness	229
5	Polymer Gels as Scaffolds for Cell Cultivation	231
	5.1 Cell Growth on Various Gels	232
	5.2 Effect of Charge on Cell Growth	234
	5.3 Cell Proliferation on Tough Gels	235
6	Robust Gels with Low Friction: Excellent Candidates as Artificial Cartilage	236
	6.1 Wearing Properties of Robust DN Gels	237
	6.2 Robust Gels with Low Friction	238
	6.3 Biological Responses of DN Gels in Muscle and Subcutaneous Tissue	240
References		242

1 General Introduction

Biological tissues mainly consist of water and macromolecular components such as fibrous collagen and proteoglycans. Biological tissues display fascinating properties: extremely low surface sliding friction between cartilages, strong adhesion between hard bones and tendons, extraordinarily strong mechanical toughness and shock-absorbance of cartilages, tendons, muscles, etc. [1, 2]. From the viewpoint of material science, biotissues, having an elastic modulus of 10^4 Pa–10^7 Pa and containing 50–85% of water, are essentially soft and wet materials. Some of the excellent properties of biotissues, such as sustainability to large deformation, low sliding friction, shock absorption, and substance permeability, are believed to originate from the soft and wet nature of biotissues.

However, to date, almost all of the artificial organs (artificial hearts, blood vessels, hips and knees, etc.) have been constructed from hard and dry materials. These artificial organs, to some extent, have successfully served as substitutes for real organs, but are still far from satisfactory. For example, artificial hips and knees made of metals and ceramics, which are hard and dry materials, lack the shock-absorbing function and have a high frictional resistance against sliding motion [3]. Another example is artificial blood vessels. Blood blots occur in artificial blood vessels made of polymers due to protein adsorption at the surface of the blood vessel wall [4].

In order to construct artificial organs with an excellent function similar to the real biological organ, material innovation is urgently required, i.e., one should find a man-made soft and wet material as substitute for biotissue. Synthetic hydrogels, which consist of polymer networks swollen with large amounts of water (the same as biotissue), are the only candidates [5]. Different from biotissue, the water content of a hydrogel varies over a wide range, usually 50–99.9% of the total weight, depending on the chemical structure. Hydrogels are solids on the macroscopic scale: they have definite shapes and do not flow. At the same time, they behave like solutions on the molecular scale: water-soluble molecules can diffuse in hydrogels with various diffusion constants that reflect the size and shape of the molecules.

Gel is a fascinating material because of its unique stimuli–response properties, such as volume phase-transition [6] and chemomechanical behavior [7, 8], and for its possible wide application in many industrial fields. Recently, hydrogels have become especially attractive in biological fields due to their possible applications as soft artificial tissues. Hydrogels are expected to be used as scaffolds for repairing and regenerating a wide variety of tissues and organs, or even as substitutes for tissues and organs, because their three-dimensional network structure and viscoelasticity are similar to the macromolecular-based extracellular matrix (ECM) in biological tissue.

However, to be qualified for biological applications, significant problems and disadvantages of hydrogels need to be resolved, including:

1. Improving the mechanical toughness of the hydrogel. In the physiological condition, many tissues (such as blood vessels, articular cartilages, semilunar cartilages, tendons, and ligaments) exist in a severe mechanical dynamic environment. For example, the articular cartilages sustain a daily compression of 10 MPa, whereas most hydrogels are mechanically very weak and fail under sub-megapascal stresses.
2. Controlling the surface properties of hydrogel. For example, cartilage shows a low frictional resistance against sliding motion, but the resistance of hydrogels is not well understood.
3. Controlling viability to cells and compatibility to tissues. For example, a hybrid artificial blood vessel consists of a gel vessel with its inner surface covered by endothelial cells, therefore, gels on which the endothelial cells are able to form a continuous monolayer should be developed.
4. Controlling the transportation of ions, hormones and nutrients, and improving the permeability to proteins.

Therefore, to replace natural tissues with hydrogels, it is important to design and synthesize hydrogels with some crucial properties, such as high mechanical strength, low surface friction and wear, and high cellular viability. Gel scientists have strived to design hydrogels with these crucial properties by using synthetic polymers instead of natural polymers, because the chemical properties of synthetic gels are easily controllable and reproducible, and the cost of synthetic gels is relatively low.

In this chapter, recent progress on the study and development of hydrogels with tough mechanical strength, low frictional coefficient, and cellular viability will be described. Furthermore, attempts to apply such gels as substitutes for biological tissues will be introduced.

2 Tough Double Network Gels

Applications of hydrogels as mechanical devices are fairly limited due to the lack of mechanical strength of hydrogels. Reported values of the fracture energy of typical hydrogels fall within a range of 10^{-1}–10^0 J/m^2 [9–11], which are much smaller than those of the usual rubbers, i.e., $\sim 10^3$ J/m^2 [12]. Many researchers have thought that the lack of mechanical strength of a gel is unavoidable because of its solution-like nature, i.e., low density of polymer chains and low friction between the chains. Furthermore, it is well known that in hydrogels synthesized from monomer solutions, a very heterogeneous structure is formed during the gelation [13]. This is also considered to be a factor that decreases the mechanical strength. However, when we pay attention to biological systems, we find some biological hydrogels with excellent mechanical performances. For example, cartilage, containing 75 wt% of water, exhibits a large work of fracture (1 kJ/m^2) [14], which allows the cartilage to sustain a daily compression of several megapascals [15]. It is a challenging problem in modern gel science to bridge the gap between man-made gels and biological gels.

Many approaches, such as slide-ring gels [16–18] and nanocomposite hydrogels [19–22], have been used in attempts to significantly improve the mechanical properties of hydrogels by rendering the network junction points mobile. A new approach, incorporating a neutral polymer of high relative molecular mass within a swollen polyelectrolyte network, (double-network hydrogels, DN gels), was recently discovered and developed [23–31]. DN gels have the highest fracture toughness among cross-linked hydrogels. The mechanical properties of DN gels prepared from many different polymer pairs were shown to be much better than those of the individual components. Among all the polymer pairs studied so far, the gel made of poly (2-acrylamido-2-methyl-1-propanesulfonicacid) (PAMPS) polyelectrolyte and polyacrylamide (PAAm) neutral polymer stands out with unusually favorable properties, e.g., the PAMPS/PAAm DN gels, containing about 90 wt% water, possess both hardness (elastic modulus of 0.3 MPa) and toughness (compressive strength of $\sim$20 MPa, tearing fracture energy of $\sim$1 kJ/m^2).

2.1 Optimized Structure for Tough DN Gels

DN gels are synthesized via a two-step network formation: the first step is to form a rigid gel that is highly cross-linked, and the second is to form a loosely cross-linked network in the first gel [23]. Hereafter, the DN gels are referred to as $P_1 - x_1 - y_1 / P_2 - x_2 - y_2$, where P_i, x_i and y_i ($i = 1, 2$ for the first and second network, respectively) are the abbreviated polymer name, molar monomer concentration, and the percentage cross-linker concentration (in mol%) with respect to the monomer for the i-th network, respectively.

Figure 1 shows a comparison of behaviors of the PAMPS-1-4 single network (SN) gel and the PAMPS-1-4/PAAm-2-0.1 DN gel prepared at the optimized conditions under uni-axial compression. Figure 2 shows stress–strain curves for the

Soft and Wet Materials: From Hydrogels to Biotissues

Fig. 1 Images of PAMPS gel (**a**) and PAMPS/PAAm DN gel (**b**) under and after uniaxial compression. Crosslinking density (ratio of cross-linker to monomer) was 4 mol% for PAMPS and 0.1 mol% for PAAm. Fracture stress was 0.4 MPa for PAMPS gel and 17.2 MPa for PAMPS/PAAm DN gel. (Reproduced, with permission, from [23])

Fig. 2 Stress–strain curves for the PAMPS/PAAm DN gel, PAMPS gel, and PAAm gel. Crosslinking density was 4 mol% for PAMPS and 0.1 mol% for PAAm. (Reproduced, with permission, from [23])

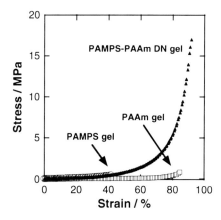

PAMPS-1-4/PAAm-2-0.1 DN, the PAMPS-1-4 SN, and the PAAm-2-0.1 SN gels. The DN gel holds up to a stress of 17.2 MPa, where the vertical strain λ is about 92%. On the other hand, the PAMPS gel and PAA gel break at stresses of 0.4 MPa ($\lambda = 41\%$) and 0.8 MPa ($\lambda = 84\%$), respectively.

The DN gels exhibit high mechanical strength when satisfying the following structure features:

1. An optimal combination is PAMPS gel as the first network and PAAm gel as the second network.
2. The first network PAMPS is tightly cross-linked and the second network PAAm is loosely cross-linked. For example, when the cross-linking density of the first network is kept constant at 4 mol% and that of the second network is changed systematically from 0 to 1.0 mol%, the mechanical strength of the DN gels changes dramatically, as shown in Fig. 3a. However, in this case, all the PAMPS/PAAm DN gels show a constant elastic modulus of 0.3 MPa, a water content of 90 wt%, and a molar ratio of the second network to the first network of 20, regardless of the change in the cross-linking density of the second network.
3. The amount of the second network is in large excess to that of the first network. For example, the strength of PAMPS/PAAm DN gel increases when the molar ratio of the second network to the first network increases, as shown in Fig. 3b.

The fracture energy G of the DN gels was measured by the tearing test as a function of the crack velocity V and showed the following: (1) The fracture energy G ranges from 10^2 to 10^3 J/m^2, which is 100–1000 times larger than that of normal PAAm gels (10^0 J/m^2) or PAMPS gels (10^{-1} J/m^2) with similar polymer concentrations as the DN gels. (2) G shows weak dependence on the crack velocity V. (3) G at a given

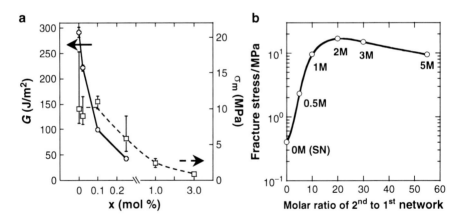

Fig. 3 Relationship between the mechanical strength of PAMPS/PAAm DN gels and their structure parameters. (**a**) Dependence of the fracture energy G at $v = 0.5 \times 10^{-2}$ m/s (*circles*) and the fracture stress σ_m (*squares*) of PAMPS-1-4/PAAm-2-y_2 on the cross-linking density y_2 of the second component, PAAm the vertical legend x in figure (a) has the same meaning with y_2 in this paper. (Reproduced, with permission, from [25].) (**b**) Effects of the molar ratio of the second network to the first network on the mechanical strength of the PAMPS-1-4/PAAm-x_2-0.1 DN gels. The different composition ratios of the second to the first network were obtained by immersing the first network gel in monomer solutions constituting the second network with various concentrations. *Numbers* in the figures denote the value of the monomer concentration x_2 for polymerizing the second network. Data at zero molar ratios denote the values of single network gels. (Reproduced, with permission, from [23])

value of *V* increases with decreasing cross-linking density of the second network, in agreement with the compressive strength (Fig. 3a).

The DN gels possess both high strength (high elastic modulus) and toughness (large fracture energy). Adjusting the compositions of the first and second networks of the gel can independently control these two quantities, which is convenient for practical application. Invention of DN gels has not only made a great breakthrough in finding wide application of gels in industry and biomedical fields, but also addresses fundamental problems in gel science.

2.2 Necking and Hysteresis of Tough DN Gels

A kind of yielding phenomenon was found in PAMPS-1-2/PAAm-2-0.02 DN gels made from relatively sparse (fragile) first networks [27]. On tensile tests, narrowing zones ("necks") appear in the sample and grow with further stretching, as shown in Fig. 4 [27]. During neck propagation, a plateau region appears in the loading curve.

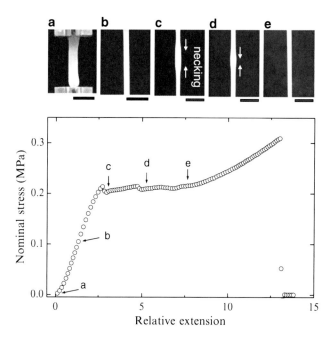

Fig. 4 (**a–e**) Pictures demonstrating how the necking process makes progress. Images correspond to the *arrowed data points* in the graph. *Scale bars*: 10 mm. Width of the undeformed gel in (**a**) corresponds to the thickness of the sample (∼4 mm). In (**c**) and (**d**), the *upper and lower parts* (necked regions) of the gel are slightly narrowed compared with the *middle part* (unnecked region). The necked regions grow up with the extension of the sample (*opposing arrows* in the pictures). *Below*: Loading curve of PAMPS-1-2/PAAm-2-0.02 DN gel under uniaxial elongation at an elongation velocity of 500 mm/min. (Reproduced, with permission, from [27])

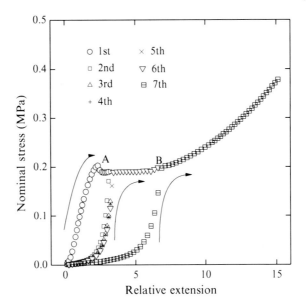

Fig. 5 Loading curves of PAMPS-1-2/PAAm-2-0.02 DN gel under uniaxial elongation at an elongation velocity of 250 mm/min. A and B mark the plateau region. (Reproduced, with permission, from [27])

The plateau value of the tensile stress is almost independent of the stretching rate. After neck propagation, the gel becomes fairly soft and sustains large elongation, up to an elongation strain of around 20.

The large deformation of the softened DN gels produced by necking was almost reversible. Figure 5 shows loading curves of repeated elongation tests for PAMPS-1-2/PAAm-2-0.02 DN gel at an elongation velocity of 250 mm/min. It should be noticed that the first curve, the plateau of the sixth curve (from A to B in Fig. 5), and the latter part of the seventh curve (beyond B) are smoothly connected, composing a curve that seems to be obtained by a single loading. The observations on the softened gels after the tensile test demonstrate that an irreversible structural change takes place inside the gels, although their appearance is almost unchanged.

From the above experimental findings and the fact that the PAMPS SN gels are quite brittle and break into small pieces at small extensions, it seems that in the course of necking deformation, the first PAMPS network fragments into small clusters; and the clusters play the role of cross-linker in long PAAm chains (Fig. 6), because the second network is loosely cross-linked.

The fragmentation should involve dissipation, which is reflected in the hysteresis of the loading curves in Fig. 5. Since the work done to the system is consumed to create the cluster, the cluster size is in the order of hundreds of nanometers. This value may reflect quenched heterogeneity of the first network of DN gels found in our study [30, 31].

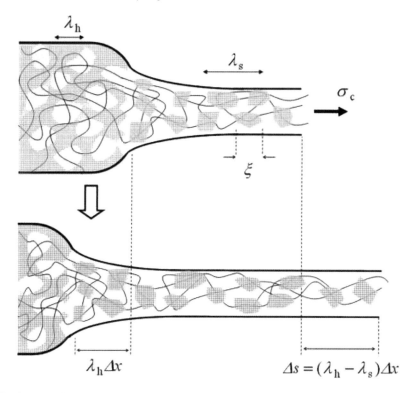

Fig. 6 Illustrations of the necking propagation and a model of the network structure of the softened DN gel after the necking. ξ is the typical size of the PAMPS clusters, σ_c is the critical stress for yielding. Δs is the stretched distance and Δx is the corresponding portion of length (Δx is the length of the portion at the reference (stress-free) state) converted from the un-necked to the necked regions; λ_s and $\lambda_h (\ll \lambda_s)$ are elongation ratios for the necked (softened) and un-necked (hard) regions, respectively. (Reproduced, with permission, from [27])

From the viewpoint of polymer physics, the softened gel after necking is a fascinating system. First, the PAMPS clusters, playing the role of cross-linker, are not chemically bonded to the PAAm chains; thus, the "molecular weight between cross-linking" can be adjusted by chain sliding in response to deformation. The capability for chain sliding is held in common with the slide-ring gels, which are well known as being ductile and tough hydrogels [16–18]. Second, the PAMPS cluster can be redivided; when a cluster in a deformed sample suffers a larger stress than other clusters, the inequality of stress can be corrected by redivision. Owing to these effects, the resultant network structure has fewer mechanical defects. The excellent mechanical properties of the softened gel can be attributed to the above structural features.

Systematic loading and unloading experiments in uniaxial tension and uniaxial compression have also been performed on DN gels that do not show necking. These DN gels also show a significant hysteresis during the first loading cycle, increasing very strongly with the applied maximum deformation (Fig. 7) [29]. Such a large hysteresis was never observed during a second loading cycle and was attributed to

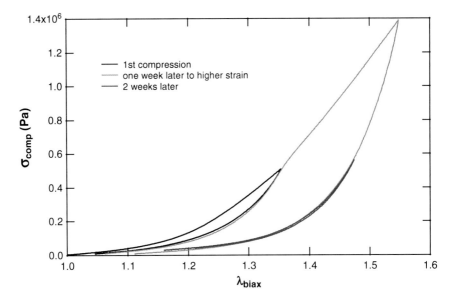

Fig. 7 Successive loading cycles in compression of the same DN gel sample. λ_{biax} is the biaxial extension ratio. It relates to the compression ratio λ_{comp} as $\lambda_{biax} = \frac{1}{\sqrt{\lambda_{comp}}}$. (Reproduced, with permission, from [29])

the fracture of covalent bonds in the first network. Assuming that the entire energy dissipated during the hysteresis cycle can be attributed to a fracture of the first network strands by a Lake–Thomas mechanism, the results suggest that fracture and unloading of only 1% of the bonds within the network leads to a decrease of up to 50% of the number of strands. These results demonstrate the very large degree of heterogeneity of the network. If such a dissipative mechanism is active at the crack tip, it is likely to greatly increase the energy necessary to propagate a macroscopic crack.

2.3 Proposed DN Gel Toughening Mechanism: Local Damage Zone Models

On the basis of the necking phenomena and the large hysteresis of DN gels, Brown [32] and Tanaka [33] have independently proposed a similar theoretical model for explaining the extraordinary high fracture energy of DN gels. The model assumes that on the strongly stretched region in front of the crack tip, the material first yields to transform into a very soft material with an intrinsic fracture energy G_0, and then the crack tip passes through the softened (damaged) zone (Fig. 8). Using the energy balance concept of fracture mechanics, a scaling level expression has been proposed, in which the effective fracture energy G is expressed in terms of the yielding

Soft and Wet Materials: From Hydrogels to Biotissues 213

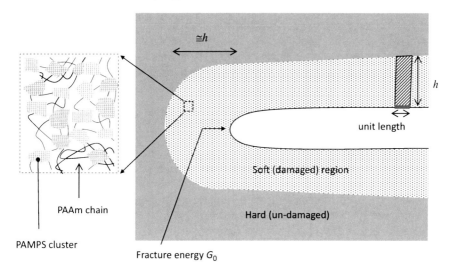

Fig. 8 Local damage zone model to explain the extraordinarily large fracture energy of DN gels. (Reproduced, with permission, from [33])

stress σ_c, the size of the softened zone h, the intrinsic fracture energy G_0, and the strain where the necking finishes ε_c [33]:

$$G = G_0 + \sigma_c \varepsilon_c h.$$

An order estimation of G using the numerical data from the necking gel predicts that the necking zone is in the order of hundreds of micrometers [33], which will be verified by experimental observation.

2.4 Structure of DN Gels Characterized by Small-Angle Neutral Scattering

It is still not clear from the local damage zone model how the stress can be transferred from the second network to the first network via a distance as large as hundreds of micrometers. Small-angle neutron scattering (SANS) was performed to elucidate the molecular origin of the superior mechanical properties of PAMPS/PAAm DN gels [30, 31]. A strong heterogeneity of the PAMPS first network has been observed [31], which is in agreement with the prediction obtained from the hysteresis results [29]. Furthermore, measurements on PAMPS/PAAm DN gels and their solution blend counterparts indicate that the two polymers interact favorably with each other while in water. This favorable PAMPS–PAAm interaction given by the condition $\chi_{PE-NP} \ll \chi_{PE-water} < \chi_{NP-water}$, where χ is the Flory–Huggins interaction parameter, is consistent with some of the salient features of the DN structure revealed by SANS, and it might also contribute to the ultimate mechanical properties of DN gels. This observation suggests that the attractive enthalpic interactions between two components might play a role in the toughening of DN gels. The SANS measurements further indicate that deformation of DN gels results in periodic and mesoscale ($\sim 1.5\,\mu m$) compositional fluctuations in both PAMPS and PAAm.

On the basis of the SANS results, a molecular mechanism has been recently proposed for the toughness enhancement of DN gels [34]. This mechanism rationalizes the changes in molecular structure of the DN gel constituents observed via in-situ neutron scattering measurements, the composition dependence of the solution viscosity, and the thermodynamic interaction parameters of PAMPS and PAAm molecules obtained previously from neutron scattering studies. More specifically, this proposed mechanism provides an explanation for the observed periodic compositional fluctuations in the micrometer range induced by large strain deformation.

2.5 Tough DN Gels from Neutral Polymers

The discovery of a hydrogel with good mechanical performance should enable the hydrogel to find a wide application in industry, such as in fuel cell membranes, load-bearing water absorbents, separation membranes, optical devices, low friction gel machines, and in the printing industry and biomechanical fields, such as for artificial cartilages, tendons, blood vessels, and other biotissues.

To realize the biomedical applications, it is important to apply the concept of the high performance DN gels to various polymer combinations, including biocompatible polymers. Hydrogels derived from natural polymers frequently demonstrate adequate biocompatibility and have been widely used in tissue engineering [35]. On the basis of the design principle for a tough DN gel, some progress has been made recently by combining bacterial cellulose (BC) and gelatin [36].

BC is a kind of extracellular cellulose produced by bacteria of genus *Acetobacter* and consisting of a hydrophobic ultrafine fiber network stacked in a stratified structure [37]. Scanning electron microscopy (SEM) images shows that BC gel consists of alternating dense and sparse cellulose layers with a period of $10\,\mu m$, which contributes to the anisotropic mechanical properties. It has a high tensile modulus (2.9 MPa) along the fiber-layer direction but a low compressive modulus (0.007 MPa) perpendicular to the stratified direction. Due to its poor water-retaining ability, water is easily squeezed out from the BC gel network (although an as-prepared BC contains 90% water) and there is no recovery of the swelling property because of hydrogen-bond formation between cellulose fibers.

Gelatin is a polypeptide derived from an ECM structural protein, collagen. Gelatin gel can retain water and recovers from repeated compression. Due to its poor mechanical strength, a gelatin gel is easily broken into fragments under a modest compression of 0.12 MPa. By using the DN gel method, it is possible to create a strong gel consisting of BC and gelatin that has both biocompatibility and high mechanical strength [36].

BC/gelatin DN gel is synthesized by immersing BC substrate in gelatin solution (30 wt%), then the gelatin is cross-linked by 1-ethyl-3-(3-dimethylaminopropyl) carbodiimide hydrochloride (EDC) (1 M). Wide-angle X-ray diffraction (WAXD) shows that the crystal pattern of BC is well maintained, and SEM shows that homogeneous gelatin textures fill in the cellulose-poor layers in BC/gelatin gel. The structure of BC/gelatin indicates that the anisotropic mechanical strength of BC is maintained after combination with gelatin. Therefore, measurements of mechanical strength against compression and elongation were performed perpendicular to and parallel to the stratified direction of BC/gelatin DN gels, respectively. Typical compressive and elongation stress–strain curves are shown in Fig. 9. BC/gelatin DN gel shows a compressive elastic modulus of 1.7 MPa (Fig. 9a), which is more than 240 times higher than that of BC gel (0.007 MPa) and 11 times higher than that of the gelatin gel (0.16 MPa). The fracture strength of BC/gelatin DN gel against compression in the direction perpendicular to the stratified structure reached 3.7 MPa. The strength is in the range of articular cartilage (1.9–14.4 MPa). This value is about 31 times higher than that of gelatin gel. Moreover, the BC/gelatin DN gel recovers well under a repeated compressive stress up to 30% strain. The elongation stress–strain curve (Fig. 9b) shows that BC/gelatin DN gel can sustain nearly 3 MPa elongation stress with 23 MPa elastic modulus, which is 112 times greater than that of the gelatin gel.

Similar improvement in the mechanical strength was also observed for the combination of BC with polysaccharides such as sodium alginate, gellan gum, and ι-carrageenan. For example, the compressive elastic modulus of BC-ι-carrageenan DN gel is 0.12 MPa, which is 17 and 13 times higher than that of individual BC and ι-carrageenan gels, respectively. In summary, mechanical properties of synthetic

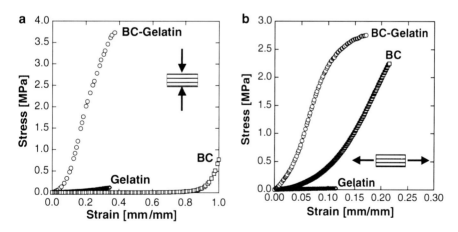

Fig. 9 Comparison of compressive (**a**) and elongational (**b**) stress–strain curves of BC-gelatin DN, gelatin, and BC gels. The compression and elongation were performed perpendicular to and along with the stratified direction of BC and BC-gelatin DN gels, respectively. Concentration of gelatin in feed was 30 w% and EDC concentration 1 M. (Reproduced, with permission, from [36])

and natural gels can be effectively improved by the inducing the double network structure. The composition of two kinds of networks can be adjusted according to different applications.

3 Friction and Lubrication of Gels: An Approach to Low Friction Materials

Industrial or environmental problems caused by high frictional surfaces of materials exist in our daily life. The search for materials with a low surface friction has been one of the classical and everlasting research topics for material scientists and engineers. Despite many efforts, it has been shown that surface modification or adding of lubricant are not very effective in reducing the steady-state sliding friction between two solids, which show a frictional coefficient $\mu \sim 10^{-1}$ even in the presence of lubricant [38].

In contrast to man-made systems, fascinating surface behavior of bio-organs can be observed. For example, the cartilages of animal joints have a friction coefficient in the range 0.001–0.03, remarkably low even for hydrodynamically lubricated journal bearings [39–46]. It is not well understood why the cartilage friction of the joints is so low, even in conditions where the pressure between the bone surfaces reaches as high as 3–18 MPa and the sliding velocity is never greater than a few centimeters per second [40]. Under such conditions, the lubricating liquid layer cannot be sustained between two solid surfaces and the hydrodynamic lubrication does not work.

The fascinating biotribological properties originate from the soft and wet nature of soft tissue. For example, cartilage consists of a three-dimensional collagen network filled with a synovial fluid [41–46]. Cartilage cells synthesize a complex ECM; the weight bearing and lubrication properties of cartilage are associated primarily with this matrix and its high water content. The main macromolecular constituents of ECM are the proteoglycan, aggrecan, and the cross-linked network of collagen fibrils.

Therefore, soft tissue existing in a "gel state" should account for the specific tribological behavior in biological systems. In an effort to find a way to reduce the sliding friction of man-made systems by elucidating the low-friction secrets of biological surfaces from the viewpoint of hyodrogels, the friction and lubrication of hydrogels has been extensively investigated over the past 10 years [47–67].

3.1 Rich and Complex Friction Behavior of Hydrogels

3.1.1 Load Dependence

The friction between solids obeys Amonton's law (1699), $F = \mu W$, which says that the frictional force F is linearly proportional to load W, and does not depend

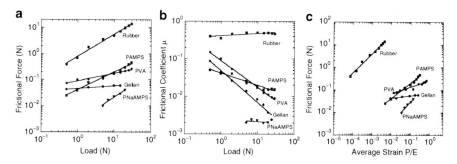

Fig. 10 Dependence of frictional force F on load W (**a**), the frictional coefficient on load (**b**), and the frictional force on the average strain (**c**) for various kinds of hydrogels sliding on glass substrate. Sliding velocity 7 mm/min. Sample sizes for PVA, gellan, and rubber $3 \times 3\,\text{cm}^2$; PAMPS and PNaAMPS $2 \times 2\,\text{cm}^2$. Compressive modulus E: PVA 0.014 MPa; gellan 0.06 MPa; PAMPS 0.25 MPa; PNaAMPS 0.35 MPa; rubber 7.5 MPa. Degree of swelling q: PVA 17; gellan 33; PAMPS 21; PNaAMPS 15. The measurements were performed by a tribometer in air. (Reproduced from [49])

on the apparent contact area A of two solid surfaces nor on the sliding velocity v [38]. However, gel friction does not simply obey Amonton's law, but shows rich and complex features. Gel friction strongly depends on the properties of gels that are determined by their chemical structure, such as hydrophilicity, charge density, cross-linking density, water content, and elasticity [47, 49]; on the surface properties of opposing substrates, such as surface charges, hydrophobicity [49, 50, 64], and roughness [65]; and on the measurement conditions, such as the normal load and sliding velocity [47, 49–51].

Figure 10 shows the friction behavior of hydrogels with different chemical structures: PVA, gellan, PAMPS, and its sodium salt PNaAMPS [49]. These hydrogels were slid on a smooth glass substrate at a sliding velocity of 7 mm/min, using a tribometer. Figure 10a shows the relationship between the normal load (W) and the frictional force (F), which obeys a power law, $F \propto W^\alpha$ where the scaling exponent α lies in a range of 0–1.0, depending on the chemical structure of the gels.

The frictional coefficient μ, which is defined as the ratio of the frictional force to applied load, is shown in Fig. 10b. The frictional coefficient μ of these gels accordingly shows unique load dependencies, which are quite different from those of solids. The μ of PVA, gellan, and PAMPS gels decreases with an increase of load. On the other hand, the μ of the PNaAMPS gel is constant over a change of load (similar to that of rubber) but its value is as low as 0.002, which is two orders of magnitude lower than those of solids. PAMPS and PNaAMPS gels are only different in their counterions, but they show a striking difference in frictional behavior on the glass surface.

So, the frictional behavior of gels strongly depends on their chemical structure, and the frictional force of gels is two or three orders of magnitude lower than that of a rubber. The behavior can be continuously observed for a few hours.

3.1.2 Sample Area Dependence

The linear dependence of friction on load established in solid friction, $F = \mu W$, is explained in terms of the yielding mechanism i.e., the solid surface is not molecularly flat and the real contact area between two surfaces increases with an increase of load due to yielding. Thus, the friction has no dependence on the apparent contact area of the two solid surfaces, and Amonton's law holds [38].

To elucidate the feature of interface contact between a gel and the opposing plate, the frictional force of various kinds of gels were measured by varying the contact area of gel A under a constant load W [49]. It was found that F also shows a power law relation with A, which can be denoted as $F \propto A^\beta$. Combining the results of F on W and A, one has $F \propto W^\alpha A^\beta$ As shown in Fig. 11, the correlation between α and β was found to be $\beta \approx 1 - \alpha$. Therefore, $F \propto AP^\alpha$, where $P = W/A$ is the average normal pressure and α lies between 0 and 1, depending on the chemical structure of the gel. This result demonstrates that the frictional force per unit area (frictional stress) is related to the normal pressure P, instead of to the load W, by the power law. For solid friction, the relation $F \propto AP^\alpha$ is also valid, with $\alpha = 1$.

A polymer gel is easily deformable, with a typical elasticity ranging from 1 kPa to 1000 kPa, owing to the presence of a large amount of water. A small pressure would be sufficient to cause a large deformation of a gel. This favors interfacial contact with the opposing surface. As shown in Fig. 10c, the average strains, $\lambda = P/E$, of gels under the experimental load range are more than several percent higher than that of a rubber and, needless to say, much higher than that of solid. Here, E is the compressive elastic modulus of the gel. Since the surface roughness of gels synthesized on smooth surfaces such as glass plate or silicon wafer is of several nanometers, which is of the order of the network mesh size of a gel, the whole gel surface should contact with the smooth glass substrate in the mesh size scale. All the gel samples were measured under a similar strain range; nevertheless, they showed quite different pressure dependencies.

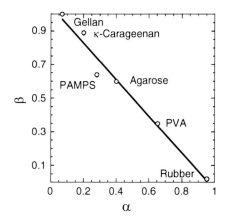

Fig. 11 Relationship between α and β, where α and β are the exponents of the relation $F \propto W^\alpha A^\beta$. α was measured at $A = 9\,\text{cm}^2$; and β was measured at $W = 0.98\,\text{N}$. Sliding velocity 180 mm/min. Degree of swelling: PVA 20; gellan 33; κ-carrageenan 33; agarose 50; PAMPS 17. The measurements were performed by a tribometer in air. (Reproduced from [49])

3.1.3 Substrate Effect

The frictional behavior of gels also depends on the opposing substrates. When the nonionic PVA gel is allowed to slide on a tetrafluoreoethylene (Teflon) plate, for example, the behavior is the same as that on a glass surface. However, the behavior of strong anionic PAMPS gel on Teflon changes greatly and becomes similar to that of PVA on glass. When a pair of polyelectrolyte gels carrying the same charges, e.g., PNaAMPS gel with PNaAMPS gel, were slid with each other, very low frictional force was observed [49]. On the other hand, when two polyelectrolyte gels carrying opposite charges were slid with each other, the adhesion between the two gels was so high that the gels were broken during the measurement [50]. The phenomenon indicates that the interfacial interaction between the gel surface and the opposing substrate is crucial in gel friction.

3.2 Proposed Gel Friction Mechanism: A Repulsion–Adsorption Model

Many studies have been performed on the friction of bulk soft cross-linked polymers [68–77]. Schallamach, Leonov, and others explain the so-called adhesive friction as a specific kind of hysteresis friction of bulk soft polymers on smooth surfaces like glass [68–72]. Persson and coworkers explain that the friction of a rubber sliding on a hard and rough surface is attributed to the internal friction of the rubber [75–77].

A gel has many features in common with other soft cross-linked polymers in terms of its cross-linked polymer structure and viscoelastic properties. On the other hand, in contrast to a rubber, a gel contains a large amount of water. The water content in biological gels such as cartilage and other soft tissues is about 70–80 wt%, and in synthetic gels the water content can be as high as 99.9 wt%. Water in gels is strongly solvated to the polymer network and could not be squeezed out easily like a sponge. Due to the presence of a large amount of water, the internal friction of a gel is much less than that of a cross-linked polymer melt or rubber. For a typical gel, $\tan\delta$, a characteristic parameter of the viscoelastic properties of soft materials, is in the order of 0.01, whereas that of a rubber is never less than 0.1.

When a soft gel is pressed onto a smooth hard surface in water, the water is squeezed outwards, and the stability of the water film intercalated at the interface depends on the sign of the spreading coefficient of water between the solid substrate and soft gel, i.e., $S = \gamma_{SG} - (\gamma_{SW} + \gamma_{WG})$, where γ_{SG}, γ_{SW} and γ_{WG} are interfacial energies between the substrate and gel, substrate and water, and water and gel, respectively [65, 78, 79].

If $S > 0$, corresponding to a repulsive interaction between the polymer network of the gel and substrate, water acts as a lubricant, there is no direct contact between the gel and the solid, and a thin water layer of nanometer-order thickness remains stable at the interface.

If $S < 0$, corresponding to an adhesive interaction between the polymer network of the gel and substrate, the water layer begins to dewet when its thickness becomes a few hundred nanometers, and the gel forms contacts with the substrate. These contacts grow (dewetting) only above a critical nuclei size, $R_c \approx e^2/h_0$ [78, 79]. Here, e is the water layer thickness and $h_0 = -S/E$ is the length that describes the competition between the surface energy gain $(-S)$ and gel elasticity (E). On a strongly adhesive substrate $(S \ll 0)$, the critical nuclei size is small, so many nuclei grow to form dewetting domains. At the boundary between the dewetting domains, liquid drops are trapped. Therefore, on a strongly adhesive substrate, we expect heterogeneous contact consisting of dewetting and wetting domains.

The contact heterogeneity is characterized by $h_0 = -S/E$. On a weakly-adhesive substrate on which $h_0 \leq \xi$, water drops cannot be trapped, and homogeneous contact is expected [65]. Here, ξ is the mesh size of the gel network.

Gong and coworkers have proposed a repulsion–adsorption model, from the viewpoint of interfacial interaction between polymer and solid, to describe the friction behavior of gels on a smooth substrate (Fig. 12) [48, 65]. The main argument of the adsorption–repulsion model is as follows: By considering a gel in contact with a solid wall in water to be analogous to a polymer solution, the polymer network on the surface of the gel will be repelled from the solid surface if the interface interaction is repulsive (wetting), and will be adsorbed to the solid surface if it is attractive (dewetting). In the repulsive case, the friction is due to the lubrication of the hydrated water layer of the polymer network at the interface, which predicts that the frictional stress f should be proportional to the sliding velocity. In the attractive

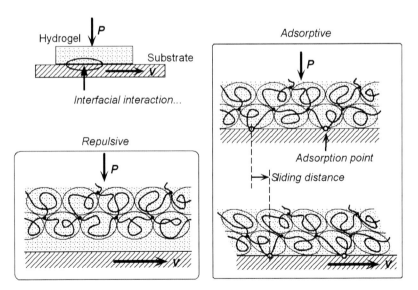

Fig. 12 Repulsion–adsorption model for the gel friction on a smooth solid substrate. P normal pressure, v sliding velocity. (Reproduced, with permission, from [67])

Soft and Wet Materials: From Hydrogels to Biotissues

case, the friction of a gel is from two contributions: (1) elastic deformation of the adsorbing polymer chain, and (2) lubrication of the hydrated layer of the polymer network. In Sects. 3.2.1 and 3.2.2, we describe the repulsion–adsorption model in detail.

3.2.1 On a Smooth but Weakly Adhesive Substrate

The molecular model is confined to the case of weak adhesion ($S < 0$) to ensure homogeneous contact [48, 65]. In this case, two sources contribute to the frictional stress of a gel: elastic deformation of an adsorbing polymer chain σ_{el} and the lubrication of the hydrated layer of the polymer network σ_{vis}, which can be represented as follows (Fig. 12):

$$\sigma = \sigma_{el} + \sigma_{vis} \qquad (2)$$

In order to estimate the first contribution from the elastic adsorption, a C^* gel picture is used, in which the gel is a collection of adjacent blobs of radius ξ that has a characteristic relaxation time $\tau_f \approx \xi^2/D_{coop}$, where $D_{coop} \approx T/6\pi\eta\xi$ is the cooperative diffusion constant of the gel, T is temperature and η is the viscosity of the solvent [80]. Each blob is associated with a partial polymer chain (the polymer chain between two next-neighboring cross-linking points). The scaling theory relates this molecular structure of the gel with its elastic modulus E by the equation $E \cong T/\xi^3$ [80].

Using this C^* picture, the elastic force produced by the stretching of a single chain is $f_{el} \cong Tv\tau_b/\xi^2$, the total number of chains per unit area is $m_0 \cong 1/\xi^2$, and among these chains, only a fraction $\tau_b/(\tau_b + \tau_f)$ are in an adsorbing state. The frictional stress due to the adsorption of polymer chains is given as [48, 65]:

$$\sigma_{el} \cong \frac{Tv\tau_f}{\xi^4} \frac{(\tau_b/\tau_f)^2}{(\tau_b/\tau_f + 1)}, \qquad (3)$$

where τ_b is the life-time of adsorption. This contribution is the same as the adhesive friction proposed by Schallamach for rubber friction [68–72]. τ_b is related to the characteristic relaxation time of the polymer gel τ_f by the expression $\tau_b/\tau_f = u^{-1}\varphi(\alpha/u)$, and it decreases with an increase in the sliding velocity v. Here, $\varphi(\alpha/u)$ is a function of α and u, and $\alpha = v\tau_f/\xi$ is the scaled sliding velocity. The dimensionless parameter u is related to the adsorption energy of one chain F_{ads} as follows:

$$u = \exp(-F_{ads}/T). \qquad (4)$$

Thus, u is related to the spreading constant S by:

$$S \cong T\xi^{-2}\ln u. \qquad (5)$$

For weak adhesion, u is less than but close to unity, which corresponds to an S value that is slightly negative. When $\alpha/u \ll 1$, $\varphi(\alpha/u) \to 1$. This corresponds to the equilibrium case of $\tau_b = \tau_f \exp(F_{ads}/T)$ at $v = 0$.

The scaling theory is related to the molecular structure of the gel with elastic modulus E by the expression $E \cong T/\xi^3$. Thus, (3) may be rewritten as:

$$\sigma_{el}/E \cong \frac{\alpha}{u} \frac{\varphi^2(\alpha/u)}{[u+\varphi(\alpha/u)]}. \tag{6}$$

When $\alpha \ll 1$, $\sigma_{el}/E \propto \alpha^1$, and if $\alpha \gg 1$, $\sigma_{el}/E \propto \alpha^{-0.7}$. At $\alpha \approx 1$, σ_{el}/E reaches a maximum, as shown by Fig. 13a [48, 65].

Friction due to the viscous flow of solvent at the vicinity of the interface also occurs. By choosing the no-slip boundary at a layer that is one blob deep from the gel surface:

$$\sigma_{vis} \approx \frac{\eta v}{\xi}. \tag{7}$$

The characteristic velocity $v_f = \xi/\tau_f$ is defined as the case when the sliding velocity $v \ll v_f$ and the frictional force is due to the elastic force of the stretched polymer chain and increases with the sliding velocity. When $v \gg v_f$ the polymer does not have sufficient time to form an adsorbing site, and the friction decreases with an increase in the velocity in this velocity region. v_f can be expressed in terms of E using the scaling relation $v_f \propto T^{1/3}E^{2/3}/\eta$.

On the other hand, the friction from the second contribution, i.e., the viscous friction, increases monotonously with the velocity. Therefore, the first contribution is dominant when $v \ll v_f$, and the second contribution dominates when $v \gg v_f$.

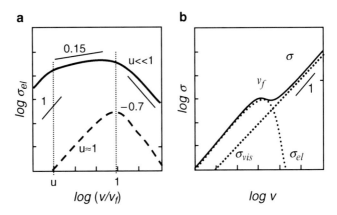

Fig. 13 Velocity dependence of frictional stress for a soft gel sliding on a smooth adhesive solid substrate. The result is based on the molecular picture in Fig. 12, which considers the thermal fluctuation of adsorption and desorption of the polymer chain. (**a**) The elastic term of the frictional stress of a gel. See text for a description of parameter u. (**b**) Summation of the elastic term and the viscous term. When $v \ll v_f$, the characteristic polymer adsorption velocity, the elastic term is dominant. At $v \gg v_f$, the viscose term is dominant. Therefore, transition from elastic friction to lubrication occurs at the sliding velocity characterized by the polymer chain dynamics. (Modified from figure 1 in [65])

At $v \approx v_f$, a transition from elastic friction to hydrated layer lubrication occurs, as shown schematically in Fig. 13b. When $v \gg v_f$, the friction $\sigma \approx \sigma_{vis} \propto \eta v/\xi$. It is estimated that $v_f \approx 2 \times 10^{-2}$ m/s if $D \approx 10^{-6}$ cm^2/s and $\xi = 5$ nm are used for PVA gel [65].

3.2.2 On a Smooth but Strongly Repulsive Substrate

In this case ($S > 0$), the sources contributing to the frictional stress of a gel are only from the lubrication of the hydrated layer of the polymer network σ_{vis} (Fig. 12) So:

$$\sigma_{vis} \approx \frac{\eta v}{h}. \tag{8}$$

Here, h is determined by the balance of the repulsion between the sliding surfaces and the normal pressure applied. For example, in the case of two similarly charged gels, the repulsive interaction originates from the osmotic pressure of the mobile counterions, and the h value might change over a range of several nanometers to several hundred nanometers, depending on the normal pressure applied [50].

3.3 Comparison of the Model with Observation

3.3.1 Gel Friction on Adhesive Surface

The effect of substrate adhesion on the frictional behavior of sliding hydrogels has been recently examined on smooth solid substrates with various hydrophobicities. The velocity dependence of frictional stress of a soft hydrogel, PVA, on adhesive smooth solid substrates in water using a strain-controlled parallel-plate rheometer is shown in Fig. 14. On a hydrophilic substrate that is weakly adhesive to gel (Fig. 14a), the friction–velocity relationship is divided into three regions: a creep region at low velocity ($v < \xi/\tau_b^0$), an elastic region at intermediate velocities ($\xi/\tau_b^0 \leq v < \xi/\tau_f$), and a lubricating region at high velocities ($v > \xi/\tau_f$), where $\tau_b^0 = \tau_f/u$, τ_f, and ξ are the polymer chain adsorption time at the thermodynamic equilibrium condition, polymer chain relaxation time, and mesh size, respectively. The friction behavior is satisfactorily described by the repulsion–adsorption model described above except for the creep region. There is no distinct water invasion by the so-called hydrodynamic effect, and the lubricating layer is formed by the polymer hydration with a thickness of the order of ξ that is velocity-independent [65].

On strongly adhesive substrates (Fig. 14b, c), friction increases with the substrate hydrophobicity in the low-velocity region, showing a weak velocity-strengthening, and a dramatic friction transition at around $v = 10^{-3}$ m/s. This value is one order lower than the characteristic velocity of the polymer chain $v_f = \xi/\tau_f$. The friction behavior is satisfactorily described by the repulsion–adsorption model below the transition region The friction transition is explained in terms of the elastic

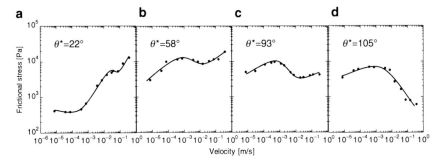

Fig. 14 Velocity dependence of dynamic frictional stress of PVA gel against solid substrates of various hydrophobicities in water: (**a**) G2, (**b**) SW, (**c**) OTS-glass, and (**d**) F-glass. Hydrophobicity of the substrates is in the order G2 < SW < OTS-glass < F-glass. Normal strain 26%; normal stress 14 kPa. Contact angles (θ^*) of substrates to water are shown. (Reproduced, with permission, from [65])

dewetting–wetting transition that is caused by the invasion of trapped water formed due to the heterogeneous dewetting.

Furthermore, when the substrate has a θ^* value higher than 105° (Fig. 14d), the friction–velocity curve exhibits a bell shape, which monotonously drops at $v > 10^{-3}$ m/s. On such a strongly hydrophobic surface, the apparent lubrication layer thickness is estimated to be in the order of several hundred nanometers. The result suggests that on such a hydrophobic surface, a large amount of water might be entrapped at the interface due to heterogeneous dewetting, which favors hydrodynamic lubrication at high velocity.

Therefore, on a weakly adhesive substrate, where the soft contact of the gel on the solid in water is homogeneous, the molecular model described above is essentially valid, and the friction at low velocity is due to elastic deformation of adsorbing polymer chain At high velocity, when the polymer chain of the gel has not enough time to form an adsorption site on the substrate, the friction is due to viscous dissipation of the lubricating layer. On a strongly adhesive substrate, the heterogeneous dewetting might lead to low friction at elevated velocity. The latter phenomenon is applicable in the design of low friction systems working at a high shear velocity.

3.3.2 Gel Friction on Repulsive Surface

According to the repulsion–adsorption model, a solvent layer is formed at the interface when the gel–substrate interaction is repulsive, and the frictional shear stress arises from the viscous flow of the solvent layer. Therefore, the frictional stress should increase with the increase in the sliding velocity, i.e., $\sigma \propto v$.

Figure 15 shows the frictional behavior of a ring-shaped PNaAMPS gel, which is negatively charged, rotated against a glass surface, which is also negatively charged in water, using a strain-controlled parallel-plate rheometer [53]. As shown in Fig. 15, the frictional stress σ increases with an increase in velocity, whereupon the profiles

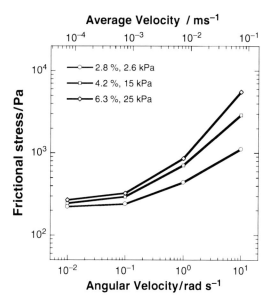

Fig. 15 Angular velocity dependence of the frictional force for PNaAMPS gel rotated on a piece of glass surface in pure water at 25 °C at different normal strains. Values of the normal compressive strains (%) and normal stresses (kPa) applied during the measurement are shown. Sample thickness 3 mm; degree of swelling 27. (Reproduced, with permission, from [53])

depend on the applied normal compressive strain, and therefore the normal stress. At low normal strain (2.8%), σ is almost constant for low velocities and then gradually increases at higher velocities. At higher normal strains, σ increases moderately in the low velocity range but increases distinctly at higher velocities. In other words, the velocity dependence is less notable at low velocities and normal strains and becomes stronger at higher velocities and normal strains.

Although the frictional stress increases monotonously with an increase in velocity, the relationship between σ and v is far more complicated than can be expected from the simple lubrication mechanism, which predicts $\sigma \propto v^{1.0}$. This discrepancy might originate from the smaller lubricating layer thickness. The real local normal stress, which is determined by the compressive stiffness of the gel (>0.8 MPa), might be too large to form a continuous lubricating layer that sustains a hydrodynamic lubrication.

4 Gels with Low Friction

4.1 Friction Reduction by Template Effect

It has been discovered that the surface structure, and therefore friction, of a gel are strongly dependent on the substrate on which the gel is synthesized [81]. For example, gels that are synthesized between a pair of glass substrates have a mirror-like smooth surface. This is the same when a hydrogel is synthesized on other hydrophilic substrates, such as mica and sapphire. However, even for the same chemical structure, a hydrogel exhibits an eel-like slim surface when it is synthesized on hydrophobic substrates such as Teflon and polystyrene (PS). The differences in the surface nature of the gels synthesized on different substrates are so obvious that they can be easily distinguished by touching with one's finger. If a hydrogel is synthesized between two plates, one of which is hydrophobic and the other hydrophilic, heterogeneous gelation occurs [82]. After the equilibrated swelling in water, the gel exhibits a significant curvature, whereby the gel surface formed on the hydrophobic Teflon surface is always the outside of the curvature and that on the glass is the inside. This is because the gel surface close to the Teflon has a higher swelling, while the one next to it has a lower value. It has been clarified that the gel has a gradient structure and that the surface formed on Teflon (or other hydrophobic substrates) has a low cross-linking density, with branched dangling polymer chains [81].

The frictional force and frictional coefficient of PAMPS gels synthesized on a glass plate and on a PS plate were measured against a glass plate in water using a rheometer (Fig. 16) [52]. The frictional stress of the gel prepared on PS substrate shows a lower value than that prepared on a glass substrate. Especially in the low velocity range, the frictional stress of the gel prepared on PS substrate can be as low as 1 Pa, a value equivalent to the shear stress on the wall of blood vessels [1] The frictional coefficient of the gel prepared on PS reaches 10^{-4}, at least two orders lower in magnitude than that of the gel synthesized on glass at the low velocity range. The reduction in friction is attributed to the presence of branched dangling chains on the gel surface prepared on the hydrophobic substrate, as revealed by the result for the PAMPS gels containing free linear PAMPS polymer chains prepared on the glass plate, which showed similar low friction coefficients (Fig. 16).

Such a substrate effect is observed in a wide variety of hydrogels prepared from water-soluble vinyl monomers (e.g., the sodium salt of styrene sulfonate, acrylic acid, and acrlyamide), and on various hydrophobic substrates, such as Teflon, polyethylene, polypropylene, PVC, and polymethyl methylacrylate (PMMA) [52]. This template effect is due to retardation of the radical polymerization near the rough and hydrophobic substrates that trap oxygen at the solid surface [82].

4.2 Friction Reduction by Dangling Chains

The effect of polymer brushes on the reduction of sliding friction was also observed for solid friction, using surface force apparatus (SFA) measurements [83–86]. For example, Klein et al. reported a massive lubrication between mica surfaces modified by repulsive polyelectrolyte brushes in water [83]. These results show that polymer dangling chains on solid or gel surfaces can dramatically reduce the surface friction if the polymer brush has a repulsive interaction with the sliding substrate.

Soft and Wet Materials: From Hydrogels to Biotissues

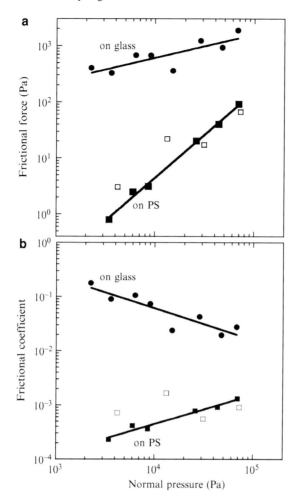

Fig. 16 Normal pressure dependence of the frictional force (**a**) and the frictional coefficient (**b**) of PAMPS gels slid against a glass plate in water at an angular velocity of 0.01 rad/s: prepared on glass, swelling degree 21 (*circles*); prepared on PS, swelling degree 27 (*closed squares*); and containing linear polymer chains, swelling degree 15 (*open squares*). (Reproduced, with permission, from [52])

The dramatic friction reduction effect of dangling chains has also been found in the negative load dependence of some physically cross-linked polysaccharide gels, such as gellan and κ-carrageenan. For example, the frictional force of gellan gel is almost independent of normal load when the pressure is lower than a critical value [51]. However, the frictional force decreases with an increase of normal load when the pressure is increased to a critical value (about 10^4 Pa). Polysaccharide gels are loosely cross-linked by salt. They undergo "sol–gel" transition reversibly and transfer to sol state under heating or pressure, and gradually dissolve in water even at room temperature. Gellan and κ-carrageenan are loosened partially or dissolved slowly from the gel network under a high pressure exceeding a critical value. The dissolved polymer chains form the dangling-like architecture on the gel surface, leading to an increased hydrodynamic layer, which contributes to the decrease in frictional force beyond the critical load.

To quantitatively investigate the relationships between the friction coefficient and the polymer brush properties (such as polymer length, density, and structure), hydrogels of poly(2-hydroxyethyl methacrylate) (PHEMA) with well-defined polyelectrolyte brushes of poly(sodium 4-styrenesulfonate) (PNaSS) of various molecular weights were synthesized, keeping the distance between the polymer brushes constant at ca. 20 nm [59]. The effect of polyelectrolyte brush length on the sliding friction against a glass plate, an electrorepulsive solid substrate, was investigated in water over a velocity range of 7.5×10^{-5} to 7.5×10^{-2} m/s. It was found that the presence of polymer brushes can dramatically reduce the friction when the polymer brushes are not very long. With an increase in the length of the polymer brushes, this drag reduction effect only works at a low sliding velocity, and the gel with long polymer brushes even shows a higher friction than that of a normal network gel at a high sliding velocity. The strong dependence on polymer length and sliding velocity indicate a dynamic mechanism of the polymer brush effect [59].

The presence of dangling chains might effectively enhance the hydrodynamic thickness of the solvent layer at the sliding interface. As discussed previously, when the interfacial interaction is repulsive between gel and the substrate on which the gel is slid, a water layer is retained at the interface even under a large normal load to give a low friction. Under the same pressure, the static solvent layer thickness should be the same for both the chemically cross-linked gel and the gel having branched dangling polymer chains on its surface. However, under a shear flow, which occurs in the process of friction measurement, the polymer brushes are deformed more easily than the cross-linked network. This would increase the effective thickness of the hydrodynamic layer, resulting in a reduction of the shear resistance. This discovery should enable the hydrogel to find wide application in many fields where low friction is required.

4.3 Friction Reduction by Dilute Polymer Solution

The friction events in a biological system mostly occur between soft and wet tissues intermediated by viscoelastic polymer fluids such as synovial fluid or mucus [1]. For example, mucus adheres to many epithelial surfaces, where it serves as a diffusion barrier against contact with noxious substances (e.g., gastric acid and smoke) and as a lubricant to minimize shear stresses; such mucus coatings are particularly prominent on the epithelia of the respiratory, gastrointestinal, and genital tracts. Mucus is also an abundant and important component of saliva, giving it virtually unparalleled lubricating properties [1].

The effect of a *nonadhesive* polymer solution on the friction of an *adhesive* gel on a glass substrate has been studied [64]. For example, friction of physically

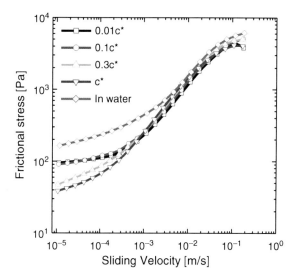

Fig. 17 Sliding velocity dependence of the frictional stress for PVA gel sliding against a piece of glass surface in dilute PEO solution with different concentrations in terms of overlap concentration C^* at 20°C. Sample thickness ~3 mm; normal pressure 14 kPa; Mw of PEO 20,000. (Reproduced, with permission, from [64])

cross-linked poly(vinyl alcohol) (PVA) gel against glass in linear poly(ethylene oxide) (PEO) solution was investigated at various PEO concentrations (Fig. 17). There is no specific interaction between these two neutral and hydrophilic polymers. At low sliding velocity ($10^{-5}, 10^{-4}$ m/s), distinct PEO polymer effects are observed: The frictional stress in PEO solutions is lower than that in pure water, decreasing with the increase in PEO concentration and reaching a minimum at the crossover concentration, c^* (Fig. 17). At fast sliding velocity ($10^{-2}, 10^{-1}$ m/s), all the friction curves in dilute PEO solution superpose with the curve in pure water, independent of PEO concentration. These results indicate that in the low sliding velocity region, where adsorption of PVA gel on glass plays the dominant role in friction, PEO chains screen the adsorption of PVA chains to the glass surface. In the fast sliding velocity region, PEO chains are either extensively stretched or form a deplete layer on the glass surface due to the high shear rate, so the liquid lubrication with a viscosity of $\eta \approx \eta_{\text{water}}$ prevails. These results give an essential idea of the role of mucins in biological systems in reducing the friction at the low sliding velocity region, i.e., they screen the interaction between two sliding surfaces and sustain water.

4.4 Friction Reduction by Substrate Roughness

The relationship between the frictional stress and velocity of a PVA hydrogel sliding on rough glass substrates in water was studied with a strain-controlled rheometer

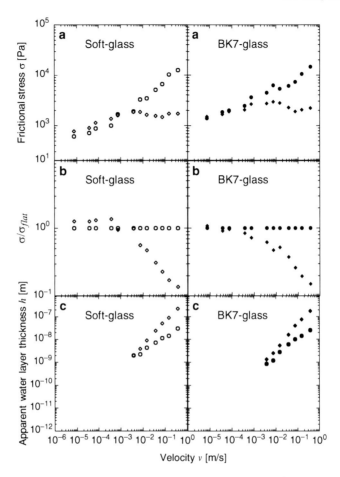

Fig. 18 Velocity dependence of dynamic frictional stress (**a**), dynamic frictional stress normalized by the value on a flat surface (G20 or BK7 1/10) (**b**), and the apparent water layer thickness, estimated by supposing a lubrication mechanism of $\sigma = \eta v/h$ (**c**), of the PVA gel sliding against soft glass (*left column, open symbols*) and BK7 glass (*right column, closed symbols*) of different roughness R_a. The measurements were performed in water at 20°C under a normal strain of 26%: G20, $R_a = 20$ nm (*open circles*); G1000, $R_a = 1000$ nm (*open diamonds*); BK7 1/10, $R_a = 20$ nm (*closed circles*); BK7 frost, $R_a = 1000$ nm (*closed diamonds*). (Reproduced, with permission, from [66])

(Fig. 18). When the substrate roughness increased, the friction in the low velocity region slightly increased, whereas it did not change remarkably at high velocity when the substrate roughness R_a increased up to 100 nm. For a rougher surface with $R_a = 1000$ nm, the friction began to decrease slightly at a sliding velocity of $v = 10^{-4}$–10^{-3} m/s and abruptly at $v \approx 10^{-2}$ m/s. When $v > 10^{-1}$ m/s, the friction showed a tendency to increase again with an increase in velocity. These behaviors are well explained in terms of contact dynamics, which features two characteristic

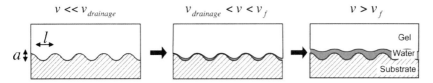

Fig. 19 Contact of a soft gel sliding on a rough solid substrate. The substrate roughness l and a are assumed to be much larger than the mesh size ξ of the gel. As the sliding velocity (v) increases, the effective contact area decreases. (Reproduced, with permission, from [66])

velocities, i.e., v_f and $v_{drainage}$, as shown in Fig. 19. Here, v_f is the characteristic velocity of the adsorption model. $v_{drainage}$ is the characteristic water drainage velocity at the asperities and is given by [66]:

$$v_{drainage} \cong \frac{\xi^2 P}{l\eta} \frac{1}{(1 - \xi^2/a^2)}.$$

Here, l and a are the characteristic wavelength and depth of asperities, respectively; ξ the mesh size of gel network; P the normal pressure applied; and η the viscosity of water. When the characteristic wavelength of asperities l and depth a is much larger than the mesh size of the gel ($l \geq a \gg \xi$), we have $v_{drainage} \ll v_f$. At $v < v_{drainage}$, the friction increases with the roughness due to the enhanced contact area between the gel and the substrate. At $v_{drainage} < v < v_f$, the gel does not have sufficient time to form complete contacts with the surface asperities and the contact area decreases, which leads to a decrease in friction with the velocity. At $v \gg v_f$, there is no contact between the gel and substrate, and a continuous lubricating layer forms at the interface. This again leads to an increase in the frictional stress with velocity.

The results show that the friction of a gel on a weak adhesive substrate can be minimized even in the high velocity region, depending on substrate roughness. This result provides some essential ideas for designing a soft gel system with low friction over a wide velocity range, which is important in bioengineering applications where low friction is required, such as in artificial articular joints and artificial hearts.

5 Polymer Gels as Scaffolds for Cell Cultivation

To mimic the macromolecular-based ECM in biological tissue, the cell adhesion and proliferation properties of hydrogels are critical parameters. However, various hydrogels that originate from natural resources, such as alginate [87], chitosan [88, 89], and hyaluronic acid [90], and that are synthetically created, such as poly (N-isopropylacrylamide) (PNIPAAm) [91], PEO [92], PVA [93], and poly(ethylene glycol) (PEG) [94], show a poor cellular viability without modification with cell adhesive proteins or peptides, such as collagen, laminin, fibronectin, and the RGD (Arg-Gly-Asp) sequence.

Recently, it has been found that bovine fetal aorta endothelial cells (BFAECs) can spread, proliferate, and reach confluency on synthetic hydrogels with negative charges, such as PNaAMPS, PNaSS, and poly(acrylic acid) (PAA), without surface modification with any cell adhesive proteins or peptides [95–98]. The artificial cell scaffolds from synthetic sources have many advantages over natural sources like collagen: the chemical properties of synthetic gels are easily controllable and reproducible; they are infection-free, withstand high-temperature sterilization, and are relatively low cost. Therefore, the success in cultivating cells on negatively charged synthetic gel surfaces without any surface modification of adhesive proteins would substantially promote the application of gels for tissue engineering. Cell behaviors on various hydrogels and the effect of gel charge density on cell growth are described in Sects. 5.1–5.3.

5.1 Cell Growth on Various Gels

The behavior of BFAECs on several kinds of negatively charged hydrogels such as PAA, PMAA, PNaSS, and PNaAMPS, and on neutral hydrogels such as PAAm and PVA with different cross-linking density have been investigated [95]. These polymers form strong hydrogels by the DN structure with proper combinations [23]. Understanding the cellular behavior of these gels helps to design gels that have cell viability as well as tough mechanical strength.

Phase-contrast micrographs of BFAECs on various kinds of gels cultured at 6 h and 120 h are shown in Fig. 20. At 6 h of culture, BFAECs could be observed adhered to or spread on the gel surface, but not proliferating. Cells do not spread on nonionic PAAm gels. Clearly, PAAm gels have a strong inhibiting effect on cell spreading. It is thought that weak interaction between the proteins and PAAm surface [99, 100] results in the reduced cell activity.

However, about 50% spreading ratio is observed on another nonionic gel, PVA, when the cross-linking concentration is 6 or 10 mol%. It decreases to about 15% when the cross-linking concentration is decreased to 2 or 4 mol%. This result indicates that cell viability on PVA gel is better than that on PAAm gel at the initial culture stage. Cells proliferate with culture time and reach subconfluence at about 168 h.

Cell viability on PAA and PMAA gels, which are weak polyelectrolytes with carboxylic acid groups on the side chains of the polymers, were also investigated. On PAA gels, cells proliferate with culture time, eventually reaching confluency at 144 h, with a cell density of about 7×10^4 cell/cm^2. In contrast to cell proliferation on PAA gel, although cells can proliferate on PMAA gel at 4 days, cells cannot proliferate after that. This indicates that cell proliferation is very sensitive to chemical structure and to the hydrophobic/hydrophilic properties of gels. The above results indicate that cell spreading and proliferation are sensitive to the cross-linking concentration of weak polyelectrolyte gels. It was reported that carboxylic acid groups of the PAA-grafted surface have a negative effect on cell adhesion, spreading, and

Soft and Wet Materials: From Hydrogels to Biotissues

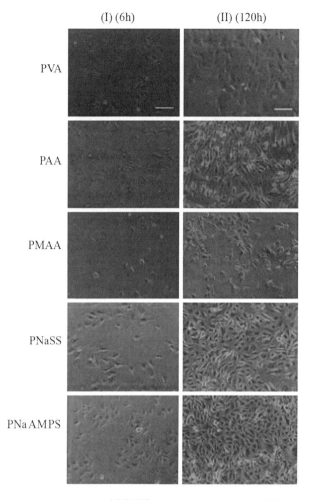

Fig. 20 Phase-contrast micrographs of BFAECs on various gels at the initial stage (6 h, *column I*) and at a prolonged culture time (120 h, *column II*). Crosslinker concentration of gels: PVA 6, PAA 2, PMAA 1, PNaSS 10, and PNaAMPS 6 mol%. Original magnification 10×; *scale bar*: 100 μm. (Reproduced, with permission, from [95])

growth [101]. Another research study showed that there were only a few cells attached onto PAA-grafted surface, and that cell proliferation was very slow [102]. Our results indicate that if PAA is cross-linked under suitable conditions, PAA hydrogel is also able to act as a cell scaffold.

On PNaSS and PNaAMPS gels, which are strong polyelectrolytes because both have sulfonate functional groups, distinct cell behaviors are observed as compared with PAA and PMAA gels. Cells proliferate to confluency on PNaSS and PNaAMPS gels, with a cell density higher than 1.1×10^5 cell/cm^2 at 144 h. Cell proliferation behaviors on PNaSS and PNaAMPS gels are nearly the same and are independent of cross-linking concentration.

Many chemical and physical properties of the substrate affect cell compatibility, such as charge density [103, 104], interfacial energy, hydrophobicity/hydrophilicity balance, mobility of the polymer chain, and morphology of the material [105]. The results on gels suggested that the protein absorption from serum-containing medium onto a PNaSS and PNaAMPS gel surface might occur to facilitate cell adhesion, spreading, and proliferation. The study demonstrated that BFAECs can grow on PVA, PAA, PMAA, PNaSS, and PNaAMPS gels without modification with any proteins or peptides, and that the cells exhibit a high level of spreading and proliferation.

5.2 Effect of Charge on Cell Growth

It was reported that cell–substrate interaction is closely related to substrate surface charge [105]. The behavior of cells on gels also showed similar results to those described in the previous section. A more quantitative study on the relationship between gel charge density and cell behavior has been performed by using copolymer gels consisting of a neutral polymer, poly(N, N'-dimethylacrylamide) (PDMAAm), on which the cell does not show growth, and a negatively charged polyelectrolyte, PNaAMPS, on which the cell shows confluent growth.

As shown in Fig. 21a, the degree of swelling of the copolymer gels increases with an increase of the molar fraction, F, of charged the moiety due to the enhanced charge density of the copolymer gels. Taking into account the change in charge density caused by the change in copolymer composition, the relationship between the zeta potential of poly(NaAMPS-co-DMAAm) gels and cell behavior is

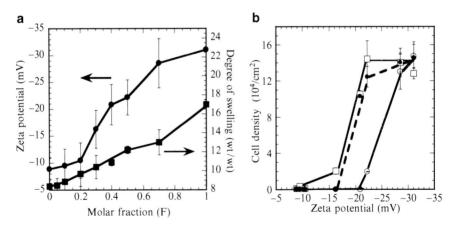

Fig. 21 (a) Zeta potential and degree of swelling of poly(NaAMPS-co-DMAAm) gels as a function of NaAMPS molar fraction F. (b) Cell density at 120 h as a function of zeta potential of poly(NaAMPS-co-DMAAm) gels. Endothelial cells used were BFAECs (*open squares*), HUVECs (*closed circles*), and RSTFCs (*open circles*). Error ranges are standard deviations over $n = 4$–8 samples. (Reproduced, with permission, from [98])

analyzed. Figure 21a shows the zeta potential of poly(NaAMPS-co-DMAAm) gels equilibrated in HEPES buffer solution as a function of F. Figure 21b shows various kinds of endothelial cell (BFAECs, HUVEC, and RSTFCs) density at 120 h as a function of the zeta potential of the copolymer gels.

Although PDMAAm is neutral, it shows a negative zeta potential in HEPES buffer solution, perhaps due to its ionic adsorption in buffer solution. With the increase of F, the zeta potentials of poly(NaAMPS-co-DMAAm) gels increase and saturate to -30 mV. Figure 21b shows that BFAEC density increases when the zeta potential is -16 mV ($F = 0.3$), and the cells proliferate to confluence when the zeta potential is -20 mV ($F = 0.3-0.4$). HUVEC density sharply increases and the cells begin to proliferate when the zeta potential is -20 mV ($F = 0.4$). RSTFC density increases when the zeta potential is -22.2 mV ($F = 0.5$), and the cells proliferate to confluence when the zeta potential is -28.5 mV ($F = 0.7$). The results demonstrated that there is a critical zeta potential for cell spreading and proliferation on the poly(NaAMPS-co-DMAAm) gels.

It is not known why cell spreading and proliferation can be manipulated by the charge density of gels. It is thought that the charge density of gels adjusts the gel adsorption of adhesive proteins from the serum-containing culture medium. To elucidate this, the effect of charge density on the amount of total adsorbed protein from serum-containing culture medium, and on the fibronectin contained in the total adsorbed protein on poly(NaAMPS-co-DMAAm) gels have been investigated [98]. The effects of gel charge density on total protein and fibronectin adsorption coincide well with cell behavior. The results indicated that gel charge density adjusts protein adsorption, i.e., more negative charge, more protein adsorption, which favors cell spreading and proliferation.

5.3 Cell Proliferation on Tough Gels

In previously studies, we have reported that DN gels show a fracture strength as high as a few tens of megapascals, although both of the individual networks are mechanically weak [23]. However, tough DN gels, such as PNaAMPS/PDMAAm and PNaAMPS/PAAm, do not facilitate cell spreading and proliferation, although many cells can adhere to the surfaces of these gels. This is because in the process of DN synthesis, the second monomer mostly polymerizes on the surface of the first network and, therefore, the surface chemical composition and the properties of DN gels are determined by the second component (i.e., the neutral component, such as PDMAAm and PAAm), which is not suitable for cell spreading and proliferation.

According to the results of critical charge effect on cell behavior, as previously described, we can design a tough triple network (TN) gel that facilitates cell spreading and proliferation, and, at the same time, preserves high mechanical strength by inducing a copolymer as the third component to DN gels.

PNaAMPS/PDMAAm/poly(NaAMPS-co-DMAAm) TN gels, with the third network having a molar fraction, $F = 0.5$, have been synthesized. These TN gels are very tough, having fracture stresses as high as 1–3 MPa, compared with single

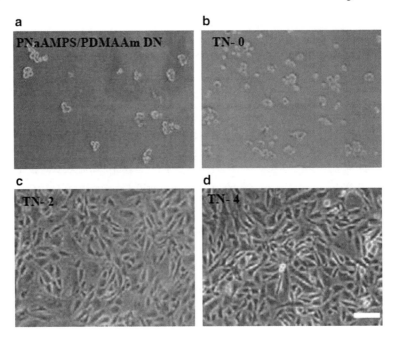

Fig. 22 Phase-contrast micrographs of BFAECs cultured on double network (DN) and triple network (TN) gels: PNaAMPS/PDMAAm DN gel (**a**), PNaAMPS/PDMAAm/poly(NaAMPS-*co*-DMAAm) TN gels (**b–d**). In the third network, NaAMPS molar fraction was $F = 0.5$. The cross-linker concentration of the third network was 0 (**a**), 2 (**b**), and 4 mol% (**c**). *Scale bar*: 100 μm. (Reproduced, with permission, from [98])

network gels PNaAMPS (0.63 MPa), and poly(NaAMPS-*co*-DMAAm) with $F = 0.5$ (0.26 MPa). BFAEC proliferation on these TN gels has been studied. Images of BFAECs on these TN gels at 120 h are shown in Fig. 22. On PAMPS/PDMAAm DN gel and a TN gel in which the third network is not cross-linked (sample TN-0), no cell spreading is observed (Fig. 22a, b). When the third network cross-linker concentration is 2 or 4 mol% (samples TN-2 and TN-4), cells proliferate and reach confluence (Fig. 22c, d). These results indicate that the surface properties of the TN gels with poly(NaAMPS-*co*-DMAAm) as the third component are the same as those of the single network poly(NaAMPS-*co*-DMAAm) gel.

6 Robust Gels with Low Friction: Excellent Candidates as Artificial Cartilage

Most biological tissues are in the gel state; for example, articular cartilage is a natural fiber-reinforced hydrogel composed of proteoglycans, type II collagen, and approximately 70% water. Normal cartilage tissue contributes highly to joint functions that involve ultralow friction, distribution of loads, and absorption of impact energy. When normal cartilage tissue is damaged, it is extremely difficult to regenerate these tissues with currently available therapeutic treatments Therefore, it is important to develop substitutes for normal cartilage tissue as a potential therapeutic option.

The important characteristic of materials used as load-bearing tissue replacements is that the mechanical properties are comparable to those of native tissue. As described above, DN hydrogels can withstand a few to tens of megapascals of mechanical strength. The value is comparable to the severe loading conditions imposed on human articular cartilage (1.9–14.4 MPa) [1], satisfying the mechanical property requirement of articular cartilage substitute.

6.1 Wearing Properties of Robust DN Gels

For application of DN gels to artificial articular cartilage, it is crucial to evaluate the wear properties, because articular joints are subjected to rapid shear force of high magnitude for millions of cycles over a lifetime. However, there are no established methods for evaluating the wear property of gel. The pin-on-flat wear testing that has been used to evaluate the wear property of ultrahigh molecular weight polyethylene (UHMWPE), which is only one established rigid and hard biomaterial used in artificial joints, was used to evaluate the wear property of DN gels. The wear property of four kinds DN gels composed of synthetic or natural polymers, PAMPS/PAAm, PAMPS/PDMAAm, BC/PDMAAm, and BC/gelatin were evaluated [106]. The properties of the four DN gels used for the wearing test are shown in Table 1.

The morphologies of DN gels under one million cycles of friction, which is equivalent to 50 km of friction (50×10^6 mm), are shown in Fig. 23. The maximum wear depth of the PAMPS/PAAm, PAMPS/PDMAAm, BC/PDMAAm, and BC/gelatin gels was 9.5, 3.2, 7.8, and 1302.4 μm, respectively. It is amazing that the maximum wear depth of PAMPS/PDMAAm DN gel is similar to the value of UHMWPE (3.33 μm). In addition, although the maximum wear depth of

Table 1 The average values of the water content, and the material properties concerning the PAMPS/PAAm DN gel, PAMPS/PDMAAm DN gel, BC/PDMAAm DN gel, and BC/Gelatin DN gel, which were used in the wearing study. (Reproduced, with permission, from the literature [106])

DN gels	PAMPS/PAAm	PAMPS/PDMAAm	BC/PDMAAm	BC/Gelatin
Water content (%)	90.0	94.0	85.0	78.0
Elastic modulus (MPa)	0.33	0.20	1.6	1.7
Stress at failure (MPa)	17.2	3.1	2.9	3.7
Strain at failure (%)	92	73	50	37

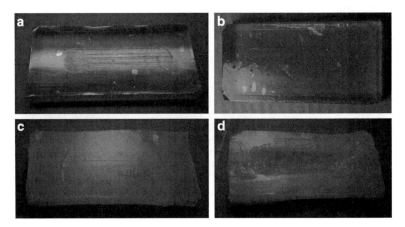

Fig. 23 Gross observations of representative samples after the wear tests: (**a**) PAMPS/PAAm DN gel, (**b**) PAMPS/PDMAAm DN gel, (**c**) BC/PDMAAm DN gel, and (**d**) BC/gelatin DN gel. (Reproduced, with permission, from [106])

PAMPS/PAAm DN gel and BC/PDMAAm DN gel is about 2–3 times higher than that of UHMWPE, these gels could bear the one million times of cyclic friction. The results demonstrate that PAMPS/PAAm, PAMPS/PDMAAm, and BC/PDMAAm DN gels are resistant to wear to a greater degree than conventional hydrogels, and that PAMPS/PDMAAm DN gel is potentially useful as a replacement material for artificial cartilage. On the other hand, BC/gelatin DN gel, which is composed of natural materials, shows extremely poor wear properties compared with other DN gels. Some reasons, such as relatively low water content, higher friction coefficient, and the ease of becoming rough by abrasion, are attributed to the lower wear properties of BC/gelatin DN gel.

For designing materials that could be potentially used as artificial cartilage, suitable viscoelasticity, high mechanical strength, durability against repetitive stress, low friction, high resistance to wear, and resistance to biodegradation within the living body are required. It is difficult to develop a gel material that satisfies even two of these requirements at the same time. However, recent developments in synthesizing mechanically strong hydrogels have broken though the conventional gel conception, and opened a new era of soft and wet materials as substitutes for articular cartilage and other tissues.

6.2 Robust Gels with Low Friction

It has been described that a hydrogel with polyelectrolyte branched dangling polymer chains on its surface can effectively reduce the surface sliding frictional coefficient to a value as low as 10^{-4}.

Table 2 Mechanical properties of DN, TN and DN-L gels. (Reproduced, with permission from the literature [61])

Gels	Water content (wt%)	Elasticity (MPa)	Fracture stress σ_{max} (MPa)	Fracture strain ε_{max} (%)
DN	84.8	0.84	4.6	65
TN	82.5	2.0	4.8	57
DN-L	84.8	2.1	9.2	70

Based on this research, the new soft and wet materials, with both low friction and high strength, are synthesized by introducing a weakly cross-linked PAMPS network (to form a TN gel) or an noncross-linked linear polymer chain (to form a DN-L gel) as a third component into the optimal tough PAMPS/PAAm DN gel [61]. The TN and DN-L gels were synthesized by UV irradiation after immersing the DN gels in a large amount of a solution of 1 M AMPS and 0.1 mol% 2-oxoglutaric acid with (to form TN) and without (to form DN-L) the presence of 0.1 mol% MBAA. The mechanical properties of the gels are summarized in Table 2. After introducing cross-linked or linear PAMPS to the DN gel, the fracture strength of TN and DN-L gels was on the order of megapascals, and the elasticity of the gels was higher than that of DN gel ($\sim$2 MPa). The fracture strength of DN-L was remarkably increased because PAMPS linear chains can effectively dissipate the fracture energy.

Figure 24a, b show the frictional forces (F) and frictional coefficient (μ) of the three kinds of gels as a function of normal pressure (P) [61]. The gels were slid on the glass plate in water. The results clearly shows that the frictional coefficient decreases in the order DN > TN > DN-L, indicating that introduction of PAMPS, especial linear PAMPS, as the third network component obviously reduces the frictional coefficient of the gels.

The DN gel has a relatively large frictional coefficient ($\sim 10^{-1}$) because the second network, nonionic PAAm, dominates the surface of the DN gel and is adsorptive to the glass substrate.

However, when the PAMPS network is introduced to the DN gel as the third component, frictional coefficient of the TN decreases to $\sim 10^{-2}$, which is two orders of magnitude lower than that of the DN gel. This is because the surface of the TN gel is dominated by PAMPS, resulting in repulsive interaction with the glass substrate and reduction of the frictional force. Furthermore, when linear PAMPS chains are introduced to the surface of the DN gel, the frictional coefficient is significantly reduced to $\sim 10^{-4}$, which is one to three orders of magnitude less than that of TN and DN gels, respectively. This demonstrates that the linear PAMPS chains on the gel surface reduce frictional force due to further repulsive interaction with the glass substrate [48, 52, 53].

It should be emphasized that the lower frictional coefficient of DN-L gel can be observed under a pressure range of 10^{-3}–10^5 Pa, which is close to the pressure exerted on articular cartilage in synovial joints. The results demonstrate that the linear polyelectrolyte chains are still effective in maintaining lubrication, even under an extremely high normal pressure.

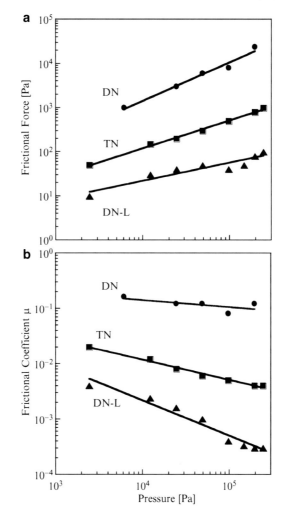

Fig. 24 Normal pressure dependence of frictional force (**a**) and frictional coefficient (**b**) of tough DN gels against a glass plate in pure water: DN (*circles*), TN (*squares*), and DN-L (*triangles*) gels. Sliding velocity 1.7×10^{-3} m/s. (Reproduced, with permission, from [61])

6.3 Biological Responses of DN Gels in Muscle and Subcutaneous Tissue

The biological reaction of four types of DN gels in muscle and subcutaneous tissues have been evaluated, using implantation tests according to the international guidelines [107, 108]. The four gels tested were PAMPS/PDMAAm DN gel, PAMPS/PAAm DN gel, BC/PDMAAm DN gel, and BC/gelatin DN gel.

Soft and Wet Materials: From Hydrogels to Biotissues

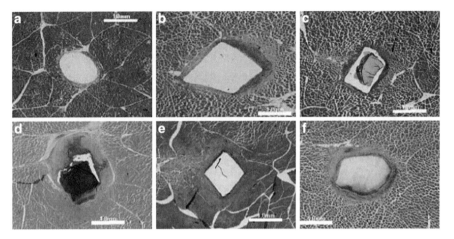

Fig. 25 Histological observations at 4 weeks (HE stain, 2×). (**a**) Negative control (high density polystyrene), (**b**) Positive control (polyurethane containing 0.75% zinc diethyldithiocarbamate), (**c**) PAMPS/PDMAAm gel, (**d**) PAMPS/PAAm gel, (**e**) BC/PDMAAm gel, (**f**) BC/gelatin gel. Scale bar is 1 mm. (Reproduced, with permission, from [108])

The implantation tests demonstrated that, although PAMPS/PDMAAm DN gel induced a mild inflammation at 1 week, the degree of the inflammation significantly decreased to the same degree as that of the negative control at 4 weeks, as shown in Fig. 25. Because the previous studies showed that the PAMPS/PDMAAm DN gel has excellent wear properties, comparable to UHMWPE in pin-on-flat-type wear testing [106], and that this gel is hardly degraded when implanted into a living body [107], we believe that this gel has the possibility to be applied as artificial cartilage in the future. However, to verify this possibility, many other factors needed for artificial cartilage repair, such as porosity, cell nutrition, changes in water content and compressive strength, capsule formation in the cartilage tissue, and so on, should be evaluated in future studies. Secondly, BC/gelatin gel showed the same degree of inflammation as that of the negative control at 1 week, and then showed a gradual absorbance at 4 and 6 weeks. This gel has the potential to be applied as an absorbable implant. The PAMPS/PAAm and BC/PDMAAm gels induced a significant inflammation at both 1 and 4 weeks. These DN gels are difficult to apply as clinical implants in their current state of development.

Acknowledgements This research was financially supported by a Grant-in-Aid for the Specially Promoted Research (No. 18002002) from the Ministry of Education, Science, Sports and Culture of Japan. The author thanks T. Kurokawa, Y. Tanaka, Y. M. Chen, H. Na, H. Furukawa, and the graduate students in LSW for their contributions to double network gel research. The author also thanks K. Yasuda, C. Creton, W. L. Wu, and H. Brown, for useful discussions and contributions to this work.

References

1. Fung YC (1993) Biomechanics: mechanical properties of living tissues, 2nd edn. Springer, New York
2. McCutchen CW (1978) Lubrication of joints, the joints and synovial fluid. Academic, New York
3. Derbyshire B, Fisher J, Dowson D, Hardaker C, Brummitt K (1994) Comparative study of the wear of UHMWPE with zirconia ceramic and stainless steel femoral heads in artificial hip joints. Med Eng Phys 16:229
4. Foley DP, Melkert R, Serruys PW (1994) Influence of coronary vessel size onrenarrowing process and late angiographic outcome after successful balloonangioplasty. Circulation 90:1239–1251
5. DeRossi D, Kajiwara K, Osada Y, Yamauchi A (1991) Polymer gels-fundamentals and biomedical applications. Plenum, NewYork
6. Tanaka Y, Nishio I, Sun ST, Ueno-Nishio S (1982) Collapse of gels in an electric field. Science 218:467–469
7. Osada Y, Okuzaki H, Hori H (1992) A polymer gel with electrically driven motility. Nature 355:242–244
8. Osada Y, Gong JP (1998) Soft and wet materials: polymer gels. Adv Mater 10:827–837
9. Zarzycki J (1988) Critical stress intensity factors of wet gels. J Noncryst Solids 100:359–363
10. Tanaka Y, Fukao K, Miyamoto Y (2000) Fracture energy of gels. Eur J Phys E 3:395–401
11. Bonn D, Kellay H, Prochnow M, Ben-Djemiaa K, Meunier J (1998) Delayed fracture of an inhomogeneous soft solid. Science 280:265–267
12. Lake GJ, Thomas AG (1967) The strength of highly elastic materials. Proc R Soc Lond 300:108–119
13. Furukawa H, Horie K, Nozaki R, Okada M (2003) Swelling-induced modulation of static and dynamic fluctuations in polymer gels observed by scanning microscope light scattering. Phys Rev E 68:031406.1–031406.14
14. Simha NK, Carlson CS, Lewis JL (2003) Evaluation of fracture toughness of cartilage by micropenetration. J Mater Sci Mater Med 14:631–639
15. Abe H, Hayashi K, Sato M (1996) Data book on mechanical properties of living cells, tissues, and organs. Springer, Tokyo
16. Okumura Y, Ito K (2001) The polyrotaxane gels: a topological gel by figure-of eight cross-links. Adv Mater 13:485–487
17. Karino T, Okumura Y, Ito K, Shibayama M, (2004) SANS studies on spatial inhomogeneities of slide-ring gels. Macromolecules 37:6177–6182
18. Karino T, Okumura Y, Zhao C, Kataoka T, Ito K, Shibayama M. (2005) SANS studies on deformation mechanism of slide-ring gel. Macromolecules 38:6161–6167
19. Haraguchi K, Takehisa T. (2002) Nanocomposite hydrogels: a unique organic-inorganic network structure with extraordinary mechanical, optical, and swelling/de-swelling properties. Adv Mater 10:1120–1124
20. Haraguchi K, Takehisa T, Simon F (2002) Effects of clay content on the properties of nanocomposite hydrogels composed of poly (N-isopropylacrlamide) and clay. Macromolecules 35:10162–10171
21. Haraguchi K, Farnworth R, Ohbayashi A, Takehisa T (2003) Compositional effects on mechanical properties of nanocomposite hydrogels composed of poly(N,N'-dimethylacrylamide) and clay. Macromolecules 36:5732–5741
22. Haraguchi K, Li HJ (2006) Mechanical properties and structure of polymer-clay nanocompoiste gels with high clay content. Macromolecules 39:1898–1905
23. Gong JP, Katsuyama Y, Kurokawa T, Osada Y (2003) Double network hydrogels with extremely high mechanical strength. Adv Mater 15:1155–1158
24. Na YH, Kurokawa T, Katsuyama Y, Tsukeshiba H, Gong JP, Osada Y, Okabe S, Karino T, Shibayama M (2004) Structural characteristics of double network gels with extremely high mechanical strength. Macromolecules 37:5370–5374

25. Tanaka Y, Kuwabara R, Na YH, Kurokawa T, Gong JP, Osada Y (2005) Determination of fracture energy of double network hydrogels. J Phys Chem B 109:11559–11562
26. Tsukeshiba H, Huang M, Na YH, Kurokawa T, Kuwabara R, Tanaka Y, Furukawa H, Osada Y, Gong JP (2005) Effect of polymer entanglement on the toughening of double network hydrogels. J Phys Chem B 109:16304–16309
27. Na YH, Tanaka Y, Kawauchi Y, Furukawa H, Sumiyoshi T, Gong JP, Osada Y (2006) Necking phenomenon of double-network gel. Macromolecules 39:4641–4645
28. Huang M, Furukawa H, Tanaka Y, Nakajima T, Osada Y, Gong JP (2007) Importance of entanglement between first and second components in high-strength double network gels. Macromolecules 40:6658–6664
29. Webber RE, Creton C, Brown HR, Gong JP (2007) Large strain hysteresis and Mullins effect of tough double-network hydrogels. Macromolecules 40:2919–2927
30. Tominaga T, Tirumala VR, Lin EK, Gong JP, Furukawa H, Osada Y Wu WL (2007) The molecular origin of enhanced toughness in double-network hydrogels: A neutron scattering study. Polymer 48:7449–7454
31. Tominaga T, Tirumala VR, Lee S, Lin EK Gong JP, Wu WL (2008) Thermodynamic interactions in double-network hydrogels. J Phys Chem B 112:3903–3909
32. Brown HR (2007) A model of the fracture of double network gels. Macromolecules 40:3815–3818
33. Tanaka Y (2007) A local damage model for anomalous high toughness of double-network gels. Europhys Lett 78:56005
34. Tirumala VR Tominaga T Lee S, Butler PD, Lin EK, Gong JP, Wu WL (2008) A molecular model for toughening in double-network hydrogels. J Phys Chem B 112:8024–8031
35. Lee KY, Mooney DJ (2001) Hydrogels for tissue engineering. Chem Rev 10:1869–1879
36. Nakayama A, Kakugo A, Gong JP, Osada Y, Takai M, Erata T, Kawano S (2004) High mechanical strength double-network hydrogel with bacterial cellulose. Adv Funct Mater 14:1124–1128
37. Hestrin S, Schramm M (1954) Synthesis of cellulose by *Acetobacter xylinum*: preparation of freeze dried cells capable of polymerizing glucose to cellulose. Biochem J 58:345
38. Presson BNJ (1998) Sliding friction: physical principles and applications, 2nd edn., NanoScience and Technology Series. Springer, Berlin
39. McCutchen CW (1962) The frictional properties of animal joints. Wear 5:1–17
40. McCutchen CW (1978) Lubrication of joints, the joints and synovial fluid. Academic, New York
41. Dowson D, Unsworth A, Wright V (1970) Analysis of "Booted lubrication" in human joints. J Mech Eng Sci 12:364–369
42. Ateshian GA, Wang HQ, Lai WM (1998) The role of intestitial fluid pressurization and surface porosities on the boundary friction of articular cartilage. J Tribol 120:241–251
43. Hodge WA, Fijian RS, Carlson KL, Burgess RG, Harris WH, Mann RW (1986) Contact pressures in the human hip joint measured in vivo. Proc Natl Acad Sci USA 83:2879–2883
44. Grodzinsky AJ (1983) Electromechanical and physicochemical properties of connective tissue. Crit Rev Biomed Eng 9:133–199
45. Buschmann MD, Grodzinsky AJ (1995) A molecular model of proteoglycan-associated electrostatic forces in cartilage mechanics. J Biomech Eng 117:179–192
46. Wojtys EM, Chan DB (2005) Meniscus structure and function. Instr Course Lect 54:323–330
47. Gong JP, Higa M, Iwasaki Y, Katsuyama Y, Osada Y (1997) Friction of gels. J Phys Chem B 101:5487–5489
48. Gong JP, Osada Y (1998) Gel friction. A model based on surface repulsion and adsorption. J Chem Phys 109:8062–8068
49. Gong JP, Iwasaki Y, Osada Y, Kurihara K, Hamai Y (1999) Friction of gels. 3. Friction on solid surfaces. J Phys Chem B 103:6001–6006
50. Gong JP, Kagata G, Osada Y (1999) Friction of gels. 4. Friction on charged gels. J Phys Chem B 103:6007–6014
51. Gong JP, Iwasaki Y, Osada Y (2000) Friction of gels. 5. negative load dependence of polysaccharide gels. J Phys Chem B 104:3423–3428

52. Gong JP, Kurokawa T, Narita T, Kagata K, Osada Y, Nishimura G, Kinjo M (2001) Synthesis of hydrogels with extremely low surface friction. J Am Chem Soc 123:5582–5583
53. Kagata G, Gong JP, Osada Y (2002) Friction of gels. 6. effects of sliding velocity and viscoelastic responses of the network. J Phys Chem B 106:4596–4601
54. Kurokawa T, Gong JP, Osada Y (2002) Substrate effect on topographical, elastic, and frictional properties of hydrogels. Macromolecules 35:8161–9166
55. Baumberger T, Caroli C, Ronsin O (2002) Self-healing slip pulses along a gel/glass interface. Phys Rev Lett 88:75509
56. Nitta Y, Haga H, Kawabata K (2002) Time dependent static friction force of agar gel-on-glass plate immersed in water. J Phys IV France 12:319–320
57. Kagata G, Gong JP, Osada Y (2003) Friction of gels. 7. Observation of static friction between like-charged gels. J Phys Chem B 107:10221–10225
58. Baumberger T, Caroli C, Ronsin O (2003) Self-healing slip pulses and the friction of gelatin gels. Eur Phys J E 11:85–93
59. Ohsedo Y, Takashina R, Gong JP, Osada Y (2004) Surface friction of hydrogels with well-defined polyelectrolyte brushes. Langmuir 20:6549–6555
60. Tada T, Kaneko D, Gong JP, Kaneko T, Osada Y (2004) Surface friction of poly(dimethyl siloxane) gel and its transition phenomenon. Tribol Lett 17:505–511
61. Kaneko D, Tada T, Kurokawa T, Gong JP, Osada Y (2004) Mechanically strong hydrogels with an ultra low friction coefficient. Adv Mater 17:535–538
62. Kurokawa T, Tominaga T, Katsuyama Y, Kuwabara R, Furukawa H, Osada Y, Gong JP (2005) Elastic-hydrodynamic transition of gel friction. Langmuir 21:8643–8648
63. Jiang Z, Tominaga T, Kamata K, Osada Y Gong JP (2006) Surface friction of gellan gels. Colloids Surf A Physicochem Eng Asp 284–285:56–60
64. Du M, Maki Y, Tominaga T, Furukawa H, Gong JP, Osada Y, Zheng Q (2007) Friction of soft gel in dilute polymer solution. Macromolecules 40:4313–4321
65. Tominaga T, Biederman H, Furukawa H, Osada Y, Gong JP (2008) Effect of substrate adhesion and hydrophobicity on hydrogel friction. Soft Matter 4:1033–1040
66. Tominaga T, Kurokawa T, Furukawa H, Osada Y, Gong JP (2008) Friction of a soft hydrogel on rough solid substrates. Soft Matter 4:1645–1652
67. Gong JP (2006) Friction and lubrication of hydrogels – its richness and complexity. Soft Matter 7:544–552
68. Schallamach A (1963) A theory of dynamic rubber friction. Wear 6:375–382
69. Chernyak YB, Leonov AI (1986) On the theory of the adhesive friction of elastomers. Wear 108:105–138
70. Savkoor AR (1965) On the friction of rubber. Wear 8:222–237
71. Ludema KC, Tabor D (1966) The friction and visco-elastic properties of polymeric solids. Wear 9:329–348
72. Vorvolakos K, Chaudhury MK (2003) The effects of molecular weight and temperature on the kinetic friction of silicone rubbers. Langmuir 19:6778–6787
73. Brown HR (1994) Chain pullout and mobility effects in friction and lubrication. Science 263:1411–1413
74. Brown HR (1994) Chain mobility and pull-out effects in lubrication and friction. Faraday Discuss 98:47–54
75. Persson BNJ (2001) Elastoplastic contact between randomly rough surfaces. Phys Rev Lett 87:116101
76. Persson BNJ (2001) Theory of rubber friction and contact mechanics. J Chem Phys 115:3840–3861
77. Persson BNJ, Volokitin A (2002) Theory of rubber friction: nonstationary sliding. Phys Rev B 65:134106
78. Martin A, Clain J, Buguin A, Brochard-Wyart F (2002) Wetting transitions at soft, sliding interfaces. Phys Rev E 65:31605
79. de Gennes PG Brochard-Wyart F, Quere D (2003) Capillarity and wetting phenomena: drops, bubbles, pearls, waves. Springer, New York

80. de Gennes PG (1979) Scaling concept in polymer physics Cornell University Press, Ithaca, New York
81. Kii A, Xu J, Gong JP, Osada Y, Zhang XM (2001) Heterogeneous polymerization of hydrogels on hydrophobic substrate. J Phys Chem B 105:4565–4571
82. Peng M, Kurokawa T, Gong JP, Osada Y, Zheng Q (2002) Effect of surface roughness of hydrophobic substrate on heterogeneous polymerization of hydrogels. J Phys Chem B 106:3073–3081
83. Klein J, Kumacheva E, Mahalu D, Perahia D, Fetters LJ (1994) Reduction of frictional forces between solid surfaces bearing polymer brushes. Nature 370:634–636
84. Grest GS (1999) Normal and shear forces between polymer brushes. In: Granick S (ed) Polymers in confined environments. Advances in Polymer Science, vol 138. Springer, Berlin, pp 149–183
85. Raviv U, Giasson S, Kampf N, Gohy JF, Jerome R, Klein J (2003) Lubrication by charged polymers. Nature 425:163–165
86. Wojtys EM, Chan DB (2005) Meniscus structure and function. Instr Course Lect 54:323–330
87. Rowley JA, Madlambayan G, Mooney DJ (1999) Alginate hydrogels as synthetic extracellular matrix materials. Biomaterials 20:45–53
88. Chang YY, Chen SJ, Liang HC, Sung HW, Lin CC, Huang RN (2004) The effect of galectin 1 on 3T3 cell proliferation on chitosan membranes. Biomater 25:3603–3611
89. Chung TW, Lu YF, Wang SS, Lin YS, Chu SH (2002) Growth of human endothelial cells on photochemically grafted Gly-Arg-Gly-Asp (GRGD) chitosans. Biomater 23:4803–4809
90. Park YD, Tirelli N, Hubbell JA (2003) Photopolymerized hyaluronic acid -based hydrogels and interpenetration networks. Biomater 24:893–900
91. Stile RA, Healy KE (2001) Thermo-responsive peptide-modified hydrogels for tissue regeneration. BioMacromolecules 2:185–194
92. West JL, Hubbell JA (1999) Polymeric biomaterials with degradation sites for proteases involved in cell migration. Macromolecules 32:241–244
93. Schmedlen RH, Masters KS, West JL (2002) Photocrosslinkable polyvinyl alcohol hydrogels that can be modified with cell adhesion peptides for use in tissue engineering. Biomaterials 23:4325–4332
94. Fittkau MH, Zilla P, Bezuidenhout D, Lutolf MP, Human P, Hubbell JA, Davies N (2005) The selective modulation of endothelial cell mobility on RGD peptide containing surfaces by YIGSR peptides. Biomaterials 26:167–174
95. Chen YM, Shiraishi N, Satokawa H, Kakugo A, Narita T, Gong JP, Osada Y, Yamamoto K, Ando J (2005) Cultivation of endothelial cells on adhesive protein free synthetic polymer gels. Biomaterials 26:4588–4596
96. Chen YM, Shen KC, Gong JP, Osada Y (2007) Selective cell spreading, proliferation, and orientation on micropatterned gel surfaces. J Nanosci Nanotechnol 7:773
97. Chen YM, Tanaka M, Gong JP, Yasuda K, Yamamoto S, Shimomura M, Osada Y (2007) Platelet adhesion to human umbilical vein endothelial cells cultured on anionic hydrogel scaffolds. Biomaterials 28:1752–1760
98. Chen YM, Gong JP, Tanaka M, Yasuda K, Yamamoto S, Shimomura M, Osada Y (2008) Tuning of cell proliferation on tough gels by critical charge effect. J Biomed Mater Res A 88:74–83, doi:10.1002/jbm.a.31869
99. Nakayama Y, Anderson JM, Matsuda T (2000) Laboratory-scale mass production of a multimicropatterned grafted surface with different polymer regions. J Biomed Mater Res 53:584–591
100. Lee JH, Lee JW, Khang G, Lee HB (1997) Interaction of cells on chargeable functional group gradient surfaces. Biomaterials 18:351–358
101. Lee SD, Husiue GH, Chang PCT, Kao CY (1996) Plasma-induced grafted polymerization acrylic acid and subsequent grafting of collagen onto polymer film as biomaterials. Biomaterials 17:1599–1608
102. Lee JH, Lee JW, Khang G, Lee HB (1997) Interaction of cells on chargeable functional group gradient surfaces. Biomaterials 18:351–358

103. Schneider GB, English A, Abraham M, Zaharias R, Stanford C, Keller J (2004) The effect of hydrogel charge density on cell attachment. Biomaterials 25:3023–3028
104. Magnani A, Priamo A, Pasqui D, Barbucci R (2003) Cell behaviour on chemically microstructured surfaces. Mater Sci Eng C 23:315–328
105. Kishida A, Iwata H, Tamada Y, Ikada Y (1991) Cell behavior on polymer surfaces grafted with non-ionic and ionic monomers. Biomaterials 12:786–792
106. Yasuda K, Gong JP, Katsuyama Y, Nakayama A, Tanabe Y, Kondo E, Ueno M, Osada Y (2005) Biomechanical properties of high toughness double network hydrogels. Biomaterials 26:4469–4475
107. Azuma C, Yasuda K, Tanabe Y, Taniguro H, Kanaya F, Nakayama A, Chen YM, Gong JP, Osada Y (2007) Biodegradation of high-toughness double network hydrogels as potential materials for artificial cartilage. J Biomed Mater Res A 81A:373–380
108. Yasuda K, Tanabe Y, Azuma C, Taniguro H, Onodera S, Suzuki A, Chen YM, Gong JP, Osada Y (2008) Biological responses of novel high-toughness double network hydrogels in muscle and the subcutaneous tissues. J Mater Sci Mater Med 19:1379–1387

Index

A

Adhesive friction 219
Adsorption–repulsion model 219
Aggrecan 216
Alginate 231
Amonton's law 216
Artificial organs 204

B

Bacterial cellulose (BC) 214
Biotissues 203
Block copolymers 6
Blood blots 204
Bovine fetal aorta endothelial cells (BFAECs) 232
Brillouin light scattering (BLS) 41
Brownian hard spheres 72
Brownian motion 2, 22, 57, 62, 71, 170, 188, 191
Brownian particles 55, 60
Brownian suspensions, non- 175, 188

C

Cage effect 74, 86
Cages 126
Carrageenan 215, 218, 227
Cartilage, artificial 203, 236
 cells 216
Cell scaffold 203
Cell spreading/proliferation 235
Cellulose 214
Centroiding algorithm 180
Chitosan 231
Clay-polymer hybrids 36
Colloidal dispersion 1, 55, 163
Colloidal gels 3
Colloidal glasses 1, 119, 165
Colloidal-like particles 120
Confocal imaging 163
Confocal microscopy 172
Core-corona 21
Core-shell particles 169
Correlated image tracking (CIT) 183
Crystallization 27
Cycloheptylbromide (CHB) 169

D

Daoud–Cotton density 6, 16
Decahydronaphthalene/tetrahydronaphthalene 169
Density correlator 70
Density fluctuations 68
Diblocks 6
Diffusion dynamics 12
Dioctadecyl-tetramethylindocarbocyanine (DiIC18) 168
Double network 203
Double-network hydrogels 205
Drug delivery/release 43, 122, 125
Dynamic light scattering (DLS) 21
Dynamics 1

E

Elasticity 120, 125, 129, 217, 220, 239
 emulsion droplets 125
Elastohydrodynamic interactions 117
Elastohydrodynamic slip 148
Emulsion droplets, elasticity 125
Emulsions 8, 43, 124, 171
Equilibrium 68
1-Ethyl-3-(3-dimethylaminopropyl) carbodiimide hydrochloride (EDC) 215
Ethylacrylate 122

F

Fast channel flow, fluids/pastes 189
Fibronectin 231
Flow curve 55
Flow geometries 173
Fluid-to-glass transition 57
Fracture energy 212
Friction 203
    reduction 225

G

Gel friction 217
Gelatin 214, 237
Gellan 217, 227
Glass transition 1, 55, 163
    shear moduli 76
Glasses 3
    colloidal 119
Grafted particles 1
    colloidal 7
Grafting density 8
Green–Kubo relations 62

H

Hairy particles 1
Hard glasses 118, 167
Hard-sphere colloids 193
Herschel–Bulkley equation 149, 190
Hookian spring constant 57
HUVEC density 235
Hyaluronic acid 231
Hydrodynamic interactions 72
Hydrodynamic slip 148
Hydrogels 203
Hydroxy stearic acid (HSA) 7
Hysteresis 209

I

Integration, transients approach 55
Integrations through transients (ITT) 59, 73
Isotropically sheared hard spheres model (ISHSM) 89

L

Laminin 231
Laponite 121
Liposomes 124
Load dependence 216
Lubricating film 145, 151
Lubrication 73, 140, 144, 151, 189, 216, 225, 239

M

MCT-ITT 59
Metal ion adsorption 122
Mica in nylon 36
Micelles 1, 21, 171
    block copolymers 6
Microgels 8, 122
    particles 8
Mode coupling closure 71
Mode coupling theory (MCT) 55, 59, 187
Model soft spheres 4
Mooney–Rivlin law 129
Mucus 228
Multilamellar vesicles 124

N

Nanoparticle-polymer hybrids 1
Nanoparticles, dispersion in polymer matrices 37
Near equilibrium 131
Necking 209
Network particles 122
Newtonian viscosity 57
7-Nitrobenzo-2-oxa-1,3-diazole-methyl methacrylate (NBD-MMA) 168
Non-equilibrium stationary state 55
Nonergodicity 118, 126

O

n-Octadecyl alcohol 7
Osmotic pressure 136

P

PAMPS/PAAm DN gels 205
Particle imaging velocimetry (PIV) 170, 178
Peclet number 63, 107, 189
PEP–PEO star-like block-copolymer micelles 21
Percus–Yervick (PY) approximation 72
Phase diagrams 1
Phospholipids 124
PMMA-coated silica particles 24
PMMA-PHS 184
PNIPAM 9, 80, 106, 122, 231
    coated PS latex 32
    particles 9
Poly(2-acrylamido-2-methyl-1-propanesulfonicacid) (PAMPS) 205
Poly(acrylic acid) (PAA) 122, 232
Poly(butadienyl)lithium 4
Poly(butyl methacrylate) 7

Poly(N,N'-dimethylacrylamide) (PDMAAm) 234
Poly(dimethyl siloxane) 7
Poly(ethylene glycol) (PEG) 7, 231
Poly(ethylene oxide) (PEO) 7, 229, 231
Poly(ethylene propylene)/poly(ethylene oxide) 6
Poly(2-hydroxyethyl methacrylate) (PHEMA) 228
Poly(N-isopropylacrylamide) (PNIPAM) 9, 80, 106, 122, 231
Poly(methyl methacrylate) (PMMA) 7, 121, 122, 167, 190, 226
Poly(NaAMPS-co-DMAAm) 234
Poly(propylene oxide) (PPO) 7
Poly(sodium 4-styrenesulfonate) (PNaSS) 228
Poly(vinyl alcohol) (PVA) 223, 229, 231
Poly(N–vinylcaprolactam) 122
Poly(2-vinylpyridine) (P2VP) 39, 122
Polybutadiene 6, 20
　stars 4, 20
Polydimethylsiloxane 121
Polydisperse dispersion, asymptotics 102
Polydispersity 5, 20
Polyelectrolyte brushes 226
Polyelectrolyte microgels 10
Polyethylene 226
Polyethylene oxide 121
Polyisoprene 6
Polylectrolyte microgels 123, 127
Polymer–colloids 117, 123
Polymer-grafted silica 7
Polypropylene 226
Polysaccharide gels 227
Polystyrene (PS) 5, 7, 121, 226
Polyurethane-PMMA 8
PS-P2VP 39
PS-poly(ethyl propylene) (PEP) 38
PVC 226

R
Refractive index (RI) 21, 40, 44, 167, 169
Relaxation 119
Reptation 29
Repulsion–adsorption 219
Rheology 1
　nonlinear 117, 163
(Rhodamine isothiocyanate) aminostyrene (RAS) 168

S
Sample area dependence 218
Self-consistent-field-theory (SCFT) 21

Shear moduli 55, 57, 76, 136
Shear rheology 148
Shear thickening 171
Shear thinning 117
Slip regimes 141
Small angle neutron scattering (SANS) 18
Small angle X-ray scattering (SAXS) 18, 213
Smoluchowski operator 61
Soft colloids 1
　glassy state 29
Soft glasses 118
Soft lubrication 144
Softness 1, 3
　tunability 11
Solvent friction 60
Solvent-particle interactions 72
Stars 1, 131
　colloidal 1, 4, 18, 120
　interactions 12
Steady shear 55
Stearyl alcohol 7, 121
Stokes–Einstein–Sutherland diffusion coefficient 61
Strength 203
Styrene sulfonate 226
Surface force apparatus (SFA) 226
Surface rheology 140, 142
Synovial fluid 228

T
Tetrafluoreoethylene (Teflon) 219, 226
Toughness 203, 206, 214
Tracking algorithms 181
Triblocks 6
Triton X-100 138

U
Ultrahigh molecular weight polyethylene (UHMWPE) 237

V
Viscoelasticity 34, 59, 80, 205, 238
　linear 55, 89, 117
Vitrification 28

W
Wall slip 117, 140

Z
Zwanzig–Mori type equation of motion 71, 95